LF

Contributions to Behavior-Genetic Analysis

THE MOUSE AS A PROTOTYPE

Contributions to Behavior-Genetic Analysis

THE MOUSE AS A PROTOTYPE

GARDNER LINDZEY
Department of Psychology
University of Texas
Austin, Texas

DELBERT D. THIESSEN
Department of Psychology
University of Texas
Austin, Texas

APPLETON-CENTURY-CROFTS
EDUCATIONAL DIVISION/MEREDITH CORPORATION
New York

PRINTED IN THE UNITED STATES OF AMERICA
390–56275–0

Contributors

JAN H. BRUELL
University of Texas, Austin, Texas

ROBERT L. COLLINS
The Jackson Laboratory, Bar Harbor, Maine

J. C. DEFRIES
University of Colorado, Boulder, Colorado

BARBARA J. GRIEK
University of Colorado, Boulder, Colorado

JAMES D. HAWKINS
San Jose State College, San Jose, California

J. P. HEGMANN
University of Colorado, Boulder, Colorado

GARDNER LINDZEY
University of Texas, Austin, Texas

MARTIN MANOSEVITZ
University of Texas, Austin, Texas

GERALD E. MCCLEARN
University of Colorado, Boulder, Colorado

THOMAS E. MCGILL
Williams College, Williamstown, Massachusetts

WILLIAM MEREDITH
University of California, Berkeley, California

KEITH OWEN
University of Texas, Austin, Texas

DAVID A. RODGERS
The Cleveland Clinic, Cleveland, Ohio

KURT SCHLESINGER
       University of Colorado, Boulder, Colorado
ROBERT K. SELANDER
       University of Texas, Austin, Texas
DELBERT D. THIESSEN
       University of Texas, Austin, Texas
MAL WHITSETT
       University of Texas, Austin, Texas
JAMES R. WILSON
       University of Colorado, Boulder, Colorado
SUH Y. YANG
       University of Texas, Austin, Texas

Contents

INTRODUCTION ix

Section I GENETIC ANALYSIS OF BEHAVIOR 1

1. THE USE OF ISOGENIC AND HETEROGENIC MOUSE
   STOCKS IN BEHAVIORAL RESEARCH 3
   Gerald E. McClearn, James R. Wilson, and
   James E. Meredith

2. GENETIC ANALYSIS OF OPEN-FIELD BEHAVIOR 23
   J. C. DeFries and J. P. Hegmann

3. GENETIC ANALYSIS OF MALE SEXUAL BEHAVIOR 57
   Thomas E. McGill

Section II GENE-ENVIRONMENTAL INTERPLAY 89

4. GENETIC VARIATION AND HOARDING 91
   Martin Manosevitz and Gardner Lindzey

5. THE SOUND OF ONE PAW CLAPPING: AN INQUIRY
   INTO THE ORIGIN OF LEFT-HANDEDNESS 115
   Robert L. Collins

Section III SINGLE GENE MODIFICATION OF
   BEHAVIOR 137

6. SINGLE GENE SUBSTITUTIONS AND BEHAVIOR 139
   James D. Hawkins

7. CHROMOSOME MAPPING OF BEHAVIORAL ACTIVITIES 161
   Delbert D. Thiessen, Keith Owen, and Mal Whitsett

vii

Section IV GENE-PHYSIOLOGIC DETERMINATION
              OF BEHAVIOR   205

   8. MECHANISM-SPECIFIC BEHAVIOR: AN EXPERIMENTAL
       ALTERNATIVE   207
       David A. Rodgers

   9. THE GENETICS AND BIOCHEMISTRY OF AUDIOGENIC
       SEIZURES   219
       Kurt Schlesinger and Barbara J. Griek

Section V THE EVOLUTION OF BEHAVIOR   259

  10. BEHAVIORAL POPULATION GENETICS AND WILD
       Mus Musculus   261
       Jan H. Bruell

  11. BIOCHEMICAL GENETICS AND BEHAVIOR IN WILD
       HOUSE MOUSE POPULATIONS   293
       Robert K. Selander and Suh Y. Yang

INDEX   335

Introduction:
Mus musculus, Mechanisms and Man

Mus musculus is, and probably will remain, the principal mammalian species for investigating fundamental relations between genes and behavior. The mouse has circled the globe with man and inhabited every niche common to mammalian species. Man and mouse easily coexist in each other's world and in some cases provide roots for each other's existence. While the rodent may live opportunistically off man, man delves into its biology and behavior and thereby enriches his own knowledge. The prey-predator symbiosis between man and mouse is one of the oldest adaptations that has evolved.

We have selected for this book prime examples of how behavior genetics has used the mouse strategically to explore the genetic foundations of behavior. Reviews of some of the most significant trends in behavior genetics are included. The various chapters identify biometric procedures that are commonly applied to many different relevant problem areas. The importance of gene-environment interaction is stressed throughout, and physiologic pathways are described for several areas of research. For the first time the use and abuse of inbred strains of mice are discussed within the framework of evolutionary adaptation, and considerable attention is devoted to the wild mouse as an appropriate species for study. Both theory and original investigations are reported, and an attempt is made to demonstrate for the reader the basic importance of mouse research and its relevance to all facets of behavioral investigation.

THE GENESIS OF MOUSE RESEARCH

Man's earliest use of the mouse can be traced to pet fanciers and breeders in antiquity (Keeler, 1931; Grüneberg, 1952). The waltzing mouse with its exaggerated circling motions was one of the first behavioral mutants to reach recorded history. It can be traced to its origin in China as early as 80 B.C. The historical chain extends to the present time. If five generations are assumed to arise each year, then approximately 10,245 generations have passed since the waltzing mouse was first described. This amounts to the equivalent of 204,900 years of human life—truly an impressive pedigree.

It is even said (Sturtevant, 1965) that Gregor Mendel originally worked out laws of genetic segregation and recombination with the mouse but suppressed the mammalian work for fear of antagonizing the Church. He, of course, eventually published his monumental paper in 1865 using common garden peas. Nevertheless von Guaita (see Grüneberg, 1952) and others demonstrated regularities of inheritance in the mouse at the close of the 19th century that, after the rediscovery of Mendel's work in 1900, could clearly be interpreted along the same lawful lines as the pea experiments. In fact, von Guaita published data on the inheritance of waltzing in mice that nicely demonstrated that circling was the result of a single genetic unit. This experiment is perhaps the first systematic study in behavior genetics. The universal application of Mendel's insights set the tone for all genetic investigations to follow.

Behavior genetics is one of the outgrowths of this knowledge, and the applications of these insights in mouse research have resulted in this animal's becoming a primary model for all mammalian systems. The use of *Mus* is a logical outgrowth of the need for a variety of genetic populations that can be manipulated easily, that resemble man biologically, and that are readily available for experimental use. Many specially bred domestic as well as wild mouse lines serve these purposes very effectively.

Taxonomically *Mus musculus* is in the order Rodentia, the family Muridae, and the subfamily, Maurinae (Simpson, 1945). Within *Mus musculus* many domestic and aboriginal forms exist. All these genetic groups provide important sources of experimental material for evaluating intrinsic laws of behavior and evaluating the adaptation of man.

Much behavior genetic work has concentrated on highly inbred strains originally developed for differential susceptibility to neoplastic diseases. The major strains and their lineage are shown in Fig. 1. The first inbred line, now known as DBA, was begun at Harvard in 1909 by Clarence Cook Little. Within 15 years after the origin of the first inbred line most of the other strains used in cancer research had been established. In 1921 at Cold Spring Harbor, Leonell C. Strong mated an albino mouse obtained from Halsey J. Bagg with one from an albino stock from Little's colony and began inbreeding. This resulted in the A strain, now an important line in behavior genetic investigations. In 1920, Strong made a series of crosses between Bagg albinos and the *DBA* line and developed the *C3H, CBA, C, CHI,* and *C121* inbred strains. The *C3H* line was later separated into several sublines.

Another strain often used in behavioral research, the *C57,* also dates from 1921. Little obtained two littermates from Miss A. E. C. Lathrop, a mouse fancier in Granby, Massachusetts, and mated these. The offspring segregated as black and brown, which, with inbreeding, led to the *C57BL* and *C57BR* strains. J. M. Murray later developed the *C57L* strain from a coat-color mutant found in a *C57BR* subline.

Charleton MacDowell, also at Cold Spring Harbor, originated the *C58* line from descendants of Miss Lathrop's progenitors of the *C57* lines. MacDowell also inbred the Bagg albino stock to develop the well-known BALB/c inbred strain.

Other strains were soon to follow. Jacob Furth of the Henry Phipps

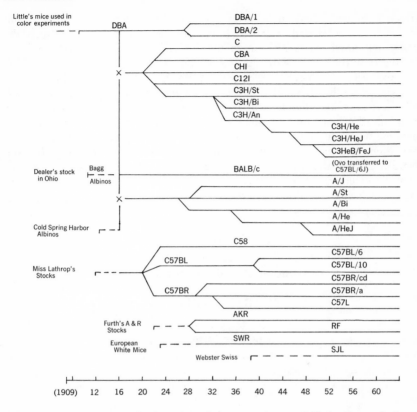

FIG. 1. Origins of the older inbred strains. (From Staats, 1966. In *Green, E. L., ed.* Biology of the Laboratory Mouse. *Courtesy of McGraw-Hill Book Co.)*

Institute in Philadelphia began inbreeding mice in 1928 to yield the *AK* and *RF* strains. Clara J. Lynch of the Rockefeller Institute inbred within the common Swiss albino mice to obtain such lines as the *SWR/J* and *SJL/J*. Altogether over 200 lines have appeared (Staats, 1966), usually inbred by brother-sister matings for a minimum of 20 generations. In addition, special stocks are maintained that carry single genes of interest (Lane, 1964), and at least 80 neurologic mutants, of unusual value for behavioral investigations, are available (Sidman et al., 1965).

A great deal of information about special strains of mice can be found in edited volumes of the Federation Procedures (1960, 1963), a *Catalogue of Uniform Strains of Laboratory Animals Maintained in Great Britain* (1958), *Biology Data Book* (Altman and Dittmer, 1964), general reference texts (Green, 1966; Lane-Petter, 1963), and a periodical devoted to the mouse (*Mouse News Letter,* edited by Mary F. Lyon). Standardized nomenclature of inbred mice can be found in a publication edited by Staats (1964), and bibliographies of behavioral studies using inbred mice are also found in publications by Staats (1958, 1963).

Wild, commensal, and domesticated mice are only beginning to appear attractive to researchers in behavior genetics. Their heritage is without the systematic control found in inbred lines; because of this, however, their diversity and untended adaptation to natural environments can offer insights into the evolutionary background of behavior.

According to Schwarz and Schwarz (1943), wild or aboriginal mice form four separate subspecies: *Mus musculus wagneri, Mus musculus spicilegus, Mus musculus manchu,* and *Mus musculus spretus.* All four lack direct communication with man; hence, the populations offer uncontaminated gene pools for study. In addition, the first three species have developed commensal populations that have followed man around the world and consequently have diverged genetically. The range of dispersal is wide, especially for *M. m. wagneri,* including groups of mice specifically adapted to local ecologies. The three main centers of dispersal are shown in Table 1.

TABLE 1

Major Commensal Radiation of Three Stocks of Wild *Mus musculus*

First Center of Dispersal

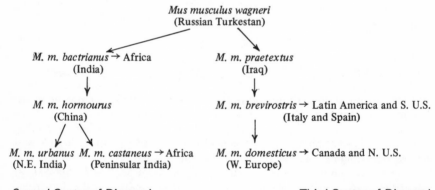

Second Center of Dispersal

Mus musculus spicilegus
(Southern Russia)
↓
M. m. musculus
(into Siberia)

Third Center of Dispersal

Mus musculus manchu
(Japan)
↓
M. m. molossinus
(Japan and Korea)

Wild mice are classified into genetic groups on the basis of origin, distribution, body weight, tail length, head size, coat color, interbreeding capacity, and dimorphic factors. The biologic separation between stocks is not always complete, giving gradations, such as in Iran where areas of *wagneri, bactrianus,* and *praetextus* meet; and hybrids, such as in Kenya where ranges of *bactrianus* and *castaneus* overlap; and the whole of the Norwegian coast where *domesticus* meets *M. m. musculus.*

Interpopulation and intrapopulation variations prevent the subspecies from being accurately classified. On the other hand, differences in gene frequency, polymorphisms, and continuous phenotypic variation provide the raw material for the study of adaptation to specific locales. Inbred strains, with their redundant and restricted genotypes, offer few of these advantages. In this volume we contrast the relative advantages of inbred and wild mice for behavioral work. Each has its place and the choice depends on the experimental question posed.

ORGANIZATION OF CHAPTERS

This volume is comprised of five sections: (1) Genetic Analysis of Behavior, (2) Gene-Environmental Interplay, (3) Single Gene Modification of Behavior, (4) Gene-Physiologic Determination of Behavior, and (5) the Evolution of Behavior. Each portion contains chapters of theoretical interest as well as chapters presenting experimental results, so that in total the material provides a fair representation of the major contemporary interests within behavior genetics. Other more general presentations of behavior genetics are found in chapters, books, and reviews by Broadhurst (1960), Bruell (1962), Fuller (1960, 1964), Fuller and Thompson (1960), Hall (1951), Manosevitz et al. (1969), McClearn (1962), McClearn and Meredith (1966), Parsons (1967), Rodgers (1966), Thiessen (1966, 1969), and Yerkes (1907).

It is appropriate that McClearn, Meredith, and Wilson initiate the section on genetic analysis of behavior. They provide a valuable commentary on the use of inbred strains of mice and their genetic derivations, and they describe common methodologic practices of many investigators, including those writing in this volume. Inbred and crossbred strains provide homogeneity and heterogeneity for special uses and allow much control over gene pools. With inbred strains one can capitalize on a minimal amount of genetic variability within a line and maximal variability between lines. Crossbred lines give more or less reproducible gene pools of high genetic variability.

McClearn and his associates clearly demonstrate the unique advantages associated with different genetic populations. Among their several contributions is an extensive genetic analysis of an eight-way cross (derived from genetic interchange among eight inbred strains) on several separate activity tasks providing over 100 different measures. The use of a systematic battery of behavioral tests and specified genotypes is a fairly recent development in behavior genetics and is sure to have wide applicability. McClearn and his colleagues give the reader an excellent model of sophistication in task design, sensitivity to behavioral measurement, and elegance of data analysis.

DeFries and Hegmann continue this section by adding what is probably the most extensive evaluation of mouse activity in the entire behavior genetic literature. The investigators make use of over 4,820 mice of the *BALB/cJ*, *C57BL/6J* strains and their cross generations. Almost every major technique of genetic analysis is described and applied to the data. Matherian analyses, parent-offspring regressions, and half-sib correlations are compared and evaluated. Five generations of selection for high and low activity are presented, and data for replicate lines as well as controls are given. Among their interesting observations is the fact that the albino gene accounts for approximately 12 percent of the additive variance in their activity measure and increases or decreases in frequency as the population is selected for low or high activity. Apparently a single gene can regulate a sizable proportion of the population variance and, in part, determines the response to selection.

The section that follows, on the interaction between gene activity and the environment, provides a complement to the first section, which emphasizes genetic determination. Clearly genes do not act *in vaccuo* but develop their expression during ontogeny and within the context in which measurements are made. McGill makes this obvious, but often overlooked, point in his detailed analyses of sex behavior in the inbred mouse. He discusses some basic biometric techniques and explains their use and limitation in the analysis of physiologic pathways of behavior. Depending on the measure of sexual responsiveness, and there are many, genes may express themselves as dominants, recessives, intermediates, or some partial form of these. Moreover, the situation in which the sexual response occurs dramatically modifies the mode of inheritance. Mechanisms underlying behavior are highly plastic. Obviously care must be taken to represent accurately the operative mechanisms and the range of environmental modifiability.

Manosevitz and Lindzey further emphasize that genetic determination is dependent upon environmental circumstances. The phenotype considered is hoarding behavior among inbred and crossbred strains of mice. Although genetic variation has a clear effect upon the hoarding response, the early developmental history or character of the proximal stimuli can also determine whether a given strain will be high or low in the hoarding response. In other words, the environment can either be expressive or suppressive for certain genes. Available evidence indicates that in *Mus musculus* hoarding is polygenically determined and heritability may range from 25 to 50 percent depending on the genotype and environment.

Collins rounds out this section with a review of hand preference in man and mouse and puts forward a remarkable environmental interpretation of what has long been considered a genetic effect. His careful review of the relevant evidence is followed by exploration of four genetic models and leads him to the conclusion that there is no substantial evidence that hand preference in humans is genetically determined. Studies with inbred mice further support this conclusion. In all probability the mouse can provide an analog for the study of prenatal and postnatal influences on handedness. Genetically specified lines are often as valuable in environmental research as they are in genetic observations.

The third section changes focus from complexly regulated behaviors, often polygenically determined, to single gene effects. Hawkins spells out goals and strategies in behavior genetics using the single gene approach and explains the often devious route of gene expression. Examples of pleiotropic effects of the agouti locus are given for sex and aggressive behavior. It is plain from Hawkins' arguments and his illustrations that genes typically operate indirectly through complex pathways and that a single gene approach to the study of behavior provides no assurance that the underlying mechanisms will be simple.

Thiessen, Owen, and Whitsett enlarge on the presentation of single gene substitutions. by presenting a thorough rationale for this experimental procedure and reviewing the available data for single gene effects in *Drosophila,* mice, man, and other species. Theoretical issues are supplemented with an extensive analysis of the albino locus in the mouse on a standardized battery of eleven tasks including over forty measures. In addition, 14 additional coat-color genes segregating on the *C57BL/6J* background are evaluated on four behavioral tasks. These analyses seem to indicate a general behavioral deficit for the albino gene and behavior-specific effects for the other genes tested. They conclude that single gene substitutions can be used to map chromosomes using behavioral phenotypes in the same way that morphologic and biochemical traits have been mapped.

The next section moves to a discussion of even more molecular problems in behavioral analyses, the gene-physiologic determination of behavior. Rodgers points out that behavior genetics has traditionally been interested in categorizing behavior in terms of gene correlates without reference to the biologic organism. Moreover, there has been a steady particularization of behavior from polygenic phenotypes to those regulated by a single gene. Nevertheless, the emphasis on categorization and correlation has not changed, it has only been cast in more minute terms. Generality tends to be lost in a maze of unrelated details. As a solution, Rodgers stresses an earlier argument (Thiessen and Rodgers, 1967) that generality across genotypes is possible by investigating mechanism-specific rather than species- or gene-specific behaviors. By that he means describing behavior in terms of underlying mechanisms without exaggerated consideration of the genetic details underlying the mechanisms. Gene influence is not, however, disregarded; it is used to advantage in manipulating the character and quantity of the mechanism of interest.

An interesting series of studies of mechanism-specific behavior is presented by Schlesinger and Griek in their chapter on audiogenic seizure susceptibility of mice. In an elegantly conceived sequence of studies, Schlesinger and Griek arrive at the conclusion that strain differences in seizure proneness may be based on biochemical differences in the brain biochemicals serotonin (5-HT), norepinephrine (NE), and gamma-aminobutyric acid (GABA). Correlational procedures are used to establish the genetic relation of seizure susceptibility and 5-HT, and NE, and pharmacologic and dietary attacks confirm the direct relation of biochemical status to convulsive disorder. The generalization that appears is that 5-HT, NE, and GABA influence neural excitation and overexcitation of critical brain areas related to seizures. Ironic as it may seem, these investigators have begun to explain a genetic variation without undue concern for the particular genes involved.

The last section of this volume considers the adaptive nature of gene pools and individuals and suggests how wild mice are best suited for evolutionary study. Certainly the wild mouse, with its natural heritage, has too long been overlooked in favor of the radically artificial, inbred mouse. Neither source of animals—the field or the laboratory—is apt to replace the other, but the wild mouse must secure its rightful place in experimental research.

Bruell addresses himself to the theoretical issues surrounding the use of wild mice in the study of behavioral population genetics. He defines basic concepts like gene pool, selective value of genes, and coadapted gene pools, and indicates how wild mice provide us with naturally selected units for study. Importantly, Bruell objects to the common arguments that species are genetically permanent and uniform and that there are species-typical behaviors that are seen among all members. Typologic classifications, especially at the level of the genotype, are useless, as Rodgers points out in another context. Bruell also adds important notes on the distribution and character of aboriginal, commensal, and feral mice. The contrast of these stocks with laboratory inbred mice is striking.

In the last chapter, Selander discusses the systematics of the house mouse and the "typology" problem. He emphasizes the genetic variation of wild populations and identifies the types of polymorphisms usually studied. His own work cited in this chapter is of special interest. Blood esterases and hemoglobin are used as genetic markers to study geographic variations in polymorphisms. Significantly, Selander finds that even within single barns without physical barriers, demes of mice are isolated breeding populations with characteristic gene frequencies. Territorial and aggressive behavior, especially by the male, are seen as potent isolating devices. As the demes are small in population size, genetic drift can determine major characteristics of the population.

There is no doubt that the rapidly developing specialty of behavior genetics has relied more heavily upon data from the mouse than any other data source. It is also clear that most or all of the major laboratories in which such data are being collected and analyzed are represented in the present volume. Thus it seems a reasonable premise that the reader will find here a valid picture of the problems, the premises, and the accomplishments of contemporary thought and research dealing with behavior genetics of the mouse.

REFERENCES

Altman, P. L., and D. S. Dittmer. 1964. Biology Data Book, Washington, Federation American Society for Experimental Biology.

Broadhurst, P. L. 1960. Experiments in psychogenetics: Applications of biometrical genetics to the inheritance of behavior. *In* Eysenck, H. J., ed. Experiments in Personality. London, Routledge and Kegan Paul, pp. 3–102.

Bruell, J. H. 1962. Dominance and segregation in the inheritance of quantitative behavior in mice. *In* Bliss, E. L., ed. Roots of Behavior. New York, Harper and Row, Pubs., pp. 48–67.

Catalogue of Uniform Strains of Laboratory Animals Maintained in Great Britain. 1958. Charshalton, Surrey, England, Laboratory Animal Centre, M.R.C. Laboratories.

Fuller, J. L. 1960. Behavior genetics. Ann. Rev. Psychol., 11:41–70.
—— 1962. The genetics of behavior. *In* Hafez, E. S. E., ed. The Behavior of Domestic Animals. Baltimore, The William and Wilkins Co., pp. 57–81.
—— 1964. Physiological and population aspects of behavior genetics. Amer. Zool., 4:101–109.
—— and W. R. Thompson. 1960. Behavior Genetics. New York, Wiley.
Ginsburg, B. E. 1958. Genetics as a tool in the study of behavior. Perspec. Biol. Med., 1:397–424.
Green, E. L., ed. 1966. Biology of the Laboratory Mouse. New York, Mc-Graw-Hill.
Grüneberg, H. 1952. The Genetics of the Mouse, 2d ed. The Hague, Nijhoff.
A Guide to Production, Care and Use of Laboratory Animals: An Annotated Bibliography. 1960. Fed. Proc., 19:1–196.
A Guide to Production, Care and Use of Laboratory Animals: An annotated Bibliography. 1963. Fed. Proc., 22:1–250.
Hall, C. S. 1951. The genetics of behavior. *In* Stevens, S. S., ed. Handbook of Experimental Psychology. New York, John Wiley and Sons, Inc., pp. 304–329.
Hirsch, J., ed. 1967. Behavior-Genetic Analysis. New York, McGraw-Hill.
Keeler, C. E. 1931. The Laboratory Mouse: Its Origin, Heredity, and Culture. Cambridge, Harvard University Press.
Lane, P. W., ed. 1964. Lists of Mutant Genes and Mutant-Bearing Stocks of the Mouse. Bar Harbor, The Jackson Laboratory.
Lane-Petter, W., ed. 1963. Animals for Research. New York, Academic Press.
Lindzey, G., and D. D. Thiessen, eds. 1970. Contributions to Behavior-Genetic Analysis: The Mouse as a Prototype. New York, Appleton-Century-Crofts.
Lyon, M. F., Collator. Mouse News Letter. Harwell, Didiot, Berkshire, England, M.R.C. Radiobiological Research Unit.
Manosevitz, M., G. Lindzey, and D. D. Thiessen, eds. 1970. Behavioral Genetics: Method and Theory. New York, Appleton-Century-Crofts.
McClearn, G. E. 1962. The inheritance of behavior. *In* Postman, L., ed. Psychology in the Making. New York, Alfred A. Knopf, Inc., pp. 144–252.
—— and W. Meredith. 1966. Behavior genetics. Ann. Rev. Psychol., 17:515–550.
Parsons, P. A. 1967. The Genetic Analysis of Behavior. Suffolk, England, Methuen.
Rodgers, D. A. 1966. Factors underlying differences in alcohol preference among inbred strains of mice. Psychosom. Med., 28:498–513.
Schwarz, E., and H. K. Schwarz. 1943. The wild and commensal stocks of the house mouse, *Mus musculus,* Linnaeus. J. Mammal., 24:59–72.
Sidman, R. L., M. C. Green, and S. H. Appel. 1965. Catalog of the Neurological Mutants of the Mouse. Cambridge, Harvard University Press.
Simpson, G. G. 1945. Principals of classification and a classification of mammals. Bull. Amer. Mus. Nat. Hist., 85.
Spuhler, J. N., ed. 1967. Genetic Diversity and Human Behavior. Chicago, Aldine Publishing Co.
Staats, J. 1958. Behaviour studies on inbred mice: A selected bibliography. Anim. Behav., 6:77–84.

——— 1963. Behaviour studies on inbred mice: A selected bibliography, II. Anim. Behav., 11:484–490.

——— ed. 1964. Standardized nomenclature for inbred strains of mice: Third listing. Cancer Res., 24:147–168.

——— 1966. The laboratory mouse. *In* Green, E. L., ed. Biology of the Laboratory Mouse. New York, McGraw-Hill Book Co., pp. 1–9.

Sturtevant, A. H. 1965. A History of Genetics. New York, Harper and Row Pubs.

Thiessen, D. D. 1966. Pleiotropism in behavior genetics: The mouse as a research instrument. Percept. Motor Skills, 23:901–902.

——— and Rodgers. 1967. Behavior and genetics as the study of mechanical-specialized behavior. *In* Spuhler J. N. ed. Genetic Diversity and Human Behavior. Chicago, Aldine Publishing Co., pp. 61–71.

——— 1970. Gene Organization and Behavior. New York, Random House, Inc.

Yerkes, R. M. 1907. The Dancing Mouse. New York, The Macmillan Co.

Contributions to
Behavior-Genetic Analysis

THE MOUSE AS A PROTOTYPE

SECTION I

Genetic Analysis of Behavior

1

The Use of Isogenic and Heterogenic Mouse Stocks in Behavioral Research[*]

GERALD E. MCCLEARN AND JAMES R. WILSON
Institute for Behavioral Genetics,
University of Colorado, Boulder

WILLIAM MEREDITH
University of California, Berkeley

INTRODUCTION

When a renewed interest in behavioral genetics developed in the 1950's, one of the principal research techniques employed in research with mice was that of comparison of inbred strains and derived generations. The many demonstrations of strain differences in an extraordinary variety of behavioral traits and, of even greater interest, strain differences in response to experimental manipulation, have resulted in the gradual acceptance of genetic control as an important aspect of experimental procedures in animal research that has no direct interest in behavioral genetics per se. Investigators have turned more and more to the consistent use of a given inbred strain of animals in elucidating their substantive problems, or to the use of several strains simultaneously, thus introducing genotype as an independent variable and making possible an assessment of the degree of generality of the results obtained. These usages of inbred strains have been discussed previously (McClearn, 1967) and are dealt with in detail elsewhere in this volume.

Although the use of inbred strains proceeds apace, increasing attention is being given to other approaches to the study of quantitative behavioral genetics of the mouse. Such techniques as selective breeding, parent-offspring regression analysis, sibling correlation analysis, and so on are appearing more and more frequently (e.g., see DeFries, 1967; DeFries and Hegmann, this

[*] The studies reported herein were supported in part by NIGMS Grants No. GM 14547 and GM 10735.

volume), and the power of these techniques assures that their use will continue and grow. Inevitably, some of these approaches will be utilized as control procedures in behavioral research.

This type of genetic analysis requires genetically heterogeneous populations, and the application to general behavioral research will also. The purpose of this paper is to describe the types of situations in which such controls will be useful, to discuss some of the criteria of a "good" genetically heterogeneous population, to describe the derivation of one such population, and to present a few examples of relevant research with this population.

Limitations of the Inbred Strain

The principal advantage of an inbred strain is the stability of its mean. The absence, or near absence, of genetic variance makes for stability over a period of time, because without genetic variability there can be no unsystematic changes of gene frequency due to sampling drift or systematic changes due to selective breeding. Thus, an investigator may have confidence in the high relative stability of his animal subjects in consecutive experiments. This relative temporal stability also suggests that strains from a common origin, maintained in different laboratories, will accumulate genetic differences only slowly, and will thus remain comparable for long periods of time.

When the focus of attention is upon the mean, these characteristics are sterling virtues. When the situation calls for analysis of variances or covariances, however, these same features become crippling handicaps. The inappropriateness of an inbred strain for assessing variance is suggested by reference to the formula for partitioning of phenotypic variance in a heterogeneous population. Making the usual assumptions of lack of genotype environment interaction and independence of genetic and environmental effects the formula is:

$$V_P = V_G + V_E$$

where V_P is the phenotypic variance, V_G is the genotypic variance, and V_E is the environmental variance (see Falconer, 1960; Roberts, 1967).

The process of inbreeding serves to decrease V_G and in the widely used, highly inbred strains of mice its value can be assumed to have approached zero. Thus, the expression describing phenotypic variance *within* a given strain reduces to:

$$V_P = V_E$$

Inbreeding has thus eliminated the entire set of genetic determinants of variability, leaving only those attributable to environmental sources.

With respect to correlation, it can be shown (Falconer, 1960) that the phenotypic covariation and correlation between traits can be represented as follows:

$$\text{Cov}_P = \text{Cov}_A + \text{Cov}_E$$
$$r_P = h_X h_Y r_A + e_X e_Y r_E$$

The terms h_X, h_Y, and r_A all rely upon the existence of additive genetic variance and consequently are eliminated by inbreeding. In the remaining term e_X is defined as $(1-h^2{}_X)^{1/2}$ and e_Y is $(1-h^2{}_Y)^{1/2}$. These terms approach 1 as inbreeding reduces genetic variance. Thus, in a highly inbred strain, $r_P = r_E$, and the only elements contributing to a correlation between two traits will be those environmental factors that have an influence on both traits. All of the covariance caused by a gene having an influence on both traits is eliminated.

Occasionally, it may be of special interest to ascertain only the environmental determinants of variance or covariance, but the meaningfulness of such an assessment is limited unless it can be compared to that due to *both* environmental *and* genetic agencies. An inbred strain is thus of circumscribed utility for describing or testing hypotheses about the range of expression of a behavioral trait, correlations between behavioral traits, multivariate analyses, correlations between behavioral and physiologic or anatomic traits, or apparatus reliability.

To be sure, animal behavior research traditionally has not been much concerned with individual differences and correlations among traits. The argument can be made, however, that an increased interest along these lines would be advantageous. In human psychology, advances in understanding of intellectual functioning and personality dynamics have been dependent to a very considerable extent upon techniques and procedures relating to heterogeneous normative groups. Indeed, the concepts of percentiles, standardized scores, reliability, validity, mental age, factor structure, and so on are so essential a part of psychometrics that it is difficult to imagine how one could go about "mental measurement" without them. The fact that such concepts are so little used in animal research where the greater opportunities for manipulation and control would render them even more powerful is lamentable in the extreme. At the very least, the application of this general perspective should serve to provide us with better measurement devices and behavioral taxonomy; it might provide significant major breakthroughs in our understanding of animal behavior.

Criteria for the Heterogeneous Population

If this is granted, what should be the nature of the normative population? It might be thought that the ideal would be a group representative of

Mus musculus at large. Such a group is not attainable in practice, however. Even if one were to begin by a large scale wild trapping operation, the sample would be biased from the outset because of differential trappability of different individual animals. Furthermore, natural selection for ability to adapt to the laboratory environment would begin at once through differential fertility, disease resistance, ability to tolerate confinement, and so on.

Purists might insist that those animals surviving such selection for laboratory life are not representative of *Mus* in a natural habitat, but such an argument overlooks the fact that for laboratory animals the laboratory *is* the natural habitat.

It appears therefore that laboratory-adaptable animals will inevitably be the basis for our heterogeneous stock. If this is so, the stocks that have seen such widespread use in psychologic research might be put forward as suitable, since they have not been deliberately inbred. However, inbreeding need not be systematic in order for reduction in genetic variance to occur, and most traditionally used commercial stocks and private colonies have been characterized by sporadic inbreeding through choice of exceptional sires and dams as well as by the general inbreeding that results inevitably from small breeding populations. The genetic situation of these groups is therefore unknown and for the purposes described, they are nearly useless.

A *systematic* genetically heterogeneous group can be made by mating of F_1s obtained from inbred strains as described by DeFries and Hegmann in this volume. Such F_2 populations have certain peculiarities: there can be at most only two alleles at any locus; the allelic frequencies for any alleles in the F_2 populations can only have values of 0, $\frac{1}{2}$, or 1; there can be no lethals; and so on. Genetically more heterogeneous groups can be obtained by intercrossing more than 2 inbred strains with more possible alleles, and more possible values of allelic frequency, although they will also be without lethals. In a sense, the larger number of inbred strains employed the better, although the real world of animal husbandry imposes some practical limitations on the number that can be used. A useful compromise appears to be an eight-way cross maintained after initial establishment by random mating in a relatively large breeding population. An eight-way cross founding population would appear to offer scope for adequate genetic variability and yet makes possible the approximate reconstitution of the group, if desirable, at some other time or place. The mode of breeding assures reasonable stability of the gene frequencies (although not, of course, of genotypes of individuals) from one generation to the next.

A heterogeneous stock was begun in the Behavioral Genetics Laboratory of the University of California, Berkeley, and is now maintained at the Institute for Behavioral Genetics, University of Colorado, Boulder. An account of this stock follows.

Development of the Genetically Heterogeneous Stock (HS)

The animals used to form the heterogeneous stock were obtained from the Cancer Research Genetics Laboratory, University of California, Berkeley, and consisted of the following strains: *A, AKR, BALB/c, C3H/2, C57BL. DBA/2, Is/Bi,* and *RIII.*

F_1 animals from different parental strains were mated to obtain four-way cross offspring. At this time it might be convenient to introduce the laboratory shorthand which evolved to permit convenient designations of strains and crosses upon a small cage-tag.

$$
\begin{array}{ll}
A = A & DBA/2 = D \\
BALB/c = B & AKR = K \\
C3H/2 = 3 & Is/Bi = I \\
C57BL = C & RIII = R
\end{array}
$$

Our convention is to name females first. Thus a $D \times A$ mating is of a *DBA/2* female with an *A* male. The F_1 offspring are described as *DA.* Similarly a mating of $AB \times 3C$ is of an *AB* female with a *3C* male. The four-way cross offspring are described by *AB3C.* Four-way cross animals with no parent strains in common were then mated to form eight-way crosses.

Throughout, matings were made so as to assure that the Y chromosomes of the eight original strains were represented with approximately equal frequency in the eightway cross generation. Although the Y chromosome is not thought to carry much genetic information other than male determiners, it is the only mouse chromosome which can be followed through the generations with no crossing over. The desirability of having approximately equal representation of the Y chromosomes in the stock seemed easily worth the effort, therefore. For example, a *RA3K* female mated to a *CDBI* male resulted in *RA3KCDBI* offspring the males of which have an *Is/Bi* Y chromosome; a *RA3K* female mated to an *IBDC* male resulted in *RA3KIBDC* offspring, the males of which have a *C57BL* Y chromosome. From the eight-way cross animals the foundation generation of the HS stock was constituted. This generation consisted of 40 mating pairs, with 5 males carrying a Y chromosome from the *A* strain, 6 from *BALB/c,* 5 from *C3H/2,* 5 from *C57BL,* 5 from *DBA/2,* 5 from *Is/Bi,* 5 from *RIII,* and 4 from *AKR.* No further attempt was made to equalize Y chromosome representation in subsequent generations, although the breeding records allow determination of Y chromosome for any male.

Eight generations of HS animals have now been bred. Each generation, consisting of 39 or 40 mating pairs, is maintained by randomly mating animals with no common grandparental ancestry. This number of mating pairs insures that inbreeding will accumulate very slowly (see Falconer, 1960). A recapitulation of part of the lineage for one family is given in Fig. 1. One animal, numbered 714, was chosen as the reference animal from generation 7. His paternal great great great great grandfather was number 110, and so on.

These animals have proved to be hardy laboratory animals, easy to

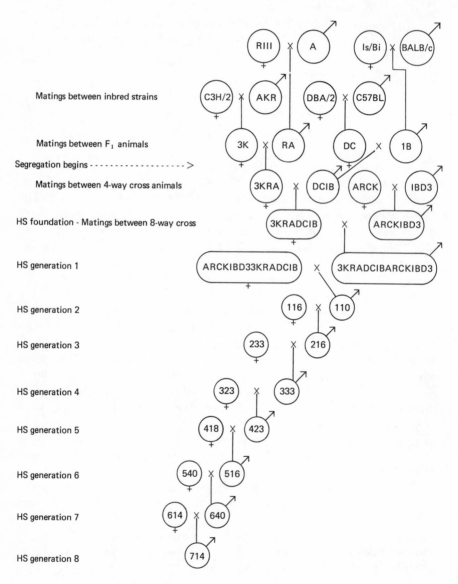

FIG. 1. Sample pedigree of an HS mouse.

handle, prolific, and fast-growing. Their reproductive success is striking, as shown in Table 1. Even the results tabulated do not do them full justice, since the second, third, and fourth litters not included in this tabulation were somewhat larger. Most females become reimpregnated at postpartum estrus and are therefore incubating a subsequent litter while nursing a prior. It should be noted that the drop in fertility and in litter size at weaning seen in generations two, three, and four may be associated with known environmental events: the pairs constituting generation two were placed in the care of a new animal caretaker and also were exposed to a flare-up of a form of infantile diarrhea which was endemic in the laboratory; animals which were to constitute generation three were flown from Berkeley to Boulder where

TABLE 1

Fertility and Litter Size in a Genetically Heterogeneous Stock of Mice Compared with Results from Several Inbred Strains[a]

Group or Strain	Lab No.	No. of Pairs	% Fertile Matings	Mean Litter Size	
				At Birth	At Weaning
HS Foundation	1	40	98	8.8	8.1
HS Generation 1	1	40	92	9.3	8.3
2	1	39	87	8.5	4.8
3	2	39	90	7.8	6.0
4	2	40	98	8.9	6.2
5	3	40	100	9.6	9.0
6	3	40	98	9.9	9.3
7	3	40	100	10.8	10.5
A	3	19	89	5.8	5.2
BALB/c Crgl	1	9	67	5.5	2.7
BALB/c An,Pi	3	7	86	6.2	5.7
BALB/c	3	20	100	7.7	6.6
C3H/2	3	16	84	5.4	3.3
C57BL Crgl	1	34	41	6.0	2.1
C57BL	3	18	72	5.9	5.1
DBA/2 Crgl	1	18	61	6.0	1.3
DBA/2 De,tr	3	21	95	5.1	4.2
Weighted Means:					
HS		318	95	9.3	7.8
Isogenic		165	75	6.0	4.2
Comparison of Laboratories:					
Berkeley–(Lab 1)		180	78	8.2	6.0
Boulder–Conventional Lab (Lab 2)		79	94	8.4	6.1
Boulder–SPF Lab (Lab 3)		224	94	8.3	7.6

[a]A pair was considered infertile if it produced no litter within 60 days of pairing. Mean litter sizes were computed using size of *first* litter produced by each pair.

they (and subsequently generation four animals) were maintained in temporary quarters pending completion of the specific pathogen free (SPF) facility now in use. A comparison of breeding results from the Berkeley laboratory, the temporary Boulder laboratory, and the present Boulder laboratory, is included in Table 1. Although the barrier system and associated procedures of the SPF laboratory were adopted mainly to afford long-term protection for valuable, and in some cases irreplaceable mouse stocks, it seems evident from the breeding results that a significant fertility gain has been a side effect.

As mentioned above, young animals from this stock develop rapidly. To investigate some of their developmental functions more fully, we measured 93 HS pups from birth through 15 days of age on body weight and several reflexes—simple behaviors modified from Fox (1965).* Results from 22

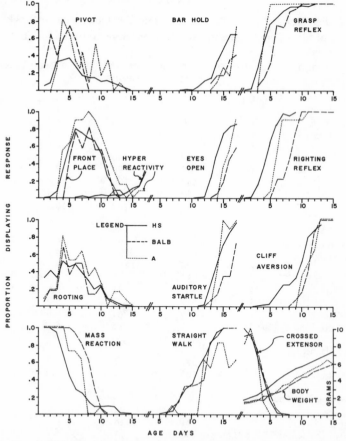

FIG. 2. Developmental functions. Only the strongest level of each response is shown.

* We thank Mr. John Belknap for technical assistance in the collection of these data.

BALB/c and 11 *A* mice are included for comparison in Fig. 2. A description of the measures employed is given below.

Pivoting. (one-minute observation period) Circular locomotion caused by side motion of front legs with hind legs essentially inactive.
   0—response not present
   1—weak attempts to pivot, typically of short duration
   2—moderate pivoting movements
   3—pivoting movements pronounced; pivoting occurs throughout most of the observation period

Straight walking. (one-minute observation period) Locomotion in an approximate straight line with all four legs involved in adult fashion.
   0—response not present
   1—straight walking for a distance at least equal to body length
   2—more than half the observation period spent in straight walking

Righting reflex. (best of two trials) Turning to rest in normal position on all four legs after being placed on side by experimenter.
   0—response not present
   1—more than one second required to attain upright posture
   2—upright posture attained almost immediately

Back righting. (average of two trials) Turning to rest in the normal position on all four legs after being placed on back by experimenter.
   0—response not present
   1—righting within 5 seconds
   2—righting within 1 second

Rooting reflex. (20-second test period) Pushing forward after bilateral stimulation by experimenter's finger on face area.
   0—response not present
   1—1 or 2 brief responses
   2—1 sustained response

Cliff drop aversion. Withdrawing from edge of surface when forepaws are placed over the edge.
   0—response not present
   1—backing within 10 seconds
   2—backing within 3 seconds
   3—backing almost immediately, within 1 second

Grasp reflex. Foot flexion with grasping of metal rod when the plantar surface is stroked by the rod.
   0—response not present
   1—contraction slight with response barely detectable
   2—complete envelopment of rod
   3—immediate and complete envelopment of rod

Front placing. Raising and placing foot on table top when dorsum of foot is placed in contact on edge of table top (animal suspended by loose skin at the back of the neck during test).

Crossed extensor. Flexion of hind limb when pinched and extension of opposite hind limb.

Bar Holding. (best of three trials) Grasping a wooden pencil with front legs and paws and supporting own weight.

    0—response not present
    1—momentary support of body weight
    2—support of body weight for 5 seconds or longer

Auditory startle. Immediate flight or "freezing reaction" following hand clap of experimenter.

Eyes open. Time of opening of eyes.
    0—both eyes closed
    1—at least one eye partially open
    2—both eyes fully open

Hyperreactivity. Overreaction or exaggerated responsiveness to novel stimuli (the "popcorn stage") with exaggerated freezing or jumping being elicited by ordinary noises and movements.

Mass reaction. Exaggerated squirming and rolling with occasional whole body convulsions following tail pinch.
    0—response not present
    1—reaction restricted to tail region
    2—reaction involving tail and hind quarters
    3—reaction involving entire body with writhing, squirming, and rolling.

Body weight. Body weight measured to the nearest tenth of a gram.

For pivoting, front placing, and rooting, the developmental function involves an increase in the proportion of animals displaying the response followed by a subsequent decrease to zero. Mass reaction and crossed extensor responses show an initially high incidence with subsequent decrease to zero. The other measures show increasing developmental functions.

For most of the measures, the development of the inbred strains lags behind that of the HS animals as might be expected on general grounds from the literature on heterosis. An exception to this is crossed extensor; the picture is somewhat obscure in the case of grasp reflex and rooting. In front placing, the *BALB/c* animals have a developmental curve very similar to that of the HS, but the *A*'s appear to be precocious in the development of the response, but retarded in the elimination of it. Pivoting is particularly interesting because of the much greater incidence of this behavior in the inbred than in the heterogeneous normative group.

The age at which sexual maturity is achieved is another developmental character of interest. Although we have not yet collected normative data on this character, we can report that HS females may become fertile very early. We found to our chagrin that some females weaned at 30 days and separated from their parents and brothers at that time gave birth to a litter about 20 days later—a litter presumably sired by the adult parent since motile sperm are not found in the tracts of young males before age 40 days (Hawtrey, 1967). These unwanted pregnancies have made it necessary for us to standardize weaning time for this stock at 21 days of age.

Activity Studies

One of the principal reasons for the development of the HS was for use in a large-scale study of mouse activity. This study was planned to combine the mathematics of quantitative genetics and the mathematics of multivariate analysis in the behavioral domain of exploratory activity and emotionality in novel situations. Very briefly, it involved collections of over 100 behavioral and elimination measures from a sample which included 1,254 isogenic mice (seven inbred strains and all reciprocal F_1 crosses) and 256 HS mice, deriving the genetic correlations among these measures and submitting this matrix to multivariate analyses in an effort to describe the obtained correlations in terms of a small number of dimensions or "factors". The phenotypic correlation matrix obtained from intercorrelating all measures taken can be partitioned into genetic and environmental components by capitalizing on the lack of genetic variance within isogenic groups, with all variation within isogenic groups attributable to environmental events. By subtracting this environmental variance from the phenotypic variance of HS animals, an estimate of genetic variance can be obtained.

For variance analyses, of course, several assumptions must be met. One assumption, homogeneity of variance among subgroups, has proven particularly difficult to meet. Using the untransformed data, a substantial mean variance relationship is evident on a majority of the measures taken, with correlations between the mean and the variance of the 98 isogenic groups (sexes plotted separately) ranging up to $+0.94$. The search for suitable transformations is continuing, and it is planned that this study will be presented in a separate monograph when the scaling problems are overcome. In the meantime, in order to present some results representative of the heterogeneous stock, the total array of results from the battery was scrutinized and a small subset was chosen consisting of measures having the least pronounced mean-variance relationships. Results on these measures are presented for males and females separately from the seven inbred strains in comparison to the HS animals. The measures chosen were obtained as follows: (1) *Emergency Latency* (E LAT). The cage of each animal was placed into a rectangular hole cut into a red opaque platform so that the top edge of the cage was even with the platform surface. The wire mesh top of the cage was removed and the animal was scored for latency to climb from within its cage onto the platform. A maximum score of 5 minutes was assigned if the animal did not emerge sooner. (2) *Sum of Crossings in Y-maze* (Y CROSS). Each animal was introduced from above into the center of a Y-maze consisting of three dead-end alleys 2 inches wide and 10 inches long interconnected and symmetrically arranged with arms joining at 120° angles. The animal was scored for the number of times it traversed the length of any alley during a 3-minute period. (3) *Rearings in the Y-maze* (Y REAR). The number of times during the 3-minute test that an animal reared so that its front paws were more than 1 inch from the floor. (4) *Barrier Latency* (B LAT). Each animal was introduced into the near compartment of a box 12 inches square divided into four triangular compartments by insertion of an X-shaped barrier. Although the insert fit diagonally from the corners of the box it only reached to the top in the central portion, being cut down to form a 1¼-inch vertical barrier near each corner. The latency to cross a barrier into an adjacent compartment was recorded. A maximum score of 3 minutes was assigned if an animal did not cross sooner. (5) *Sum of Hole-in-Wall Entries* (HW ENT). Each animal was placed into the near compartment of a box similar to the barrier apparatus, except that the X-shaped insert along the diagonals reached from bottom to top. Communication among the four wedge-shaped compartments was via 1⅜-inch diameter round holes, placed ¼ inch from the floor in the center of each common wall. The total number of entries during the 3-minute test period constituted this score. (6) *Sum of Stairs Climbed* (STAIRS). Each animal was introduced at the bottom of an enclosed set of eight wooden stairs 1½ inches wide and 3 inches long. The number of stairs traversed up and down, during the 3-minute test yielded this score. (7) *Total Urination* (URINE). After completition of the daily test on each apparatus, each animal was scored for presence or absence of urine in the apparatus. All such urination scores were summed to yield this measure. (8) *Total Defecation* (FECES). After each test each animal was scored for number of fecal boli left in the apparatus. These scores were summed to yield this measure.

The results obtained from these groups on the traits measured are summarized in Tables 2–5.

Consideration of group means (Table 2) reveals a tendency for HS mice to be somewhat more active, and to have shorter latencies, than the inbred strain mice. This general tendency is by no means true of every comparison; there are very obvious strain × apparatus interactions which lead to markedly higher scores in some inbred groups (see, e.g., *Is/Bi* STAIRS score). Blanket application of the notions of heterosis and hybrid vigor to activity traits such as those reported is obviously not appropriate. The comparison of particular interest is the relative size of trait variance in isogenic

TABLE 2

Mean Activity and Elimination Scores from Seven Inbred Strains and a Genetically Heterogeneous Stock

Strain	Sex	N	E LAT	Y-REAR	B LAT	HW ENT	Y-CROSS	STAIRS	URINE	FECES
A	♀♀	13	4.84	10.4	2.51	9.7	29.1	37.6	0.54	18.4
	♂♂	13	5.00	4.8	2.92	7.4	19.1	30.5	2.15	20.6
BALB/c	♀♀	13	4.92	9.9	1.69	15.7	41.5	63.7	2.77	26.5
	♂♂	12	4.58	9.0	2.27	21.6	54.3	59.6	2.92	30.8
C3H/2	♀♀	14	5.00	7.9	2.47	12.6	28.1	37.9	1.36	10.9
	♂♂	16	4.80	2.0	2.78	5.8	18.3	27.2	1.88	12.2
C57BL	♀♀	13	3.13	10.2	1.26	17.5	45.5	37.2	0.54	8.5
	♂♂	15	2.44	8.1	1.99	15.4	36.9	30.7	2.00	12.8
DBA/2	♀♀	13	4.49	12.4	2.09	14.8	40.2	36.1	2.38	12.9
	♂♂	13	4.93	10.4	2.74	11.1	44.3	32.9	2.54	9.6
Is/Bi	♀♀	14	4.07	7.4	2.38	16.8	46.6	99.1	5.43	20.1
	♂♂	13	3.14	11.6	1.41	22.8	45.3	89.1	5.54	12.2
RIII	♀♀	13	3.66	11.9	2.01	28.6	52.8	80.4	0.46	11.8
	♂♂	14	3.83	7.8	2.02	19.1	35.9	61.1	1.36	10.0
Isogenic	♀♀	89	4.30	9.96	2.07	16.5	40.5	56.2	1.96	15.6
	♂♂	94	4.07	7.49	2.31	14.4	35.6	46.3	2.57	15.1
HS	♀♀	136	3.77	22.4	1.31	18.8	44.0	50.4	3.38	14.4
	♂♂	120	3.84	20.6	1.43	19.2	41.5	49.3	5.70	15.4

and heterogenic stocks (Table 3). There is a clear tendency for the HS variance to be larger, but again, this is not true in every case. Comparing variance on each trait (HS females versus each inbred strain female; similarly for males), we find HS variance to be numerically larger in 99 cases, and smaller in 13. In no case is the variance of an inbred strain significantly larger than that of the HS comparison groups, so the ostensibly larger inbred variances are likely due to sampling error. Pooling the inbred variances by taking the arithmetic means, by sex, we obtain the variances shown in Table 3 as isogenic variances. As shown, the HS variance exceeds the isogenic variance in every case, and is significantly larger in 11 of the 16 comparisons ($p \leqslant 0.05$).

Proceeding now to a consideration of covariance within inbred and heterogenic groups, we find there is a tendency for covariances between traits to be higher in the heterogenic groups, but not in every case. As discussed in an earlier section, we typically expect the phenotypic covariance to be the sum of the additive genetic covariance and the environmental covariance,

$$\mathrm{Cov}_P = \mathrm{Cov}_A + \mathrm{Cov}_E$$

so that phenotypic covariance for the inbred strains is simply equal to the environmental covariance, Cov_E. A first expectation, therefore, would be for there to be higher covariances between traits measured in HS mice, since to the Cov_E term obtained from inbred groups is added the genetic covariance term, Cov_A. However, as can be seen in Table 4, this is not true in general. Reminded by the fact that some of the covariances are negative, we realize that there is no a priori reason why the genetic and environmental covariance should be equivalent in sign. Anytime they differ, the obtained phenotypic covariance may be smaller in absolute magnitude than the environmental covariance.

When the covariances are expressed as correlations,

$$r_p = h_x h_y r_A + \sqrt{(1 - h^2_x)\ (1 - h^2_y)}\ \ r_E$$

the situation becomes yet more complicated. If the heritability of one of the traits (say x) approaches zero, the first term ($h_x h_y r_A$) vanishes and the phenotypic correlation r_p approaches $r_E\ (1 - h^2_y)^{1/2}$. This of course results in phenotypic correlations obtained from heterogenic groups being smaller than r_p estimate from inbred groups. This is not the only condition in which this is true.* It may be shown that estimates from heterogenic groups will be smaller than those from isogenic groups, whenever

$$r_E > \frac{h_x h_y r_A}{1 - \sqrt{(1 - h^2_x)\ (1 - h^2_y)}}$$

Evaluating this equation for several values of trait heritabilities and genetic correlations, we obtain the results shown in Table 5. Whenever the genetic

---

* We thank Dr. John DeFries for development of this section.

TABLE 3

Variances on Selected Characters Measured in Seven Inbred Strains and a Genetically Heterogeneous Stock

		N	E LAT	Y-REAR	B LAT	HW ENT	Y-CROSS	STAIRS	URINE	FECES
A	♀♀	13	0.232	53.93	0.708	46.21	89.15	103.92	0.248	25.01
	♂♂	13	0.000	12.79	0.081	27.93	41.30	389.79	3.515	33.31
BALB/c	♀♀	13	0.075	55.15	1.541	101.90	310.25	444.21	2.331	85.72
	♂♂	12	0.963	25.17	0.903	140.41	274.56	391.08	2.576	185.47
C3H/2	♀♀	14	0.000	27.98	0.901	190.67	75.84	199.12	1.801	39.78
	♂♂	16	0.286	4.25	0.337	37.56	89.34	186.28	3.360	21.78
C57BL	♀♀	13	1.204	31.41	0.948	106.71	196.25	419.98	0.402	13.02
	♂♂	15	3.437	48.25	1.082	181.57	263.40	208.73	3.867	51.36
DBA/2	♀♀	13	1.198	90.39	1.374	170.49	239.67	501.61	4.237	37.46
	♂♂	13	0.060	31.93	0.204	101.15	346.83	375.92	4.864	48.08
Is/Bi	♀♀	14	1.781	28.80	0.847	172.17	90.24	865.92	2.674	57.27
	♂♂	13	2.590	47.93	0.845	62.95	278.83	561.46	2.248	52.90
RIII	♀♀	13	1.910	24.99	1.240	185.62	171.87	560.54	0.248	46.49
	♂♂	14	2.307	40.03	1.132	97.92	252.50	548.98	3.372	51.00
Isogenic	♀♀	89	1.026	44.66	1.080	139.11	167.61	442.19	1.706	44.88
	♂♂	94	1.378	30.05	0.655	92.78	220.96	380.32	3.400	63.41
HS	♀♀	136	2.510	112.18	1.250	189.73	264.46	561.60	5.118	69.90
	♂♂	120	2.185	93.81	1.137	167.71	277.73	709.91	6.676	82.49

17

TABLE 4
Comparison of Isogenic and HS Mice on Covariances Between Traits

E LAT Versus: (Y-REAR through FECES) and **Y-REAR Versus:** (B LAT, HW ENT, Y-CROSS)

		Y-REAR	B LAT	HW ENT	Y-CROSS	STAIR	URINE	FECES	B LAT	HW ENT	Y-CROSS
Isogenic	♀♀	-0.933	0.369	-2.294	-2.260	-2.537	0.114	0.027	-2.608	20.706	58.242
	♂♂	-1.453	-0.001	-0.704	-0.500	2.312	-0.076	-0.878	-1.232	9.268	45.082
HS	♀♀	-0.375	0.296	0.286	-2.298	-9.599	0.347	1.145	-3.208	63.100	109.717
	♂♂	-2.569	0.428	-0.964	-5.199	-7.244	-0.676	-0.098	-3.942	35.558	114.438

B LAT Versus: (HW ENT through FECES) and **HW ENT Versus:** (Y-CROSS, STAIR)

		HW ENT	Y-CROSS	STAIR	URINE	FECES	Y-CROSS	STAIR
Isogenic	♀♀	-5.383	-5.194	-6.638	-0.183	-0.656	51.241	95.924
	♂♂	-1.751	-1.860	-2.163	-0.163	0.027	54.749	83.915
HS	♀♀	-4.304	-6.112	-9.154	-0.070	1.240	86.128	79.633
	♂♂	-5.788	-7.772	-10.535	-1.037	-1.007	79.716	138.646

Y-REAR Versus: (STAIR, URINE, FECES), **HW ENT Versus:** (URINE, FECES), **Y-CROSS Versus:** (STAIR, URINE, FECES), **Stair Versus:** (URINE, FECES), **Urine Versus:** (FECES)

		STAIR	URINE	FECES	URINE	FECES	STAIR	URINE	FECES	URINE	FECES	FECES
Isogenic	♀♀	51.044	0.824	-2.876	1.212	-15.124	96.317	0.968	-10.082	8.449	-6.610	1.879
	♂♂	11.624	0.340	-2.988	2.256	2.962	77.148	7.190	6.904	4.633	-13.236	4.657
HS	♀♀	57.943	0.042	-2.385	5.074	2.953	138.003	0.407	-19.799	4.027	-37.816	9.770
	♂♂	86.088	8.292	-6.933	6.126	4.415	185.434	14.653	3.390	21.399	7.985	7.442

18

correlation is low or trait heritabilities are widely discrepant, then low or moderate values of environmental correlation will lead to larger phenotypic correlations in inbred strains than in heterogeneous populations. Examination of the correlations in Table 6 reveals many instances in which this is the case. The relatively low number of subjects in the inbred groups requires correlations on the order of ± 0.70 for statistical significance, and not many of those tabulated are of this magnitude. Accordingly, the correlations for the inbred groups were pooled, using Fisher z-transformations, and are presented as isogenic correlations. Averaging the correlations in this way would tend to mask any existing strain $\times$ apparatus interactions, but genotype $\times$ environment interactions were disregarded previously. In any case, we have no way of estimating this interaction, since the environment was not explicitly manipulated in this study.

From standard psychometric considerations we have been taught to expect a reduction in the size of a correlation coefficient if the range of measurement of either of the two variables is restricted. However, restriction of genetic variation is not the same thing as restriction of measurement and leads to quite different expectations under certain specifiable circumstances. As developed above, such genetic restriction may quite easily lead to larger correlation coefficients. It seems clear that we cannot form reasonable expectations about the genetic, environmental, and phenotypic correlations of any specified group until we know some of the important genetic parameters. This stricture will make it logically impossible to predict relative coefficient sizes until the data are in; realistically though, we will be able to do better

TABLE 5

Table of Environmental Correlations[a]

		r_A			
h_x^2	h_y^2	0.2	0.4	0.6	0.8
0.2	0.2	0.20	0.40	0.60	0.80
	0.4	0.18	0.37	0.55	0.74
	0.6	0.16	0.32	0.48	0.64
	0.8	0.13	0.27	0.40	0.53
0.4	0.4	0.20	0.40	0.60	0.80
	0.6	0.19	0.38	0.58	0.77
	0.8	0.17	0.35	0.52	0.69
0.6	0.6	0.20	0.40	0.60	0.80
	0.8	0.19	0.39	0.58	0.77
0.8	0.8	0.20	0.40	0.60	0.80

[a]When the environmental correlation exceeds the value shown in a given cell and the two characters have the heritabilities and genetic correlation value shown for that cell, the phenotypic correlation between the characters will be larger in isogenic groups than in heterogenic groups.

TABLE 6

Correlations between Traits Computed *within* Inbred Strains and Compared to Correlations on the Same Traits Using a Genetically Heterogeneous Stock

Strain	Sex	E LAT Versus:							Y-REAR Versus:						B LAT Versus:	
		Y-REAR	B LAT	HW ENT	Y-CROSS	STAIR	URINE	FECES	B LAT	HW ENT	Y-CROSS	STAIR	URINE	FECES	HW ENT	Y-CROSS
A	♀♀	-.44	-.16	-.06	-.36	-.19	-.33	-.52	-.38	.60	.72	.07	.22	.12	-.56	-.20
	♂♂	.00	.00	.00	.00	.00	.00	.00	.14	.34	.45	.26	-.19	.36	-.14	.14
BALB/c	♀♀	.22	.09	-.32	.19	-.31	.18	-.15	-.69	.21	.67	.36	.22	-.02	-.40	-.52
	♂♂	.28	-.31	-.37	-.17	-.18	-.63	-.25	-.01	.28	.80	.19	-.32	-.26	-.14	.15
C3H/2	♀♀	.00	.00	.00	.00	.00	.00	.00	-.02	.26	.75	.07	.24	.05	-.22	.18
	♂♂	.04	-.15	-.30	.15	.30	-.30	.25	-.03	.14	.04	.25	-.02	.55	.00	.16
C57BL	♀♀	.21	.14	-.15	.04	.44	.36	.20	-.03	.35	.81	.55	-.12	-.45	-.29	-.13
	♂♂	-.58	.16	-.32	-.45	-.21	-.02	-.19	-.58	.20	.79	.28	.28	.19	-.11	-.56
DBA/2	♀♀	-.50	.69	-.38	-.44	-.36	.02	-.39	-.88	.54	.91	.90	.15	-.20	-.62	-.75
	♂♂	-.29	.78	.12	-.06	.24	-.19	-.43	-.46	.08	.59	-.13	.29	.54	-.33	-.27
Is/Bi	♀♀	.03	.65	-.02	-.10	-.22	.18	.36	-.23	.07	.61	.30	-.24	-.09	-.35	-.30
	♂♂	-.47	.38	.56	.33	.40	.41	-.04	-.40	-.07	.10	-.26	-.06	-.26	-.43	-.12
RIII	♀♀	-.19	.52	-.40	.39	-.24	.17	.00	.29	-.17	.12	-.09	.11	-.55	-.65	-.47
	♂♂	.16	-.38	.06	.13	-.01	-.04	.02	-.12	.33	.75	.45	-.04	-.41	-.40	-.18
Isogenic	♀♀	-.05	.31	-.20	-.06	-.13	.09	.08	-.36	.28	.71	.38	.08	.18	-.45	.35
	♂♂	-.14	.11	-.03	-.01	.08	-.12	-.10	-.23	.19	.57	.16	.01	.00	-.23	-.11
HS	♀♀	-.02	.17	.01	-.09	-.26	.10	.09	-.27	.43	.65	.24	.00	-.03	-.28	-.34
	♂♂	-.18	.27	-.05	-.21	-.19	-.18	-.01	-.39	.28	.71	.33	.33	-.08	-.42	-.45

TABLE 6 continued

		B LAT Versus:			HW ENT Versus:				Y-CROSS Versus:			STAIR Versus:		URINE Versus:
		STAIR	URINE	FECES	Y-CROSS	STAIR	URINE	FECES	STAIR	URINE	FECES	URINE	FECES	FECES
A	♀♀	.03	.24	.44	.41	-.32	-.25	-.39	-.17	.07	.34	.63	.08	.35
	♂♂	-.56	-.75	-.07	.26	.63	-.09	-.49	.16	-.40	-.56	.18	-.24	.29
BALB/c	♀♀	-.46	-.42	-.39	.32	.46	-.07	-.09	.41	.30	-.09	.45	.17	.35
	♂♂	.21	.29	-.09	.56	.32	-.12	.31	.10	.02	.13	-.11	.25	.34
C3H/2	♀♀	-.02	-.03	-.63	-.16	.16	.10	-.34	.06	.08	.23	.25	-.25	-.12
	♂♂	-.35	-.10	.20	.28	.02	-.23	-.20	.54	.10	.01	.16	.00	-.25
C57BL	♀♀	-.23	.04	-.21	.64	.25	-.01	.30	.63	-.03	-.14	.59	-.15	-.01
	♂♂	-.19	-.10	.20	.29	.35	.31	.12	.43	.47	.21	.00	.05	.62
DBA/2	♀♀	-.87	-.25	-.01	.52	.72	.06	-.19	.81	.21	-.08	.15	-.32	.42
	♂♂	-.09	-.28	-.26	.17	.57	-.03	-.34	.21	.49	.42	.30	-.50	.44
Is/Bi	♀♀	-.45	-.05	.41	.16	.82	.46	-.16	.22	-.42	-.58	.41	-.15	.24
	♂♂	-.28	-.03	.11	.58	.78	.60	-.22	.55	.51	-.32	.40	-.24	.34
RIII	♀♀	.20	.21	-.01	.43	-.05	-.37	-.36	.28	-.37	-.30	.22	.14	-.06
	♂♂	-.14	-.29	-.07	.47	.50	.37	.22	.07	.17	-.01	-.04	-.17	.39
Isogenic	♀♀	-.32	-.04	-.07	.35	.35	-.01	-.18	.37	-.03	-.10	.40	-.07	.18
	♂♂	-.21	-.21	.00	.39	.49	.13	-.09	.31	.21	-.03	.13	-.13	.33
HS	♀♀	-.35	.03	.13	.39	.25	.16	.03	.36	.01	-.15	.08	-.19	.52
	♂♂	-.37	-.38	-.10	.37	.40	.18	.04	.42	.34	.02	.31	.03	.32

21

than this as data accumulate about the genetic parameters of behavioral traits measured in similar situations. In the meantime, collection of correlation data on the same traits using both isogenic and heterogenic stocks is in order since, by screening for cases when the HS correlations are higher, traits with a large genetic correlation can be identified. When r_A is large, a very large r_E is required for the phenotypic correlation in an isogenic population to exceed that in a heterogenic population. To enjoin the corollary, use of inbred and F_1 stocks for assessments of environmental effects seems straightforward, but does lead to new considerations. Which, for example, of the inbred strains or F_1 crosses should be used for a given study, since strain effects are likely to be quite specific? Prudence would probably require use and comparison of both isogenic and heterogenic stocks in environmental, as well as genetic, studies.

Genetically heterogeneous mouse stocks are often needed for selective breeding studies and for other experimental designs requiring genetic variation. Such stocks maintained on a breeding program designed to minimize inbreeding have been relatively unavailable, but several examples now exist, e.g., Falconer's Q stock and the HS group presently under discussion. Random-bred stocks are not generally equivalent, since in practice different "random" mating procedures and different effective mating numbers are employed by suppliers, leading to varying amounts of inbreeding. The need for inbred strains is not lessened as heterogeneous stocks become available. Their joint use is complementary, and the availablity of heterogeneous stock as a comparison group appears to provide a useful research tool.

REFERENCES

DeFries, J. C. 1967. Quantitative genetics and behavior: Overview and perspective. *In* Hirsch, J., ed. Behavior-Genetic Analysis. New York, McGraw-Hill Book Co., pp. 322–339.

Falconer, D. S. 1960. Introduction to Quantitative Genetics. Edinburgh, Oliver and Boyd.

Fox, W. M. 1965. Reflex-ontogeny and behavioral development of the mouse. Anim. Behav., 13:234–241.

McClearn, G. E. 1967. Genes, generality, and behavioral research. *In* Hirsch, J., ed. Behavior-Genetic Analysis. New York, McGraw-Hill Book Co., pp. 307–321.

Roberts, R. C. 1967. Some concepts and methods in quantitative genetics. *In* Hirsch, J., ed. Behavior-Genetic Analysis. New York, McGraw-Hill Book Co., pp. 214–257.

2

Genetic Analysis of
Open-field Behavior[*]

J. C. DeFries and J. P. Hegmann
Institute for Behavioral Genetics,
University of Colorado, Boulder

INTRODUCTION

Calvin S. Hall devised the open-field test to provide a valid and objective measure of individual differences in emotionality and, from the results of several experiments with rats (Hall, 1934; 1936a), concluded that defecation and urination in a strange open-field situation are valid indices of emotionality.[†] By the method of selective breeding, Hall (1951) conclusively demonstrated the importance of heredity as a cause of individual differences in open-field behavior. The foundation population was a "heterogeneous group" of 145 rats which were placed individually in a brightly lighted, circular enclosure (7 feet in diameter) for 2 minutes a day on each of 12 days. The resulting scores ranged from 0 (*S* neither defecated nor urinated on any of the 12 trials) to 12 (*S* either defecated or urinated during each trial). Males and females with high scores were mated to produce the first generation of the high emotional strain and those with low scores served as parents of the nonemotional strain. In succeeding generations, subjects with high scores within the emotional strain were selected, whereas in the nonemotional strain, only animals with low scores were mated. During 12 generations of such selection, a relatively consistent response was observed

[*] This investigation was supported in part by Public Health Service Research Grants GM-12486 and GM-15679, from the National Institute of General Medical Sciences. We thank E. A. Thomas for his assistance throughout the conduct of this experiment and Drs. P. S. Dawson, R. C. Roberts, and D. W. Fulker for commenting on a draft of the manuscript.
[†] For a more recent discussion of the validity of the open-field test as a measure of emotional response, see Broadhurst (1960).

in the high emotional strain. In contrast, maximum response was realized in the nonemotional strain after only one generation of selection.

Broadhurst (1960, 1967) has also utilized selective breeding as a method of genetic analysis in the study of open-field behavior in rats. However, the criterion of selection was different (total number of fecal boluses dropped during 2 minutes on each of four daily tests), and the apparatus was considerably smaller (32.75 inches in diameter). Nevertheless, a considerable divergence in the defecation scores of the "reactive" and "non-reactive" strains occurred during the first 10 generations of selection. In addition, a divergence in ambulation scores was also observed, indicating a negative association between defecation and activity in the open field. A negative correlation between ambulation and defecation in the open field had been previously reported by Hall (1936b). A critique of these and other selection experiments involving behavioral traits has been presented elsewhere (DeFries, 1967).

Heritable differences in behavior of mice in novel situations have been indicated by the simpler, but less informative, method of strain comparison. Striking differences in activity level among 15 strains of mice were found by Thompson (1953) when subjects were tested in a square (30 × 30 inches) arena. The floor of this apparatus was divided into 25 squares, and barriers were placed at the base of every other square. The number of squares traversed by a subject during a 10-minute test period was used as the activity score. Five of the 15 strains which were found to differ widely in this behavior were later tested for activity level in both a Y-maze and the arena (Thompson, 1956). With regard to arena activity, the same rank order of the strains was observed and, with only one exception, the rank order of the strains in the Y-maze corresponded to that in the arena.

McClearn (1959) has also tested for the generality of strain differences in activity level. The activity level of six inbred strains of mice and an F_1 hybrid was observed in four different kinds of apparatus: an arena similar to that employed by Thompson (1953), an open-field, a barrier apparatus, and a "hole in wall" apparatus. In general, wide differences were found among the strains and a high consistency in strain rank across the four tests was observed. McClearn (1961) has also obtained arena activity data on members of two inbred strains of mice (*C57BL/Crgl* and *A/Crgl*) which differ widely in activity level, as well as on animals of their derived F_1, F_2, and backcross generations. In general, it was concluded that perhaps more than half of the F_2 variance was due to heritable differences. The results of this experiment are in close agreement to those reported by Fuller and Thompson (1960) in which similar strains and apparatus were employed.

Although somewhat different test procedures were employed in the studies cited above, it is clear that individual differences in level of activity and defecation in an unfamiliar situation are a function of heritable differences.

In addition, open-field behavior can be objectively and efficiently measured, permitting a large sample size and thereby satisfying an important requisite of genetic analysis.

Because of the paucity of information currently available regarding the inheritance of behavioral traits and because of the importance of a character which may be related to temperament or emotionality, a genetic analysis of open-field behavior was undertaken. The primary objective of the study to be reported in this chapter was to conduct a detailed genetic analysis of differences in open-field behavior among two inbred strains of mice and among members of their derived generations. Emphasis in this study was placed on open-field activity, rather than defecation. Nevertheless, evidence for genetic and environmental associations between these behaviors will be reported, as well as evidence for a major gene effect on both characters.

QUANTITATIVE GENETIC ANALYSIS

Concepts and Methodology

Quantitative genetics is concerned primarily with the inheritance of characters whose expression is influenced by many genes, the effects of which are not readily discernible. Although the individual effects of these "polygenes" are not usually measurable, their aggregate importance may be measured by their relative contribution to the phenotypic variability in a population. Since most behavioral traits are continuously variable in expression and are probably influenced by genes at many loci, the methods of quantitative genetics are particularly applicable to the analysis of individual differences in behavior.

PARTITIONING THE PHENOTYPIC VARIANCE—The phenotype of an individual is a function of both its genotype and the environment in which it develops, i.e., $P = f(G,E)$. In quantitative genetic theory, the following linear mathematical model may be assumed:

$$P = G + E$$

where P is the phenotypic value of some trait measured on an individual, G is the genotypic value expressed in metric units and is due to the genotype of the individual, and E is an environmental deviation, i.e., a positive or negative deviation caused by environmental variations such that the mean environmental deviation for a trait in a population is zero. Therefore, assuming that G and E are uncorrelated (see Roberts, 1967, for a discussion of the validity of this assumption), the phenotypic variance (V_P) may be simply expressed in terms of the genotypic variance (V_G) and the environmental variance (V_E) as follows:

$$V_P = V_G + V_E$$

The genotypic variance may also be partitioned into components attributable to different causes. The gene, not the genotype, is the unit of transmission. Therefore, the resemblance between relatives is due to the average effects of the genes that influence the character under study. The sum of the average effects of the genes carried by an individual is defined as the additive genetic value (A) or breeding value. Nonlinear interactions may occur between alleles at a given locus causing the genotypic value to deviate from this additive genetic value. Deviations caused by such intralocus interactions are defined as dominance deviations (D). In a similar manner, nonlinear interactions may occur between alleles at different loci giving rise to epistatic interactions (I). Therefore, in principle, each allele has an average effect for each character and, when summed, these average effects result in an expected or additive genetic value. Dominance and epistasis, however, may cause the genotypic value to deviate from this expected value. Symbolically,

$$G = A + D + I$$
$$\text{and } V_G = V_A + V_D + V_I$$

where V_A, V_D, and V_I represent the additive, dominance, and epistatic variance, respectively. The additive genetic variance is thus due to the average effects of genes in a population, whereas dominance and epistasis give rise to nonadditive genetic variance. (See Lush, 1948; Falconer, 1960; for a more detailed presentation of the principles underlying the partitioning of the genotypic variance.)

One source of variation which has been disregarded in the above discussion is that due to genotype-environment interactions, i.e., the differential effects of the environment on different genotypes. Except under special conditions (Falconer, 1960), variance due to genotype-environment interactions cannot be separately measured.

HERITABILITY.—Partitioning the phenotypic variance provides a solution to the nature-nurture problem. Although the vigor of the nature-nurture controversy has declined since the 1920's (Fuller and Thompson, 1960), the problem still exists. Dobzhansky has made this point most lucidly in his recent discussion of *Heredity and the Nature of Man*:

"Furthermore, it gradually came to be recognized that the question whether the nature or the nurture, the genotype or the environment, is more important in shaping man's physique and his personality is simply fallacious and misleading. The genotype and the environment are equally important, because both are indispensable. . . . The nature-nurture problem is nevertheless far from meaningless. Asking right questions is, in science, often a large step toward obtaining right answers. The question about the roles of the genotype and the environment in human development must be posed thus: To what extent are the *differences* observed among people conditioned by the differences of their genotypes and by the differences between the environments in which people were born, grew and were brought up?" (Dobzhansky, 1964, p. 55).

Or, as Roberts (1967) has recently stated, "We need to know how much of the total variation (in a population) is due to various genetic causes, for it is axiomatic that the importance of a source of variation is proportional to the contribution it makes to the total variation."

Heritability (symbolized h^2) was first defined by Lush (1940) "as the fraction of the observed variance which was caused by differences in heredity." More recently, heritability has been defined (cf. Falconer, 1960) as the ratio of the additive genetic variance to the phenotypic variance, i.e., $h^2 = V_A/V_P$. Heritability is thus an index of the extent to which individual differences are caused by differences among the additive genetic values of members of a population. It is important to note that heritability is a population parameter and not a constant. Heritability may vary from population to population for the same trait, from one trait to another in the same population, or from time to time for the same trait in the same population (Lush, 1940).

The heritability of a character is one of the most important population parameters that may be estimated by the methods of quantitative genetics. In addition to its descriptive role, heritability also indicates the reliability of the phenotype as an index of the additive genetic value and thus may be used to predict the response to selection. Heritability is predictive, as well as descriptive, since it is equivalent to the regression of the additive genetic value of an individual on its phenotypic value (see Roberts, 1967). Nevertheless, relatively few estimates of the heritability of behavioral traits are available: "In terms of application of current genetic theory and procedure, behavioral genetics lags behind. . . . Further development of behavioral genetics will require the precise estimation of the heritabilities of a broad range of behavior patterns" (McClearn, 1963, p. 234).

Various methods are available for estimating heritability, some of which have been employed in behavioral genetics. If the additive genetic value of a character were known for each individual in a population, the additive genetic variance and hence, the heritability, could be calculated directly. However, less direct methods are necessary for the estimation of the heritabilities of polygenic characters. For example, Hirsch and Boudreau (1958) and Hadler (1964) compared the variances in isogenic and segregating populations to estimate the heritability of phototaxis in *Drosophila*. The methods of Mather (1949) which employ data from parental, F_1, F_2 and backcross generations have been used by Broadhurst and Jinks (1961) to estimate the heritabilities of several behavioral traits in a reanalysis of previously published data. However, it is important to note that the two definitions of heritability employed by Broadhurst and Jinks (1961) differ from that employed in

this chapter. Their first definition $(\frac{D}{D + E_1})$ estimates $2V_A/(2V_A + V_E)$, whereas their second definition $(\frac{\frac{1}{2} D + \frac{1}{4} H}{\frac{1}{2} D + \frac{1}{4} H + E_1})$ estimates $\frac{V_A + V_D}{V_A + V_D + V_E}$.

Sib correlations have been employed by Willham et al. (1963) in their study of the genetic variation in a measure of avoidance learning in swine. Recently, the diallel cross method has been used to estimate the heritabilities of various behavioral traits in both rats and *Drosophila* (Jinks and Broadhurst, 1965; Fulker, 1966).

Lush (1940) has suggested that the resemblance between parents and offspring is generally the most useful method for estimating heritability. Jinks and Broadhurst (1965) have utilized parent-offspring correlations to estimate the heritability of open-field ambulation and defecation in rats; however, the sample size was very small (70 male and 64 female offspring and only 19 parents) and, thus, the resulting estimates fluctuated rather widely.

The estimation of heritability by the regression of offspring on parent has several advantages over the method of parent-offspring correlation. For example, if the parents represent a selected group or if assortative mating has occurred, the correlation of parent and offspring scores will yield a biased estimate of heritability, whereas estimates from the regression of offspring on parent will not be systematically biased by such conditions (see Falconer, 1960). In addition, when progeny means or the average of male and female parents are used, the regression of offspring on parent is a simple function of heritability, whereas the parent-offspring correlation is not. However, this method and others which utilize the comparison of relatives will yield biased estimates of heritability if environmental deviations of the relatives are correlated (see DeFries, 1967, for a discussion of these problems). Although the regression of offspring on parent has been widely used in animal genetic research for the estimation of heritability, this technique has been infrequently employed in behavioral genetics. However, Connolly (1966) has recently used this method to estimate the heritability of locomotor activity in *Drosophila*.

Just as heritability may be used to predict the response to selection, the results of selection may be used to estimate the "realized heritability." Manning (1961) has utilized this technique to estimate the heritability of mating speed in *Drosophila* and Craig et al. (1965) and Siegel (1965) have used this method to estimate the heritabilities of various behaviors in chickens.

Unfortunately, the comparative approach in which estimates of heritability are obtained by several methods from the data of one population has rarely been employed. Jinks and Broadhurst (1965) have compared several methods (correlation of littermates, parent-offspring correlation, a nested design which permits an estimate of half-sib and full-sib correlations, and diallel analyses) for the estimation of the heritabilities of open-field ambulation and defecation in rats. However, standard errors were not attached to the resulting estimates so that the precision of the estimates was not apparent. In addition (although a somewhat different population was involved), estimates for the heritability of defecation were not compared to the realized heritability which could have been obtained from the results of Broadhurst's (1960, 1967) long-term selection experiment. A model for the comparative approach has been provided by Dawson (1965) in his extensive analysis of the components of the phenotypic variance of developmental rate in *Tribolium*.

GENETIC CORRELATION.—The correlation of the phenotypic values of two different characters on the same individuals in a population also has both hereditary and environmental causes. For example, assume two characters, X and Y, have additive genetic values A_x and A_y. The correlation of these additive genetic values is referred to as the "genetic correlation" and may be symbolized r_A. Environmental deviations associated with these characters, symbolized E_x and E_y, respectively, may for the sake of simplicity be assumed to also contain any deviations caused by nonadditive genetic effects. The correlation which may exist among these environmental deviations is symbolized r_E. The phenotypic correlation (r_P) of two such characters in a population is a function of the genetic and environmental correlations as follows:

$$r_P = h_x h_y r_A + e_x e_y r_E$$

where h_x and h_y are the square roots of the heritabilities of characters X and Y, $e_x = (1 - h_x^2)^{1/2}$, and $e_y = (1 - h_y^2)^{1/2}$. (See Falconer, 1960, for a derivation of this formula.)

A genetic correlation between two characters in a population may be due to pleiotropy, i.e., the manifold effects of genes, or to a linkage disequilibrium among genes at different loci which influence different traits. However, the effects of linkage which cause such a correlation should be relatively temporary. Genetic correlations may be estimated by methods analogous to those used to estimate heritability (see Falconer, 1960). Although most of these methods have not been employed in behavioral genetics, Siegel (1965) has used the correlated response to selection to estimate genetic correlations between "cumulative number of completed matings" and various other traits of chickens.

Preliminary Analysis

Open-field behavioral data were obtained on members of two inbred parental strains of mice, as well as their derived F_1, backcross, F_2, and F_3 generations. From the data of 2,641 Ss which were included in this preliminary analysis, the heritabilities of open-field activity and defecation were estimated by the following methods: (1) a "classic" analysis which employed the data of the parental, backcross, and F_2 generations; (2) regression of scores of F_3 Ss on their F_2 parents; and (3) half-sib correlations among members of the F_3 generation. The genetic correlations among these behaviors and body weight were estimated by the "cross-covariance" of scores of F_3 Ss and their F_2 parents and by analysis of covariance of scores among half-sib families in the F_3 generation. The resulting estimates will be compared to those obtained from the response to selection which will be described in a later section.

It should be emphasized that the results of these analyses will be relative only to the population under study, i.e., a random mating population of mice derived from an initial cross of two particular inbred strains. The results may not be generalized to some hypothetical population of "mice in general," especially since the strains were not selected at random. However, the results should be of interest since the estimates of heritability and genetic correlations obtained by various methods all refer to the same population so that a comparison of the validity of the methods will be possible. Only additional research will indicate the generality of these results.

SUBJECTS.—Two inbred strains of mice (*BALB/cJ* and *C57BL/6J*) which were known from previous work (Weir and DeFries, 1964; DeFries et al., 1967b) to differ widely in open-field behavior were chosen as the parental strains for this experiment. Ten males and 20 females from each strain were mated (two females per male) within strain to produce two groups of inbred progeny, *BALB/cJ* and *C57BL/6J* (symbolized *B* and *C*, respectively). In addition, 20 males and 40 females from each strain were crossed to produce reciprocal F_1 hybrid progeny (symbolized *BC* and *CB*, where *BC* refers to hybrid progeny produced by a *B* male and a *C* female parent and *CB* refers to hybrid offspring resulting from the reciprocal cross). Thus, data were obtained from members of both inbred parental strains which were reared contemporaneously with members of the F_1 generation.

The procedure of mating two females per male was continued throughout the various generations included in the preliminary analysis. Females were removed from mating cages when they were obviously pregnant and were caged individually. Resulting offspring were weaned at 35 days of age.

In order to produce Ss representing the next generation, 20 males and

40 females from each of the F_1 groups (BC and CB) were mated in all possible combinations at approximately 60 days of age, resulting in four groups of F_2 animals (symbolized $BC \times BC$, $BC \times CB$, $CB \times BC$, and $CB \times CB$, where the male parent is listed first, e.g., $BC \times CB$ refers to F_2 Ss resulting from mating BC males with CB females). In addition, 20 females from each F_1 group were mated to inbred $BALB$ males to produce two groups of Ss within the B_1 backcross generation (symbolized $B \times BC$ and $B \times CB$). In a similar manner, 20 females from each hybrid group were mated to inbred $C57BL$ males, resulting in two groups of B_2 backcross progeny ($C \times BC$ and $C \times CB$).

Seventy-two F_2 litters were obtained which were later tested in the open field. Where possible, one male and two females were chosen at random from each of these F_2 litters and then mated at random to produce Ss representing the F_3 generation. When two females or a male were not present in a litter, a substitute was selected at random from among the other litters of that group. Since each of the F_2 males was mated to two females, the resulting 128 litters in the F_3 generation consisted of a number of half-sib and full-sib families. Fifty half-sib families (progeny of a given male parent) and 119 full-sib families (litters) which contained at least one female were available in the F_3 generation, and 49 half-sib families and 119 full-sib families were available which contained at least one male. All analyses were conducted on an intrasex basis, i.e., separately for male and female progeny.

OPEN-FIELD TESTING.—All Ss in the preliminary analysis and the selection experiment were tested in an automated open field and almost always by the same experimenter (E. A. Thomas). The field is square (36 by 36 inches), and is of white, painted Plexiglas. The walls are 8 inches high with holes 5 mm in diameter drilled at 6-inch intervals at a height of 15 mm. Two sets of five light sources each are located on adjacent sides of the field and beamed through infrared filters and holes to photo-conductive cells on the opposite side, effectively dividing the floor into 36 squares, 6 by 6 inches each. Interruption of light beams activates counters which are located to the side of the field. Illumination during testing is provided by two 20-watt fluorescent tubes mounted 37 inches above the floor of the field. Illumination readings are similar in both the center and corners of the field (approximately 48 ftc., incident light).

Beginning at 40 ± 5 days of age, each S in each generation and group was tested for 3 minutes on each of 2 successive days. Each S was placed in a clear Plexiglas cylinder in one corner of the field and then released at the start of the test period. The system was dependent upon a timer which inactivated the counters after a period of 3 minutes. The total number of light beams interrupted during the two 3-minute test periods was used as each S's activity score. In addition, the total number of fecal boluses dropped was also recorded. Immediately following the second test period, each S was weighed to the nearest 0.5 gram.

Due to the presence of large heterogeneous group variances in the raw data, open-field activity and defecation scores were subjected to a square root transformation for subsequent analysis. Because mean defecation scores were low, 0.5 was added to each S's defecation score and a square root transformation was then applied to the resulting value.

GROUP MEANS.—The means and variances of the transformed open-field behavioral scores and N's of male and female Ss in each of the 13 groups included in the preliminary analysis are presented in Table 1. Large strain differences in both activity and defecation are immediately apparent. These differences between the strains are quite consistent with those found in our previous studies. Females are somewhat more active than males, especially in the segregating F_2 and F_3 generations. In contrast, males have consistently higher defecation scores than females as may be noted in each of the 13 possible comparisons of the sexes in Table 1.

By comparing the means of the F_1 groups with those of the parental groups, evidence for directional dominance may be observed in activity, but not in defecation scores. In addition, by comparing the means of the reciprocal cross F_1 groups (BC versus CB) the presence of a small, but significant, maternal effect is found: hybrid offspring produced by B mothers have lower activity ($t = 3.53$, d.f. $= 411$, $p < 0.001$) and higher defecation scores ($t = 3.49$, d.f. $= 411$, $p < 0.001$) than those produced by mothers of the C strain. Since the magnitude of these differences is approximately the same in both male and female hybrids, sex linkage may be discounted as a possible cause of this difference. With regard to activity, this maternal effect persists in the backcross and F_2 generations, i.e., backcross and F_2 Ss which have CB mothers have lower activity scores than those from BC mothers, although the differences become quite small. With regard to defecation, however, no such effect was noted in the backcross and F_2 generations.

In a separate experiment (DeFries et al., 1967a), the role of the maternal environment on the open-field behavior of inbred progeny of these two strains was assessed. Ovaries from donors of each of the two inbred strains were transplanted into hybrid recipients and, by appropriate matings, inbred offspring were obtained. In addition, inbred offspring were obtained by within-strain matings. Therefore, it was possible to compare the behavior of inbred offspring of the same strain which were carried by mothers of different strains (inbred versus hybrid) and to compare inbred offspring of different strains which were carried by mothers of the same strain (hybrid). In general, differences in the maternal environment accounted for a relatively small proportion of the variance in open-field activity and defecation, but accounted for a relatively large proportion of the variance in body weight.

TABLE 1

Means and Variances of Transformed Open-field Behavioral Scores

Generation	Group[a]	Activity[b]				Defecation[c]				N	
		Means		Variances		Means		Variances			
		Males	Females	Males	Females	Males	Females	Males	Females	Males	Females
Parental	B	4.67	4.25	6.33	4.27	2.99	2.79	0.33	0.42	30	34
	C	15.73	16.40	4.86	6.56	1.10	1.05	0.29	0.19	53	59
F_1	BC	15.09	15.04	6.70	10.37	2.16	1.78	0.47	0.38	134	128
	CB	13.98	13.66	15.80	21.06	2.36	2.05	0.54	0.48	65	88
B_1	BXBC	9.93	9.62	11.97	15.28	2.46	2.33	0.45	0.41	53	65
	BXCB	8.33	8.44	15.38	17.74	2.52	2.33	0.49	0.55	74	65
B_2	CXBC	15.48	15.34	5.03	7.39	1.38	1.14	0.36	0.27	77	80
	CXCB	14.93	14.70	12.12	6.77	1.40	1.19	0.52	0.38	83	75
F_2	BCXBC	11.88	12.63	17.96	9.23	1.99	1.50	0.62	0.43	75	68
	BCXCB	11.76	12.30	16.56	13.68	1.93	1.62	0.52	0.52	90	80
	CBXBC	12.66	12.86	15.87	16.44	1.88	1.80	0.60	0.52	82	73
	CBXCB	11.58	11.84	14.22	16.28	2.12	1.63	0.62	0.55	83	86
F_3	F_3	11.58	12.13	21.57	19.78	2.08	1.89	0.62	0.53	407	434

[a]B, C, BC, and CB refer to BALB/cJ, C57BL/6J, hybrid F_1 produced by B male and C female parents, and reciprocal cross F_1 Ss, respectively. In subsequent groups, male parent is symbolized first.

[b]Mean activity scores were obtained from transformed data, where each S's score is the square root of the sum of the activity over the 2-day test period.

[c]Mean defecation scores were obtained from transformed data, where $X = (\text{Total Boluses} + 1/2)^{1/2}$.

33

GROUP VARIANCES.—The variances of both the transformed activity and defecation scores are homogeneous across sex and across groups within the parental, B_1, B_2, F_2, and F_3 generations. However variances of the B_1 backcross groups are not homogeneous with those of the B_2 backcross groups: the pooled variances of the B_1 backcross groups for both activity and defecation are larger than those of the B_2 groups, indicating that directional dominance may be associated with high activity and low defecation scores. Evidence for directional dominance in activity was previously found by comparing the means of the parental and F_1 generations.

Variances associated with the two F_1 reciprocal cross groups are inexplicably heterogeneous. Although this unexpected result may be indicative of a scalar problem, a square root transformation has been required in this and previous studies to give homogeneous variances of scores of the two inbred parental strains, the means of which are markedly different. Variances associated with the parental and F_1 generations are also heterogeneous, suggesting the possibility of a genotype-environment interaction. Variances in activity scores of the F_3 generation are somewhat larger than those of the F_2 generation. Since mothers of F_3 Ss are members of the F_2 segregating generation, whereas mothers of F_2 Ss are isogenic, this greater variance in the F_3 generation may be due to a greater maternal variation.

A comparison of the variances of scores of the F_1 and F_2 Ss is most interesting. When the variances associated with groups within the F_1 and those within the F_2 generations are pooled, the resulting estimates of the variance of activity scores in the F_1 and F_2 generations are 12.31 and 15.20, respectively. The pooled estimates of the variance of defecation scores in the F_1 and F_2 generations are 0.452 and 0.548, respectively. For both activity and defecation scores, the variances of the F_2 population are significantly greater than those of the F_1 ($p < 0.05$, d.f. $= 411$; 629).

"TRYON EFFECT."—The above results provide no evidence for the presence of the "Tryon effect" recently discussed by Hirsch:

> "Tryon bred rats selectively on the basis of their error scores in learning a multiple T-maze. Three times between the eleventh and twenty-second generations of selection the reproductively isolated 'bright' and 'dull' strains were test-crossed to produce F_1 and F_2 progeny. According to (naive) Mendelian theory, because of segregation in the F_2, its variance should be detectably larger than that of the F_1. All three times this expected result failed to appear. This failure to obtain an increase in variance from the F_1 to the F_2 generation has been called the 'Tryon effect.' It has happened often, though not invariably, in behavior-genetic studies" (Hirsch, 1967, pp. 417–418).

Hirsch has attributed the failure to achieve an increase in variance from the F_1 to F_2 generation to insufficient sample size: "Any experiment intended to sample the spectrum of possible genotypes must be planned so that there is a statistically sufficient number of replications of the appropriate genotype matrix" (p. 419). However, an unbiased estimate of the variance may be obtained by a random sample drawn from a population, i.e., it is not necessary to replicate an entire genotype matrix in order to obtain an unbiased estimate of the variance; thus, the failure to demonstrate an increase in variance from the F_1 to the F_2 generation must have an alternative explanation.

It may be shown that the increase in variance from F_1 to F_2 is inversely proportional to the number of loci that influence the character under investigation. In other words, for any fixed range between the means of the parental strains, the larger the number of loci, the smaller will be the difference between the F_1 and F_2 variances. For simplicity, consider two parental strains, with means which differ by some value R. Let us assume that this range (R) is due to equal effects at each of N unlinked loci and that all of the genes which add to the phenotypic value are fixed in one parental strain, whereas their corresponding alleles are fixed in the other strain. If one considers only additive effects, it may be shown that the increase in the F_2 variance over that of the F_1 is as follows:

$$V_{F_2} - V_{F_1} = \frac{R^2}{8N}$$

This is merely a variation of the expression used by Mather (1949) and others to estimate the number of loci which influence a trait. It may also be shown that if complete dominance exists at each of these N loci, this increase in variance is $3R^2/(16N)$. Both expressions indicate that the increase in variance from F_1 to F_2 is inversely proportional to the number of loci which influence the trait. Thus, when traits which are influenced by a large number of loci are under investigation, a large sample size is needed to demonstrate a significant increase in variance from the F_1 to the F_2 generation: a large sample size is required to demonstrate a significant increase because the magnitude of the difference is small, not because a large sample will yield replication of the entire genotype matrix. This concept has previously been discussed in some detail by Bruell (1962).

HERITABILITY ESTIMATION.—(1) *Classic analysis.* The A, B, and C scalar tests of Mather (1949) for the presence of interactions between alleles at different loci (epistasis) were applied to the data of the parental, F_1, B_1, B_2, and F_2 generations. The results of these tests indicate that epistasis is not present in the transformed activity scores. However, the B and C tests applied to the defecation data were significant, indicating that epistatic interactions may be present in these transformed data. Nevertheless, since the square root transformation results in homogeneity of the variances of the parental scores, this transformation was employed in subsequent analyses. The difficulty of fulfilling all scalar criteria has been previously discussed by Dawson (1965).

When the scalar tests are satisfied and when no genotype-environment interactions are present in the data, the phenotypic variance may be subdivided into its additive, dominance and environmental components. The method outlined by Rasmuson (1961) was applied to the data of this experiment. This reference is recommended to the neophyte since the symbolism employed conforms closely to that utilized by Falconer (1960).

Variances were pooled across groups within the B_1, B_2, and F_2 generations and, since the variances of the reciprocal cross F_1 groups were heterogeneous, the environmental variance (V_E) was estimated from the pooled parental variance. The variance of the F_2 groups will reflect all the genetic and environmental sources of variation, i.e.,

$$V_{F_2} = V_A + V_D + V_E$$

It may be shown that the sum of the B_1 and B_2 variances is as follows:

$$V_{B_1} + V_{B_2} = V_A + 2V_D + 2V_E$$

Thus,

$$V_A = 2V_{F_2} - (V_{B_1} + V_{B_2})$$

and

$$V_D = V_{F_2} - V_A - V_E$$

These genetic and environmental components of variance were estimated from the activity and defecation data of the present study and were used to estimate both the heritability (V_A/V_{F_2}) and the coefficient of genetic determination $(V_A + V_D)/V_{F_2}$. The resulting estimates for both males and females are presented in Table 2. Standard errors (Osborne and Paterson, 1952) were estimated using the number of litters, rather than the number of Ss, in calculations of the variance of sample variances. The coefficient of genetic determination associated with defecation scores of males is smaller than heritability because the estimate of dominance variance was negative.

TABLE 2

Heritability Estimates from Classical Analysis

Open-field Behavior	Heritability		Coefficient of Genetic Determination	
	Males	Females	Males	Females
Activity	0.58 ± 0.06	0.28 ± 0.04	0.63 ± 0.06	0.49 ± 0.06
Defecation	0.42 ± 0.07	0.36 ± 0.06	0.39 ± 0.06	0.38 ± 0.06

TABLE 3

Heritability Estimates from Regression of Offspring on Midparent Values

Open-field Behavior	Males on Midparent	Females on Midparent	Pooled Estimate
Activity	0.24 ± 0.12	0.19 ± 0.12	0.22 ± 0.09
Defecation	0.04 ± 0.09	0.17 ± 0.08	0.11 ± 0.06

(2) *Regression analysis*. The covariance of offspring and parent contains one-half of the additive genetic variance plus small parts of the epistatic variance (see Falconer, 1960); thus, with a random mating population, heritability may be estimated from the regression of offspring on midparent score or by doubling the regression of offspring on male or female parents. In the present analysis, the heritabilities of open-field activity and defecation were estimated from the regression of scores of male and female F_3 Ss on midparent values and from the pooled, within-sex regression. These regressions were weighted according to the number of offspring in a litter since litter sizes were variable (see Dickerson, 1959). The resulting estimates and their standard errors are presented in Table 3. Standard errors were calculated according to Falconer (1960), using the variance of the litter means and the number of litters as N.

(3) *Sib analysis*. Analyses of variance were performed on the data of the F_3 Ss which comprise a number of half-sib and full-sib families as mentioned previously. Components of variance due to differences between progeny of different sires (σ^2_s), between progeny of dams mated to the same sire (σ^2_d) and between progeny within dams (σ^2_w), were estimated (see Dickerson, 1959; Falconer, 1960; for discussion of this hierarchical analysis). The component of variance between sires (σ^2_s), is an estimate of the covariance of half-sibs (Cov HS) and contains $\frac{1}{4}$ of the additive genetic variance plus small parts of the epistatic variance. Thus heritability may also be estimated as follows: $h^2 = 4(\text{Cov } HS)/\sigma^2_T$, where σ^2_T is the total variance and is an estimate of V_P. The resulting estimates for males, females, and a pooled estimate for both activity and defecation are presented in Table 4.

TABLE 4

Heritability Estimates from Half-Sib Correlations

Open-field Behavior	Males	Females	Pooled Estimate
Activity	0.50 ± 0.32	−0.25 ± 0.31	0.14 ± 0.14
Defecation	0.30 ± 0.32	−0.29 ± 0.31	0.02 ± 0.08

The intraclass correlation (t_{FS}) among littermates (full-sibs) was estimated as follows:

$$t_{FS} = \frac{\sigma^2_s + \sigma^2_{d\cdot}}{\sigma^2_T}$$

The resulting estimates, pooled across sex of offspring, were 0.37 and 0.22 for activity and defecation, respectively. The covariance of full-sibs contains $\frac{(\frac{1}{2}V_A + \frac{1}{4}V_D + V_{E_c})}{(V_P)}$, where V_{E_c} is the variance due to environmental deviations which are common to littermates. Since the phenotypic correlations of full-sibs clearly exceed estimates of $\frac{1}{2}h^2$ obtained in the same analysis, evidence for an environmental correlation among littermates is indicated; however, dominance variance may also account for some of this difference.

(4) *Comparison of h^2 estimates.* It is of interest to compare the various estimates of heritability which resulted from the three methods employed in the preliminary analysis: estimates from the classic analysis are relatively high for both activity and defecation. However, estimates obtained from the regression of offspring on parents are considerably lower. As indicated by their standard errors, both methods appear to yield precise estimates. Estimates from half-sib correlations are imprecise, but pooled estimates are low and relatively more precise.

Several assumptions underlying the classic analysis were not fulfilled in the present data. As mentioned previously, epistatic interactions may exist in the defecation data, as indicated by the B and C scalar tests. In addition, variances of the parental and F_1 generations were not homogeneous, indicating the presence of genotype-environment interactions. This heterogeneity indicates that an assumption that $V_{P_1} = V_{P_2} = V_{F_1} = V_E$ is not warranted. Since the variances of the reciprocal cross groups in the F_1 generation of the present study were heterogeneous, V_E was estimated from the pooled parental variance. However, only backcross and F_2 variances are used to estimate heritability form the classic analysis; thus, any bias in estimate of V_E will not systematically bias estimates of heritability. On the other hand, the coefficient of genetic determination may be systematically biased, since underestimates of V_E will result in overestimates of V_D.

It is possible that estimates of heritability from classic analysis may be inflated in the present study due to a linkage disequilibrium in F_2 data. The presence of such a disequilibrium may result in an overestimate of the additive genetic variance since the means of the parental strains differed widely. However, such a disequilibrium may affect estimates obtained from the other methods to a lesser extent.

Results from carefully designed selection experiments are needed to evaluate the validity of heritability estimates (Dickerson, 1959). The validity of the estimates obtained from the preliminary analysis will be assessed by comparing them to realized estimates obtained from the response to selection in a later section.

CORRELATION ESTIMATION.—Phenotypic correlations among activity, defecation, and body weight, calculated within each generation, group, and sex, are presented in Table 5. Phenotypic correlations between activity and defecation scores are consistently negative and 23 of the 26 estimates are significant. The correlations between these two open-field behavioral scores tend to be somewhat larger in the segregating populations than in the isogenic populations, suggesting the presence of a genetic correlation.

Correlations between activity and body weight are variable in sign; 14 of the 26 estimates are positive, 12 are negative, and only 4 are significant. Correlations between defecation and body weight are even more variable in sign (13 positive, 13 negative, and only one of the 26 values is significant). Thus, there is a consistently negative association between activity and defecation in the open field, but little or no evidence for an association of these behaviors with body weight.

Genetic correlations between activity, defecation and body weight were

TABLE 5

Phenotypic Correlations Among Open-field Activity, Defecation and Body Weight in Two Inbred Strains of Mice and Their Derived Generations

Generation	Group[a]	Activity and Defecation		Activity and Body Weight		Defecation and Body Weight	
		Males	Females	Males	Females	Males	Females
Parental	*B*	−0.235	−0.058	−0.015	−0.104	0.137	−0.208
	C	−0.424[b]	−0.302[c]	0.263	0.145	0.091	−0.182
F_1	*BC*	−0.370[b]	−0.480[b]	−0.145	−0.140	0.015	0.078
	CB	−0.397[b]	−0.489[b]	0.278[c]	0.070	0.297[c]	0.049
B_1	*B×BC*	−0.451[b]	−0.195	−0.062	−0.205	0.003	−0.102
	B×CB	−0.549[b]	−0.382[b]	−0.107	0.025	0.209	0.018
B_2	*C×BC*	−0.358[b]	−0.524[b]	−0.067	0.098	0.129	0.078
	C×CB	−0.458[b]	−0.297[c]	0.215	−0.063	−0.032	−0.021
F_2	*BC×BC*	−0.466[b]	−0.463[b]	0.288[b]	0.056	−0.200	0.020
	BC×CB	−0.403[b]	−0.450[b]	0.095	0.122	0.020	−0.066
	CB×BC	−0.247[c]	−0.428[b]	−0.053	−0.003	−0.207	−0.182
	CB×CB	−0.337[b]	−0.481[b]	0.131	−0.096	−0.153	−0.108
F_3	F_3	−0.467[b]	−0.497[b]	0.110[c]	−0.122[c]	−0.076	−0.046

[a](See Table 1 for explanation of group symbols.)
[b]Significant at 0.01 level of probability.
[c]Significant at 0.05 level of probability.

estimated by the method of cross-covariance (Falconer, 1960) of F_3 Ss and their F_2 parents. The estimate of the genetic correlation between activity and defecation is negative and relatively large (-0.76 ± 0.14) indicating that both behaviors are influenced by many of the same genes. The genetic correlations between activity and body weight and defecation and body weight, however, were small and relatively unreliable, -0.34 ± 0.24 and -0.14 ± 0.31, respectively.

Genetic correlations were also estimated from analyses of covariance of the data representing the half-sib and full-sib families in the F_3 generation. However, these estimates were quite inconsistent and subject to large standard errors (Hegmann, 1966).

As mentioned previously, the genetic correlation may be estimated by methods analogous to those used to estimate heritability. Therefore, an estimate of the "realized" genetic correlation may be obtained from the correlated response to selection. Estimates of the realized genetic correlation between open-field activity and defecation and between activity and body weight were obtained from the results of a selection experiment and the resulting values will be compared to those from the preliminary analysis in a later section.

As indicated earlier, the phenotypic correlation (r_P) between two characters (X and Y) is a function of their heritabilities (h^2_x and h^2_y), their genetic correlation (r_A) and their environmental correlation (r_E). Since all of these variables except r_E have been estimated from the data of the preliminary analysis, it is possible to estimate the environmental correlation between open-field activity and defecation. Given that the phenotypic correlation between open-field activity (X) and defecation (Y) is -0.4 and that $h^2_x = 0.2$, $h^2_y = 0.1$ and $r_A = -0.75$, an estimate for r_E of -0.35 is obtained. Thus, there is a negative correlation between the environmental deviations which are associated with open-field activity and defecation. This conclusion may also be reached by examining the phenotypic correlations between these behaviors in the isogenic parental and F_1 generations listed in Table 5. Since no genetic variation should exist within these populations, any phenotypic correlation observed should be due exclusively to environmental causes. The pooled phenotypic correlation between activity and defecation in these isogenic populations, calculated on a within-generation, group and sex basis, is -0.40, a value similar to that of r_E estimated above from the data of the segregating populations.

Selection Experiment

The foundation population for the selection experiment consisted of 40 litters chosen at random (with the restriction that each litter contain at least two males and two females) from the F_3 generation of the preliminary analysis. The most active male and female and the least active male and female from each of 10 litters were selected. The resulting 10 high-active males and 10 high-active females were then mated at random at approximately 60 days of age so as to produce progeny representing the first selected generation (S_1) of one high active line (H_1). The 10 low-active males and females were also mated at random to produce the S_1 generation of a low active line (L_1). It should be noted that the parents of lines H_1 and L_1 were littermates. In a similar manner, high-active and low-active males and females were selected from each of 10 other litters and mated at random within level of activity, resulting in offspring which represent generation S_1 of lines H_2 and L_2. Then, one male and one female were chosen at random from each of 10 other litters and mated at randam to produce progeny of control line C_1. In the same manner, animals were chosen at random from the remaining 10 litters and mated at random, resulting in a second control line, C_2. Thus, six closed lines were established; two selected for high activity $(H_1$ and $H_2)$, two selected for low activity $(L_1$ and $L_2)$, and two unselected controls $(C_1$ and $C_2)$. The rationale for the design employed (two-directional selection, with replicated lines and controls) has been discussed elsewhere (DeFries, 1967).

Selection each generation was on a within-litter basis, i.e., the most active male and female from each litter in the high active lines and the least active male and female from each litter in the low active lines were selected and then mated at random within line to produce Ss representing the next generation. In the control lines, one male and one female were chosen at random from each litter and then mated at random within line. All matings were accomplished within a 1-week period so that litters would be produced and tested contemporaneously.

In each line, two extra males and two extra females were chosen from among the largest litters and mated to provide extra litters. When 10 litters were not produced by the parents selected in the manner outlined above, animals were selected from these extra litters to maintain an effective mating population of 10 parental pairs per line per generation. Within-litter selection was practiced in order to reduce inbreeding. With an effective population size of 10 mating pairs and within-litter selection, the rate of increase in the coefficient of inbreeding should be less than 1.5 percent per generation.

Test procedures were identical to those employed in the preliminary analysis. Selection was based exclusively on total activity during the 2-day test period, although data regarding open-field defecation and body weight were also recorded.

Five generations of selection were completed while the authors were at the University of Illinois, Urbana. The selection experiment has been relocated and is being continued at the authors' present address. However, only data from the first five generations of selection (a total of 2,179 Ss) will be presented in this report.

DIRECT RESPONSE.—The mean transformed open-field activity scores of the six lines from generation S_0 (foundation population) through S_5 are shown in Fig 1. In addition, the mean, N and cumulative selection differential for each line are presented in Table 6. The cumulative selection differential is a measure of the total amount of selection which has been applied during the experiment. Each generation, the mean of the animals selected to be parents is compared to the mean of the litters (prior to selection) from which they were chosen. The difference between these two means is the selection differential. These differences were weighted according to the numbers of offspring produced by each mating pair and summed across generations, resulting in the cumulative selection differential. This cumulative selection differential may then be compared to the direct response, i.e., the difference between the means of the selected and control lines, since the response is a measure of the effect of the total selection applied.

As indicated in Table 6 and Fig. 1, there is a clear and consistent response to selection. The means of the high and low lines are approximately of equal distance from the controls and, thus, there is little evidence for an

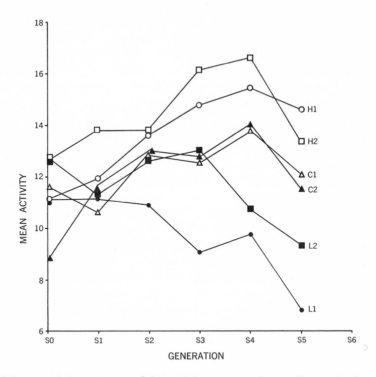

FIG. 1. Mean transformed open-field activity scores of two lines of mice selected for high activity (H_1 and H_2), two selected for low activity (L_1 and L_2), and two unselected controls (C_1 and C_2).

TABLE 6

Mean Transformed Open-field Activity Scores, N, and Cumulative Selection Differentials (CSD) of Six Lines in a Replicated, Two-directional Selection Experiment

Lines	S_0		S_1			S_2			S_3			S_4			S_5		
	Mean	N	Mean	N	CSD	Mean	N	CSD	Mean	N	CSD	Mean	N	CSD	Mean	N	CSD
H_1	11.2	75	12.0	74	3.3	13.6	72	7.1	14.8	88	9.2	15.4	71	11.7	14.5	70	14.7
H_2	12.6	90	13.8	76	3.1	13.8	69	7.0	16.1	82	9.2	16.6	79	12.6	13.2	74	15.3
C_1	11.6	93	10.6	83	–	12.7	52	–	12.5	88	–	13.8	76	–	12.0	67	–
C_2	8.9	74	11.4	62	–	13.0	54	–	12.9	71	–	14.0	72	–	11.5	77	–
L_1	11.2	75	11.2	74	3.1	10.9	63	7.8	9.0	79	10.4	9.8	73	13.6	6.8	66	18.0
L_2	12.6	90	11.2	59	3.7	12.6	70	5.0	13.0	82	7.9	10.8	85	10.8	9.3	71	14.4

asymmetrical response, at least during the first five generations of selection. However, considerable intergeneration variability exists which tends to affect all lines in the same direction. This phenomenon is often observed in selection experiments (see Falconer, 1960). The effect of this intergeneration variability may be removed by comparing the means of the selected lines to those of the controls or to each other. The difference between the means of the high and low lines represents the divergence of response to selection. The divergence between lines H_1 and L_1 and between H_2 and L_2 is plotted in Fig. 2 as a function of the cumulative selection differential. Since divergence is a measure of response which is relatively free of intergeneration variability affecting all lines in the same direction, the trend apparent in Fig. 2 is more systematic than that in Fig. 1.

CORRELATED RESPONSE.—The mean transformed defecation score and the mean body weight for each of the six lines in each generation are presented in Fig. 3 and 4. From Fig. 3 it may be seen that a clear and consistent correlated response in open-field defecation has accompanied selection for activity level. When Fig. 1 and 3 are compared, a complete reversal of rank among the high-active, control, and low-active lines is evident, again indicating a large, negative genetic correlation among these behaviors. In sharp contrast, little or no consistent trend is evident with regard to body weight.

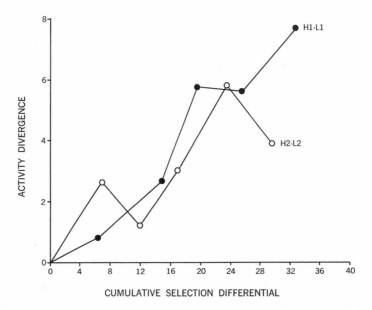

FIG. 2. Divergence of response in mean transformed open-field activity level ($H_1 - L_1$ and $H_2 - L_2$) plotted as a function of the cumulative selection differential.

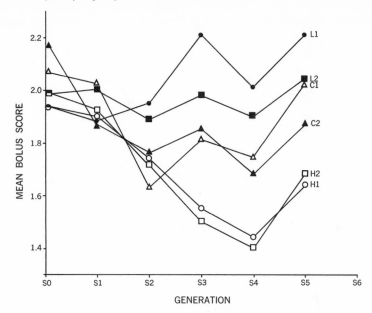

FIG. 3. Mean transformed open-field defecation scores of two lines of mice selected for high activity (H_1 and H_2), two lines selected for low activity (L_1 and L_2), and two unselected controls (C_1 and C_2).

REALIZED HERITABILITY.—When the heritability of a character and the selection differential (S) are known, the response (R) to selection may be predicted as follows:

$$R = h^2_i \; (S)$$

where h^2_i is the heritability of the particular index of selection employed. In the case of mass selection, the index is the phenotypic value of an individual without regard to the family from which it came and the heritability of that index is merely h^2. However, the present selection experiment utilized within-family selection, i.e., the index of selection was the deviation of the phenotypic value of an individual from the mean of the family or litter from which it came. Therefore, in the present case,

$$h^2_i = h^2 \; \frac{(1-r)}{(1-t)}$$

where r is the coefficient of relationship among members of the family ($r = 0.5$, since littermates are full sibs) and t is the phenotypic correlation of activity scores among litter mates, estimated previously to be 0.37.

The response to selection and the selection differential are available for each selected line in each generation of this experiment. Therefore, one

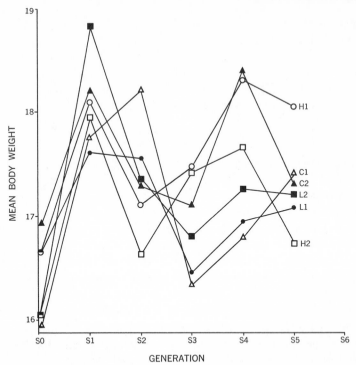

FIG. 4. Mean body weight (in grams) of two lines of mice selected for high activity (H_1 and H_2), two lines selected for low activity (L_1 and L_2), and two unselected controls (C_1 and C_2).

could estimate h^2_i from the ratio R/S for each generation of selection. However, a more reliable estimate may be obtained by utilizing the response over all generations of selection, i.e., by calculating the regression of the response on the cumulative selection differential. Since the response to selection appeared to be relatively symmetrical, h^2_i was estimated by calculating the regression of the divergence of response on the cumulative selection differentials for H_1 versus L_1, H_2 versus L_2, and from a pooled estimate. The resulting regression coefficients were then multiplied by $\dfrac{1-t}{1-r} = 1.26$, resulting in the estimates of h^2 which may be compared to those presented in the preliminary analysis. The following estimates are: H_1 versus L_1, 0.31 ± 0.04; H_2 versus L_2, 0.19 ± 0.07; and pooled, 0.26 ± 0.03.

REALIZED GENETIC CORRELATIONS.—The realized genetic correlation may be estimated from the data of each line in each generation as follows:

$$r_A = \frac{R_y h_x \sigma_x}{R_x h_y \sigma_y}$$

where r_A is the genetic correlation, R_y is the response or change observed in an unselected trait, R_x is the response in the selected trait, and h and σ are the square root of heritability and the phenotypic standard deviation of the character indicated. However, a more reliable estimate of the realized genetic correlation may be obtained from the regression of the response in the correlated trait (R_y) on the direct response (R_x) as follows:

$$r_A = b_{R_y R_x} \left(\frac{h_x \sigma_x}{h_y \sigma_y} \right)$$

Since both direct and correlated responses to selection were relatively symmetrical in the present study, the regression of the divergence of response of the correlated character on the divergence of response in activity was calculated for lines H_1 versus L_1 and H_2 versus L_2; a pooled estimate was also obtained. The resulting values were then multiplied by the appropriate ratio of the square root of the heritabilities and the phenotypic standard deviations. From the previous analyses, the heritabilities of activity, defecation and body weight were estimated to be 0.26, 0.11, and 0.20, respectively, and the phenotypic standard deviations of the three characters were 4.55, 0.75, and 2.0, respectively. The resulting estimates of the realized genetic correlations between activity and defecation and between activity and body weight are presented in Table 7. The estimates for the genetic correlation between activity and defecation are again consistently large, negative, and highly significant, whereas those between activity and body weight are consistent, but nonsignificant.

Comparison of Results.—Estimates of the heritability of open-field activity from the classic analysis were moderately large and, as indicated by their standard errors, relatively precise. However, estimates from the regression of offspring on parent were smaller, but also relatively precise. Estimates from the half-sib analysis were also small, but less precise. The realized heritability (pooled estimate) of activity was 0.26, in close agreement to that from the regression of offspring on midparent (0.22).

In addition, the realized genetic correlation of open-field activity and defecation (-0.86) is in close agreement to that estimated from the cross

TABLE 7

**Realized Genetic Correlations Obtained from the
Results of Five Generations of Selection**

Characters	H_1 vs. L_1	H_2 vs. L_2	Pooled Estimate
Activity and Defecation	-0.88 ± 0.15	-0.80 ± 0.26	-0.86 ± 0.14
Activity and Weight	0.33 ± 0.23	0.35 ± 0.44	0.34 ± 0.22

covariance of offspring and midparent scores (-0.76). The agreement between the realized genetic correlation of activity and body weight and that estimated from the cross-covariance method is less close, but as indicated by their standard errors, these estimates are relatively unreliable.

In conclusion, the results of selection are in excellent agreement to those predicted from the data of the preliminary analysis. If the validity of an estimate of heritability is defined by its agreement with the realized heritability, the results of the present study suggest that the regression of offspring on midparent provides a more valid estimate of heritability than do the other methods employed. In addition, the validity of the cross-covariance method for the estimation of the genetic correlation is demonstrated by its correspondence to the realized genetic correlation.

MAJOR GENE EFFECT

Open-field behavior is clearly influenced by genes at many loci. Minimum estimates of the number of loci (Mather, 1949) which influence open-field activity and defecation are 3.2 and 7.4, respectively (Hegmann, 1966). Since this method yields a minimum estimate of the number of loci, it may be concluded that both characters are probably polygenic. Nevertheless, it was of interest to examine the mean scores of mice with different coat colors, since most of the strains with relatively low activity in novel situations are albino, whereas those with higher activity scores are pigmented (see McClearn, 1959; Thompson, 1953).

Members of the inbred parental strain, *C57BL/6J,* are nonagouti black (*aaBBCC*), whereas *BALB/cJ* mice are albino (*AAbbcc*). Therefore, in the F_2 and later generations, five different coat colors may occur; agouti *A-B-C-;* cinnamon, *A-bbC-;* black, *aaB-C-;* brown, *aabbC-;* and albino, - - - - *cc.* The mean activity and defecation scores among the four pigmented classes did not differ greatly; however, pigmented animals had higher activity and lower defecation scores than albino animals, indicating the presence of a major gene effect on open-field behavior (DeFries et al., 1966). The mean open-field activity and defecation of albino and pigmented *Ss* in generations F_2 through F_5, pooled across generation and sex, are presented in Table 8. With regard to both activity and defecation, approximately 20 percent of the difference between the means of the two inbred parental strains was found to be due to the effect of a single gene difference at the c-locus.

Since the two parental strains differ widely in open-field behavior, it is possible that the apparent major gene effect may in fact be due to genes which are closely linked to the c-locus. However, due to the process of crossing over, this association should be relatively temporary.

With successive generations of selection for activity, the frequency of the gene for albinism should be increased in the low-active lines, decreased

TABLE 8

**Mean[a] and Variance of Activity and Defecation Scores
of Albino and Pigmented Animals in Generations
F_2 - F_5, Pooled Across Generation and Sex**

	Activity		Defecation		
	Mean ± S.E.	Variance	Mean ± S.E.	Variance	**N**
Pigmented	12.72 ± 0.18	18.65	1.78 ± 0.03	0.55	1729
Albino	10.44 ± 0.10	17.71	2.13 ± 0.02	0.49	557

[a]Unweighted mean of scores for males and females in each generation and line (generations F_4 and F_5 are generations S_1 and S_2 of the selection experiment).

in the high-active lines and remain more or less constant in the control lines (assuming, of course, equal fertility of albino and pigmented parents and equal viability of their offspring). As may be seen in Fig. 5, initial gene frequencies are approximately 0.5, but then increase in L_1 and L_2 and decrease in H_1 and H_2. The frequency of the c allele (as estimated from the square root of the percent albinos) is also increased somewhat in lines C_1 and C_2.

A major gene effect of the c-locus on open-field activity has also recently

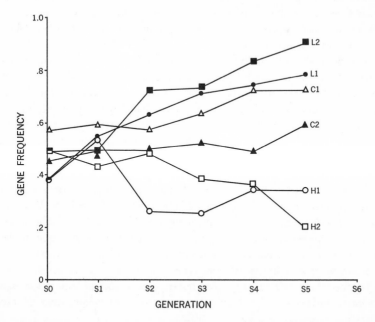

FIG. 5. Frequency of allele for albinism in two lines of mice selected for high open-field activity (H_1 and H_2), two selected for low activity (L_1 and L_2), and two unselected controls (C_1 and C_2).

been reported by investigators at two other laboratories: Henry and Schlesinger (1967) and Fuller (1967), using coisogenic lines of albino mutant and nonmutant mice of the *C57BL/6J* strain, have both reported that albino *Ss* were less active in the open field than pigmented animals. Thus, it appears that conclusive evidence for a major gene effect on open-field behavior in mice has been obtained.

Relative Importance of Major Gene Effect

Morton (1967) has recently discussed the relative merits of quantitative genetic analysis versus the characterization of major genes in the study of mental disorders. In general, Morton concludes that future genetic analysis of mental disorders should emphasize the major gene approach. In the present study, a major gene effect on open-field activity and defecation has been found and an estimate of the additive genetic variance associated with these behaviors may be obtained. From these data we shall attempt to measure the relative importance of this major gene effect versus that of as yet unidentified "polygenes" in this population.

The relevant questions are as follows: (1) What proportion of the additive genetic variance associated with open-field activity and defecation is due to the single gene effect at the c-locus? and (2) What proportion of the additive genetic covariance between these two behaviors is due to the effect at this locus? Answers to these questions should indicate the relative importance of this single gene effect on these behaviors in this population, expressed in terms which are relative to the additive genetic variance and covariance due to effects of all genes that influence the characters under study. Since the additive genetic variance and covariance are functions of gene frequency, this "relative importance" of a gene, like heritability, will vary from population to population.

When a single locus is considered, arbitrary values may be assigned to genotypes. (See Falconer, 1960, pp. 112–135, for a more complete discussion of the model.) In the present case, albinism is due to homozygosity for a completely recessive allele. Assuming that the open-field behavior of heterozygotes is no different from that of individuals homozygous for the normal allele, the arbitrarily assigned genotypic values of pigmented and albino individuals will be a and $-a$, respectively. Therefore, the difference between the mean phenotypic values of pigmented and albino animals estimates $2a$. Using the mean values presented in Table 8, the genotypic values for activity (a_x) and defecation (a_y) are 1.14 and -0.18, respectively. The value for defecation is negative, since albinos have a higher mean phenotypic value than pigmented animals.

In the F_2 and F_3 generations of this study, the frequency of both the dominant and recessive alleles should be 0.5. When complete dominance exists and when the gene frequency is 0.5, the additive genetic variance associated with a single locus is simply $\frac{1}{2}a^2$. Using this expression and the genotypic values given above, the additive genetic variance in open-field activity and defecation due to segregation at the c-locus is 0.65 and 0.0162, respectively.

From the definition of h^2, it may be seen that the additive genetic variance (V_A) due to effects of all genes is as follows:

$$V_A = h^2 V_p$$

where V_p is the phenotypic variance. Assuming that the heritabilities of activity and defecation are 0.26 and 0.11, and that the phenotypic variances as indicated in the F_3 generation are 20.72 and 0.57, V_{A_x} and V_{A_y} are estimated to be 5.39 and 0.0627, respectively.

The proportion of the additive genetic variance due to a single gene effect (assuming a gene frequency of 0.5) is simply $\frac{1}{2}a^2/V_A$. The resulting ratios for open-field activity and defecation are 0.12 and 0.26, respectively. This indicates that in the population under consideration, approximately 90 percent of the additive genetic variance in open-field activity and about 75 percent of the additive genetic variance in defecation are due to effects of genes which are as yet unidentified.

In a similar manner, the covariance between the additive genetic values for two characters which is due to the effects of a single gene may also be estimated. Under the assumptions of complete dominance and gene frequency of 0.5, the additive genetic covariance due to a single locus is $\frac{1}{2}a_x a_y$. Using the genotypic values presented above, the additive genetic covariance of open-field activity and defecation due to segregation at the c-locus is -0.103. From the definition of genetic correlation (r_A), the covariance of the additive genetic values due to effects of all loci which influence two traits is as follows:

$$Cov\ A_x A_y = r_A \left(V_{A_x} V_{A_y} \right)^{1/2}$$

Assuming that the genetic correlation of open-field activity and defecation is -0.86 and using the values of V_A presented above, Cov $A_x A_y = -0.499$.

The proportion of the covariance of the additive genetic values due to the effect of a single gene (under the above assumptions) is simply $\frac{1}{2}a_x a_y/Cov\ A_x A_y$. Upon substitution, a value of 0.21 is obtained, indicating that approximately 20 percent of the additive genetic covariance of open-field activity and defecation is due to segregation at the c-locus.

Mediation of Effect

Merrell (1965) has suggested that the study of gene substitutions, one or a few at a time, may provide information not only about the genetics of a behavioral trait, but about the trait itself. The validity of this suggestion may be illustrated by the present example, since it has been recently shown in our laboratory that the major gene effect on open-field behavior is mediated at least in part by the visual system and that albinos are more photophobic than pigmented animals.

McClearn (1960) first demonstrated that open-field activity in mice is a function of test illumination. Mice of a high-active pigmented strain (*C57BL/Crgl*) had lower activity scores when tested under red illumination than when tested under white illumination, but members of a low-active albino strain (*A/Crgl*) were more active under conditions of red light. Using a similar approach, DeFries et al. (1966) tested albino and pigmented littermates from a segregating generation in the open field under conditions of red versus white illumination. Under white illumination, pigmented animals were more active and had higher defecation scores than albinos; however, under red illumination, this difference largely disappeared. This result confirms the hypothesis of a major gene effect on open-field behavior and, in addition, indicates that this effect is mediated by the visual system. Since the open-field test has long been used as an index of emotionality, it was hypothesized that individual differences in open-field behavior are due at least in part to a visually mediated fear reaction.

In order to test this hypothesis, mice of two inbred strains (*BALB/cJ* and *C57BL/6J*) were tested under five levels of illumination (0 to 480 apparent ftc in equal log steps). Increased level of test illumination was accompanied by reduced activity in both strains, but the albino *BALB* strain was more greatly affected than the pigmented *C57BL* strain (McReynolds et al., 1967). In addition, open-field defecation scores of the BALB animals were increased as a function of increasing illumination, whereas those of the *C57BL* strain were little affected.

In a separate study (Dixon and DeFries, 1968), the effects of both level of rearing illumination and test illumination on the open-field behavior of these two inbred strains of mice were assessed. In agreement with predictions based upon the above hypothesis, *BALB Ss* had increased activity and reduced defecation scores when tested under a lower level of illumination, but had reduced activity and increased defecation scores when reared under a lower level of illumination. Effects of rearing and test illumination on the open-field behavior of *C57BL Ss*, however, were much less pronounced.

CONCLUDING REMARKS

The primary objective of the study reported in this chapter was to conduct a detailed genetic analysis of open-field behavior in mice. Open-field behavioral data were obtained on a total of 4,820 *Ss*, including members of two inbred parental strains (*BALB/cJ* and *C57BL/6J*), their derived F_1, backcross, F_2 and F_3 generations, as well as five generations of selection (two directions, with replicated lines and controls). The results of both the preliminary analysis and the selection experiment indicate that the heritability of open-field activity is approximately 0.25. In addition, a large negative genetic correlation between activity and defecation in the open field was found (approximately −0.8), indicating that individual differences in these behaviors are affected by many of the same genes. From the results of the preliminary analysis, it appears that the heritability of open-field defecation is approximately 0.1. Little or no evidence for the existence of an association of these behaviors and body weight was found.

Heritabilities were estimated by three different methods from the data of the preliminary analysis: (1) a classic analysis employing data from the parental, backcross, and F_2 generations; (2) regression of scores of F_3 Ss on their F_2 parents; and, (3) half-sib correlations among members of the F_3 generation. By comparing these values with those obtained from the results of selection, it was concluded that the regression method yielded the most valid estimate of heritability, at least under the conditions of the present study.

Although open-field behavior is clearly influenced by genes at many loci, a major gene effect was found: albino Ss have lower activity and higher defecation scores than pigmented animals. The "relative importance" of this single gene effect was measured by assessing its contribution to the additive genetic variance and covariance associated with these behaviors. Segregation at the c-locus was estimated to account for 12 percent of the additive genetic variance in open-field activity, 26 percent of the additive genetic variance of defecation, and 21 percent of the additive genetic covariance. Thus, although a major gene effect is present in these data, a relatively large proportion of residual genetic variance remains which is due to segregation at unidentified loci.

The results of several additional experiments demonstrate that this major gene effect is mediated at least in part by the visual system. These findings support the hypothesis that individual differences in open-field behavior are due to a visually mediated fear reaction and that albino Ss are more photophobic than pigmented animals.

In conclusion, it is apparent that the search for major gene effects and the methods of quantitative genetics are useful techniques which should be employed in behavioral genetics. These approaches are complementary, not mutually exclusive; both should be exploited in order to attain a more complete understanding of the genetic causes of individual differences in behavior.

REFERENCES

Broadhurst, P. L. 1960. Experiments in psychogenetics. Applications of biometrical genetics to the inheritance of behavior. *In* Eysenck, H. J., ed., Experiments in Personality, Psychogenetics and Psychopharmacology, London, Routledge and Kegan Paul, Vol. 1, pp. 1–102.

———— 1967. The biometrical analysis of behavioural inheritance. Sci. Prog., 55:123–139.

———— and J. L. Jinks, 1961. Biometrical genetics and behavior: Reanalysis of published data. Psychol. Bull., 58:337–362.

Bruell, J. H. 1962. Dominance and segregation in the inheritance of quantitative behavior in mice. *In* Bliss, E. L., ed., Roots of Behavior, New York, Harper and Row, Pubs., pp. 48–67.

Connolly, K. 1966. Locomotor activity in *Drosophila*. II. Selection for active and inactive strains. Anim. Behav., 14:444–449.

Craig, J. V., L. L. Ortman, and A. M. Guhl. 1965. Genetic selection for social dominance ability in chickens. Anim. Behav., 13:114–131.

Dawson, P. S. 1965. Estimation of components of phenotypic variance for developmental rate in *Tribolium*. Heredity, 20:403–417.

DeFries, J. C. 1967. Quantitative genetics and behavior: Overview and perspective. *In* Hirsch, J., ed., Behavior-Genetic Analysis, New York, McGraw-Hill Book Co., pp. 322–339.

———— J. P. Hegmann, and M. W. Weir. 1966. Open-field behavior in mice: Evidence for a major gene effect mediated by the visual system. Science, 154:1577–1579.

———— E. A. Thomas, J. P. Hegmann, and M. W. Weir. 1967a. Open-field behavior in mice: Analysis of maternal effects by means of ovarian transplantation. Psychon. Sci., 8:207–208.

———— M. W. Weir, and J. P. Hegmann. 1967b. Differential effects of prenatal maternal stress on offspring behavior in mice as a function of genotype and stress. J. Comp. Physiol. Psychol., 63:332–334.

Dickerson, G. E. 1959. Techniques for research in quantitative genetics. *In* Techniques and Procedures in Animal Production Research. Amer. Soc. Anim. Prod., 56–105.

Dixon, Linda K., and J. C. DeFries. 1968. Effects of illumination on open-field behavior in mice. J. Comp. Physiol. Psychol., 66:803–805.

Dobzhansky, T. 1964. Heredity and the Nature of Man. New York, Harcourt, Brace and World, Inc.

Falconer, D. S. 1960. Introduction to Quantitative Genetics, New York, The Ronald Press Co.

Fulker, D. W. 1966. Mating speed in male *Drosophila melanogaster:* A psychogenetic analysis. Science, 153:203–205.

Fuller, J. L. 1967. Effects of the albino gene upon behaviour of mice. Anim. Behav., 15:467–470.

———— and W. R. Thompson. 1960. Behavior Genetics. New York, John Wiley and Sons, Inc.

Hadler, N. M. 1964. Heritability and phototaxis in *Drosophila melanogaster*. Genetics, 50:1269–1277.

Hall, C. S. 1934. Emotional behavior in the rat. I. Defecation and urination as measures of individual differences in emotionality. J. Comp. Physiol. Psychol., 18:385–403.

———— 1936a. Emotional behavior in the rat. II. The relationship between need and emotionality. J. Comp. Physiol. Psychol., 22:61–68.

———— 1936b. Emotional behavior in the rat. III. The relationship between emotionality and ambulatory activity. J. Comp. Physiol. Psychol., 22:345–352.

———— 1951. The genetics of behavior. *In* Stevens, S. S., ed., Handbook of

Experimental Psychology. New York, John Wiley and Sons, Inc., pp. 304–329.

Hegmann, J. P. 1966. A quantitative genetic analysis of open-field behavior in two inbred strains of mice. Unpublished M.S. Thesis, Urbana, University of Illinois.

Henry, K. R., and K. Schlesinger. 1967. Effects of the albino and dilute loci on mouse behavior. J. Comp. Physiol. Psychol., 63:320–323.

Hirsch, J. 1967. Behavior-Genetic Analysis, New York, McGraw-Hill Book Co., pp. 416–435.

——— and J. C. Boudreau. 1958. Studies in experimental behavior genetics: I. The heritability of phototaxis in a population of *Drosophila melanogaster*. J. Comp. Physiol. Psychol., 51:647–651.

Jinks, J. L., and P. L. Broadhurst. 1965. The detection and estimation of heritable differences in behaviour among individuals. Heredity, 20:97–115.

Lush, J. L. 1940. Intra-sire correlations or regressions of offspring on dam as a method of estimating heritability of characteristics. Thirty-third Annual Proc. Amer. Soc. Anim. Prod., 293–301.

——— 1948. The Genetics of Populations. Ames, Iowa. Mimeographed notes.

Manning, A. 1961 The effects of artificial selection for mating speed in *Drosophila melanogaster*. Anim. Behav., 9:82–92.

Mather, K. 1949. Biometrical Genetics: The Study of Continuous Variation. London, Methuen and Co., Ltd.

McClearn, G. E. 1959. The genetics of mouse behavior in novel situations, J. Comp. Physiol. Psychol., 52:62–67.

——— 1960. Strain differences in activity of mice: influence of illumination, J. Comp. Physiol. Psychol., 53:142–143.

——— 1961. Genotype and mouse activity. J. Comp. Physiol. Psychol., 54:674–676.

——— 1963. The inheritance of behavior. *In* Postman, L., ed., Psychology in the Making, New York, Alfred A. Knopf, Inc., pp. 144–252.

McReynolds, W. E., M. W. Weir, and J. C. DeFries. 1967. Open-field behavior in mice: Effect of test illumination. Psychon. Sci., 9:277–278.

Merrell, D. J. 1965. Methodology in behavior genetics. J. Hered., 56:263–266.

Morton, N. E. 1967. Population genetics of mental illness. Eugen. Quart., 14:181–184.

Osborne, R., and W. S. B. Paterson. 1952. On the sampling variance of heritability estimates derived from variance analyses. Proc. Roy. Soc. Edinb. B., 64:456–461.

Rasmuson, M. 1961. Genetics on the Population Level. Stockholm, Svenska Bokforlaget, Bonniers.

Roberts, R. C. 1967. Some concepts and methods in quantitative genetics. *In* Hirsch J., ed., Behavior-Genetic Analysis, New York, McGraw-Hill Book Co., pp. 214–257.

Siegel, P. B. 1965. Genetics of behavior: Selection for mating ability in chickens. Genetics, 52:1269–1277.

Thompson, W. R. 1953. The inheritance of behaviour: Behavioural differences in fifteen mouse strains. Canad. J. Psychol., 7:145–155.

———— 1956. The inheritance of behavior. Activity differences in five inbred mouse strains. J. Hered., 47:147–148.

Weir, M. W., and J. C. DeFries. 1964. Prenatal maternal influence on behavior in mice: Evidence of a genetic basis. J. Comp. Physiol. Psychol., 58:412–417.

Willham, R. L., D. F. Cox, and G. G. Karas. 1963. Genetic variation in a measure of avoidance learning in swine. J. Comp. Physiol. Psychol., 56:294–297.

3

Genetic Analysis of
Male Sexual Behavior*

THOMAS E. McGILL
*Department of Psychology,
Williams College, Williamstown, Massachusetts*

INTRODUCTION

For a number of years my colleagues and I have been involved in studies concerned with the genetics and physiology of reproductive behavior in different inbred strains of mice and various crosses between different strains. This chapter reviews some of these studies in order to illustrate two important limitations to the conclusions that result from behavior-genetic analyses. A second purpose of the chapter is to present partial data from a recently completed large-scale biometric genetic analysis of the various elements of mating behavior in male mice. Space limitations, however, prevent presentation of the complete results of this latter study. Furthermore, we shall not attempt to review previous experiments on the effects of genotype on sexual behavior in other species. The interested reader will find such reviews in Broadhurst (1960), Fuller and Thompson (1960), Goy and Jakway (1962), various chapters of Hafez (1962), and Parsons (1967).

THE INITIAL STUDY: QUANTIFYING MALE SEXUAL BEHAVIOR AND ESTABLISHING STRAIN DIFFERENCES

This research program had its beginning in the academic year 1959–1960 when the author held a USPHS Postdoctoral Fellowship under the sponsor-

* This research was supported by Research Grant GM-07495 from the Institute of General Medical Sciences, U.S. Public Health Service.

ship of Professor Frank A. Beach at the University of California. Professor Beach suggested that it might be worthwhile to investigate the sexual behavior of different inbred strains of mice. There were two major reasons for initiating these investigations. In the first place, the mating behavior pattern of the male mouse had not been adequately described or measured. Second, if strain differences in elements of sexual behavior were found to exist, a number of interesting physiologic and genetic experiments could be performed.

Our first task was to induce the animals to mate under conditions that permitted observation of the behavior. Details of the methods that evolved are presented below in the discussion of our most recent genetic studies. At this point, it should suffice to note that females are brought into estrus by hormone injections and that matings are observed and scored while pairs are housed in plastic cylinders under normal room illumination.

After the development of techniques permitting the observation of sexual behavior, it was discovered that, in common with other laboratory rodents, the basic mating pattern of the male mouse consists of a series of mounts (unsuccessful attempts to penetrate the female) and intromissions (successful attempts to penetrate the female) which culminates in ejaculation. In this chapter, in accordance with recent terminology, the complete mating act including all mounts, intromissions, and the ejaculation is referred to as a "copulatory series," or simply as a "series." This nomenclature avoids confusion with regard to earlier literature where the terms "copulation" and "intromission" were sometimes synonymous.

In a typical series, a male mouse first encountering an estrous female carefully investigates her, concentrating his attention particularly on the anogenital regions. If sufficient sexual arousal occurs, the male mounts the female palpating her sides with his forepaws while simultaneously executing a series of rapid, pelvic thrusts. Quite frequently the first attempt at gaining intromission fails and the male dismounts and engages in genital cleaning. When a male is successful in gaining intromission, the rate of pelvic thrusting is greatly reduced while the amplitude is increased. The thrusts of intromission average about one-half second and are easily counted. During intromission, the male keeps one hindfoot on the floor and rests the other on the hindquarters of the female. The number of thrusts in each intromission varies from only a few to 300 or more. After an intromission, both animals generally engage in genital cleaning. This behavioral sequence of mounts, intromissions, and genital cleaning usually continues until the male ejaculates. During the ejaculatory intromission, the speed of pelvic thrusting increases, and, finally, the male quivers strongly while maintaining deep penetration of the female. At this point, he raises the hindfoot which has been resting on the floor and clutches the female with all four limbs. Most frequently, this results in both animals falling to one side. Following the male's ejaculation, both male and female again engage in genital cleaning. Specific details of the mating pattern should become clarified in the discussion of particular behavioral measures in later sections of this chapter.

Once a reliable method for observing the behavior had been developed, the next task was to divide the mating pattern into measurable elements and to search for strain differences. It is obvious from the above description that a number of different elements of the sexual behavior of male mice could be measured. In the initial study (McGill, 1962a), 16 different measures for three inbred strains were scored. Nonparametric analysis of variance revealed strain differences for 12 of these 16 measures (those listed above the line in Table 1). Mann-Whitney U-tests were applied to the three comparisons for each measure. Table 1 shows the two-tailed significance levels that resulted from the statistical analysis. Definitions of most of the measures found in Table 1 are listed in Table 3.

The most striking finding apparent from an examination of Table 1 is the frequency of statistically significant strain differences. Furthermore, later experiments using larger samples and/or different strains have indicated statistically significant strain differences for the four measures below the line that did not reach significance in the analysis of variance of this study. Similar strain differences in other inbred strains have been reported by Levine et al. (1966).

In summary, the goals of the initial study were achieved: Male mice of different inbred strains were induced to copulate under conditions permitting observation; The mating behavior was described, divided into elements, and quantified; and significant strain differences were shown to exist.

TABLE 1

Significance Levels of Strain Comparisons

Measure	BALB vs. C57	BALB vs. DBA	C57 vs. DBA
IL	.02	.02	.001
T of I	.05	.02	
PIMD		.002	.001
III	.02	.002	.001
T of M		.002	.001
No. of I	.05	.002	.002
Tot. No. Th.	.05	.002	.001
EL	.002	.02	
ED	.002	.002	
No. HM	.02	.002	.001
Rooting	.002	.002	
% Move	.02		

ML
Th. per I
No. of M.
% Bite

TABLE 2

Median Scores for the Three Strains and Significance Levels of the Three Possible Comparisons for Each Measure

Measure	Median Scores[a]			Significance Levels		
	C57BL/6J	DBA/2J	BDF$_1$	C57 vs. DBA	C57 vs. BDF$_1$	DBA vs. BDF$_1$
1. ML	42	85	42	0.02		0.002
2. Tot. No. Th.	400	129	546	0.002		0.002
3. No. of I.	17	5	18	0.02		0.02
4. % Bite	0	20	0	0.02		0.001
5. ED	23	17	19	0.02		
6. T of I	15	20	19	0.02	0.02	
7. No. HM	2	0.5	0	0.01	0.01	
8. III	28	137	42	0.002	0.001	0.002
9. T of M	2	7	3	0.002	0.002	0.02
10. PIMD	1	4	2	0.002	0.001	0.002
11. No. of M	18	16	7		0.02	
12. Th. per I	16	20	25		0.02	0.05
13. IL	107	179	93		0.02	
14. EL	1252	1376	1091			0.02

[a]All time measures are in seconds.

TWO DIFFERENT COMPARISONS OF F_1 MALES WITH INBRED MALES

Once strain differences have been established, the next logical step involves a comparison of F_1 animals with the inbred parent strains. Table 2 (McGill and Blight, 1963a; McGill, 1965) presents such a comparison; Table 3 presents definitions of the measures used in these studies. The F_1 males of Table 2 resulted from crossing *C57BL/6J* females with *DBA/2J* males.

It is apparent from Table 2 that inheritance of sexual behavior in male mice is not a simple phenomenon. Three different modes of inheritance are

TABLE 3

Definitions of Measures

ML - Mount latency. The number of seconds from the introduction of the female until the male mounts.

Tot. of Th. - The total number of thrusts with intromission preceding ejaculation.

No. of I - The number of intromissions preceding ejaculation.

% Bite - The percentage of times the male bites the female following the ejaculation duration.

ED - Ejaculation duration. The number of seconds the male spends clutching the female, and maintaining vaginal contact, following ejaculation.

T of I - Time of intromission. The number of seconds from the beginning of a mount with intromission until the male dismounts.

No. HM - The number of head mounts during a series.

III - Interintromission interval. The number of seconds from the end of one mount with intromission until the beginning of the next.

T of M - Time of mount. The length of mounts without intromission in seconds.

PIMD - Preintromission mount duration. The number of seconds from the beginning of a mount with intromission until the male's penis is inserted into the female's vagina and the first thrust of intromission occurs.

No. of M - The number of mounts without intromission per series.

Th. per I - The number of thrusts making up each intromission.

IL - Intromission latency. The number of seconds from the introduction of the female until the male gains intromission.

EL - Ejaculation latency. The number of seconds from the beginning of the first intromission until the beginning of ejaculation.

suggested by the results presented: dominance of one parental genotype or the other (measures 1 to 4 and 5 to 7), intermediate inheritance (measures 8 to 10), and heterotic inheritance (measures 11 to 13).

Our next question concerned the generality of the results reflected in Table 2. For example, for the first measure listed in Table 2 (ML = Mount Latency, defined as the number of seconds from the introduction of the female until the male first mounts), the *DBA/2J* genotype appeared recessive to the *C57BL/6J* genotype. It is possible that this result would not have been obtained if *DBA* had been crossed with a different strain. In order to test this possibility, we substituted males from the inbred strain *AKR/J* for the *C57BL/6J* males and replicated the experiment. Table 4 (McGill and Ransom, 1968) presents the genetic conclusions regarding mode of inheritance

TABLE 4

Conclusions Regarding Mode of Inheritance in Two Experiments

Measure	Experiment I DBA X C57	Experiment II DBA X AKR
ML	*DBA* recessive	No significant differences
TNT	*DBA* recessive	F_1 herotic
No. of I	*DBA* recessive	*DBA* recessive
% Bite	*DBA* recessive	*DBA* recessive
ED	*DBA* dominant	*DBA* recessive
T of I	*DBA* dominant	*DBA* recessive
No. HM	*DBA* dominant	*DBA* recessive
III	F_1 intermediate	F_1 herotic
T of M	F_1 intermediate	*DBA* recessive
PIMD	F_1 intermediate	*DBA* recessive
No. of M	F_1 herotic	F_1 intermediate
Th. per I	F_1 herotic	*DBA* recessive
IL	F_1 herotic	F_1 herotic
EL	No significant differences	No significant differences

Agreements = 4
Disagreements = 10

that resulted from the original study (*DBA* × *C57*) and from the replication (*DBA* × *AKR*).

It is obvious from Table 4 that the results of Experiment 1 have little predictive value for the different cross studied in Experiment 2. Although this may appear to be an unexpected finding, similar results are common in plant genetics. When several inbred lines of plants are crossed and the F_1 progenies studied, a "general combining ability" may be established for each of the inbred strains. However, for any particular cross, the F_1 value may deviate from the average general combining ability of the two inbred strains. Such a deviation is referred to as the "special combining ability" of the cross (Falconer, 1960, p. 281). Similar results for other aspects of the behavior of mice have been noted by Bruell (1964) and Fuller (1964).

These observations lead to an important limiting generalization (frequently made and frequently ignored) concerning the results of behavior genetic experiments: *Conclusions regarding such genetic parameters as mode of inheritance, heritability, or degree of genetic determination are specific to the populations or strains studied.*

STUDIES OF GENOTYPE AND SEX DRIVE

A second important limiting generalization may be illustrated by a series of studies on genotype and sex drive in male mice. (Results similar to those of the first study described below have been published in McGill and Blight, 1963b, and in McGill, 1965. However, these initial studies used smaller samples and fewer genotypes than were involved in the experiment shown in Fig. 1.)

In order to investigate possible effects of genotype on recovery of sex drive, sexually experienced male mice were allowed to complete a copulatory series and were then given periodic tests with receptive females until a second ejaculation occurred. The inserts in Fig. 1 show the strains studied and the size of the samples. With the obvious exception of the inbred parent strains, reciprocal crosses were studied in all cases. Since there were no statistically significant differences between any pair of reciprocal crosses, the results are combined for those genotypes.

T tests and U tests were applied to the data shown in Fig. 1. These tests revealed significant differences for all comparisons except *DBA/2J* versus F_1.

Fig. 2 presents the median results plotted after the method described by Bruell (1962). The results indicate dominance of *DBA/2J* genotype over *C57BL/6J* genotype in terms of time required to recover sex drive.

Because of the large range observed in the *C57BL/6J*, F_2, and backcross groups, the males of these strains were rested for 3 weeks and then the re-

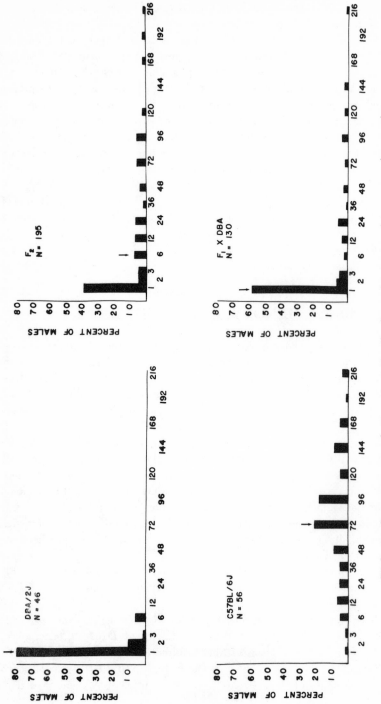

FIG. 1. Genotypes, sample sizes, and distributions of recovery time scores. Arrows indicate median scores. Note that the abscissae are not linear.

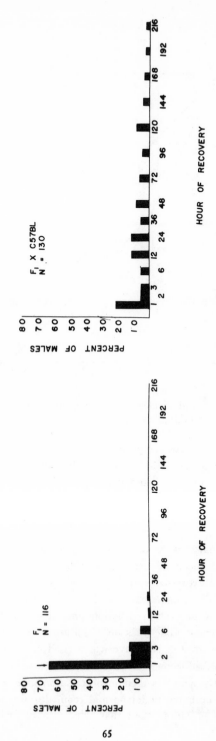

FIG. 1. Genotypes, sample sizes, and distributions of recovery time scores. Arrows indicate median scores. Note that the abscissae are not linear.

65

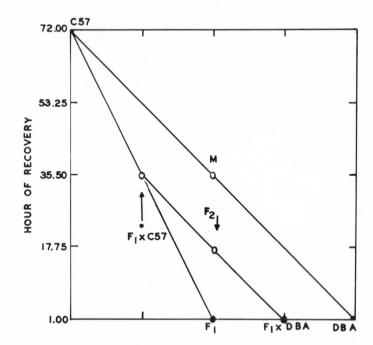

FIG. 2. Observed (solid circles) and expected (open circles) medians for the cylinder test. M = midparent.

covery of sex drive test was repeated. Test-retest correlations were carried out for all four genotypes. The correlation for the inbred strain *C57BL/6J* was not significant. The test-retest correlations for the two backcross groups and for the F_2 strain were significant. This result provides evidence for the hypothesis that genetic segregation was occurring in the F_2 and backcross groups.

When preliminary results of this experiment were presented at the Eastern Psychological Association meetings in 1963, Dr. John L. Fuller of The Jackson Laboratory expressed surprise at the results for the *C57BL/6J* inbred males. Dr. Fuller suggested that the poor performance of these males might have been due to the test situation since males of this strain are considered good breeders in The Jackson Laboratory colonies. Consequently, a second, and different, test of sex drive was carried out.

The test described above had been conducted while the animals were housed in the plastic cylinders used in observation of elements of sexual behavior. These tests occurred under normal room illumination. Obviously, this is quite a different situation from that which an animal experiences when mating in his home cage in the dark. Therefore, an experiment that more nearly duplicated the home cage situation was designed.

This second experiment made use of the fact that the semen of the male mouse forms a hard, rubbery plug in the vagina of the female (see Land and McGill, 1967; McGill et al., 1968, concerning possible functions of the plug). These copulatory or vaginal plugs are detectable by probing and provide evidence that mating and ejaculation have occurred.

Males of the strains indicated by the inserts in Fig. 3 were caged individually in plastic mouse cages with one estrous female per night for 10 consecutive nights. A search for copulatory plugs was made in the morning. Fig. 3 shows the results for this experiment. In contrast to the results shown in Fig. 1, there were no statistically significant strain differences for any of the comparisons that can be made from the data presented in Fig. 3. Of particular interest for our purposes are the results for the *C57BL/6J* males. A prediction from the data of Fig. 1 might have been that *C57BL/6J* males would mate with approximately three females during the 10 nights of testing. As Fig. 3 shows, all *C57BL/6J* males mated with at least five females and one succeeded in completing copulation with all 10. An interpretation of the discrepancies between these two studies of sex drive in male mice will be presented elsewhere. For present purposes, the studies provide an example of a second limiting generalization in behavior genetic studies: *Genetic conclusions reached in any particular experiment are specific to the total environmental conditions of that experiment.*

The two limiting generalizations stated above will probably appear shocking and disheartening to some, and obvious and inane to others. The latter group will include those deeply immersed in behavior-genetic experiments, while the former will comprise those who have looked to behavior-genetic analysis for a solution to the old nature-nurture controversy. To this group a behavior geneticist can only plead for patience and point out that such disparate items as the "laws of physiology" and the "laws of learning" have been found to differ from species to species and from situation to situation within a species.

A BIOMETRIC GENETIC ANALYSIS OF SEXUAL BEHAVIOR IN MALE MICE

With the two limitations stated above in mind, we are ready to examine partial data from a relatively large scale investigation of genetic effects on elements of the mating pattern in male mice. A classic Mendelian design involving two inbred parent strains, their F_1's, F_2's, and backcross groups was used. In the light of recent observations on the efficacy of experimental designs involving either (1) resemblance between relatives in random breeding populations, or (2) the diallel cross method of studying inbred strains and

F_1's, the Mendelian design may not have been the best one to choose. Nevertheless, as we shall see, use of this design can result in a great deal of genetic information about the population and in testable behavioral hypotheses. Table 5 shows the strains and the sample sizes for the males that met criteria in this study. Fig. 4 shows the percentage of each strain under study at particular times during the course of the experiment. Fig. 5 shows the percentage of animals of each strain under study during each month regardless of the year. Although complete balance was not achieved, possible seasonal variations cannot account for the strain differences reported below.

METHODS

General Husbandry

All males used in the experiment, including the inbred strains, were bred in the laboratory. The initial breeding stock of *C57BL/6J* and *DBA/2J* animals was purchased from The Jackson Laboratory in Bar Harbor, Maine. This stock was replenished from time to time so that our inbreds were never more than one generation removed from the Jackson strains. In all cases, breeding cages initially held one male and two females. As females became pregnant, they were removed to individual cages and given nesting material. Weaning took place at 28 days of age. Males used in data collection were not used for breeding purposes. In the segregating populations an effort was made to secure male offspring from as many different parents as possible. However, no records were kept of litters or siblings.

After weaning, males of the same genotype were caged together in groups of six in standard plastic mouse cages, $11\frac{1}{2}$ by $7\frac{1}{2}$ by 5 inches. At about 6 weeks of age, males were marked by toe clipping and recaged in groups of four.

All animals in the colony were maintained on Purina Lab Chow and tap water, both available ad libitum. Fresh water was supplied twice a week, and the cages were cleaned once a week.

The colony was maintained on a reversed light-dark cycle with both phases 12 hours in length. In general, the lights went off in the colony room at 8 A.M and came on at 8 P.M.

Treatment of Females

Females used in the tests of sexual behavior were of the strain *BALB/cJ*. All were purchased at about 6 weeks of age from The Jackson Laboratory. They were caged in groups of six and maintained in the colony room until they were approximately 10 weeks old.

BALB/cJ females were used because previous experience indicated they

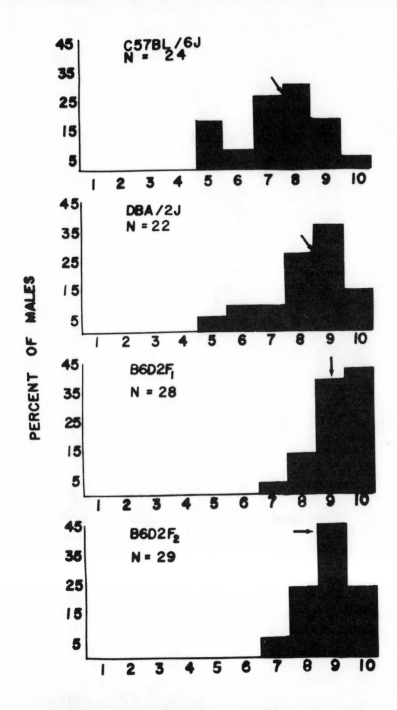

FIG. 3. Genotypes, sample sizes, and distributions of number of females plugged in the home-cage test. Arrows indicate median scores.

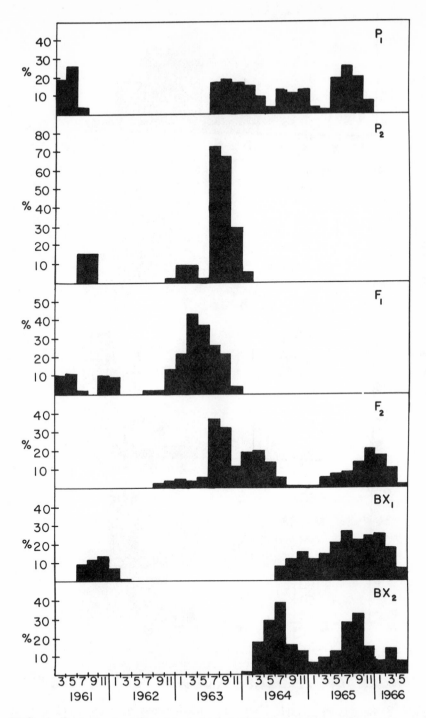

FIG. 4. Percent of each strain being studied during the course of the experiment.

70

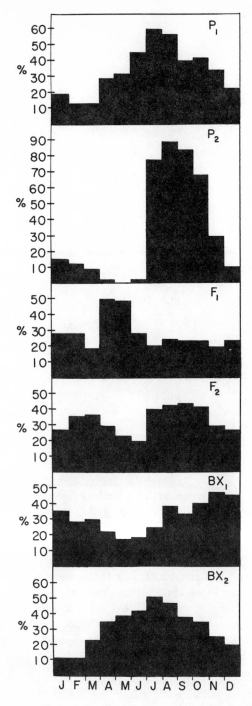

FIG. 5. Percent of each strain being studied each month from 1961 through 1966.

71

TABLE 5

The Sample

Breeding

Strain Designation		Females	Males	*n*	*N*
C57BL/6	(P_1)	Inbred			54
DBA/2	(P_2)	Inbred			52
B6D2F$_1$		P_1 ♀ X	P_2 ♂	67	
D2B6F$_1$	(F_1)	P_2 ♀ X	P_1 ♂	62	129
B6D2F$_2$		BDF_1 ♀ X	BDF_1 ♂	108	
D2B6F$_2$	(F_2)	DBF_1 ♀ X	DBF_1 ♂	116	224
$B_{x_{1a}}$		BDF_1 ♀ X	P_1 ♂	35	
$B_{x_{1b}}$		P_1 ♀ X	BDF_1 ♂	34	
$B_{x_{1c}}$	(B_{x_1})	DBF_1 ♀ X	P_1 ♂	38	141
$B_{x_{1d}}$		P_1 ♀ X	DBF_1 ♂	34	
$B_{x_{2a}}$		BDF_1 ♀ X	P_2 ♂	41	
$B_{x_{2b}}$		P_2 ♀ X	BDF_1 ♂	30	
$B_{x_{2c}}$	(B_{x_2})	DBF_1 ♀ X	P_2 ♂	38	143
$B_{x_{2d}}$		P_2 ♀ X	DBF_1 ♂	34	

could be brought into very good behavioral estrus with hormone injections. Furthermore, since they are an inbred strain, they provided control over genetically determined differences in female behavior. A strain foreign to all males under study was chosen in order to avoid the possibility that males might prefer to mate with females of their own strain; or, conversely, that they might tend to avoid females of their own strain.

At about 10 weeks of age, the females were given their first hormone injections. (See Champlin and McGill, 1967 and McGill, 1968 for details of apparatus and procedure.) 0.01 mg of estradiol benzoate was followed in 24 to 36 hours by 0.25 mg of progesterone.* In general, testing occurred from 8 to 16 hours after the progesterone injections. It had been noted that the receptivity of females tended to increase with repeated doses of the hormones. Therefore, whenever possible, females were not used in the testing situation until they had had at least two series of injections. In most cases, females were

* The hormone preparations were generously supplied by the Schering Corporation, Bloomfield, New Jersey.

rested for 2 weeks between injection series. Occasionally, when the supply of females became low, it was necessary to bring the females into heat every week. No consistent differences between heat induced every week or every 2 weeks were noted. (The injection of estrogen and progesterone is an effective contraceptive so that females can be used repeatedly.) Routinely, we continued to use a given female until she was about 7 months of age.

Testing Procedure

Testing began for individual males when they were between 8 and 10 weeks of age. It was conducted in a room adjacent to the colony room under normal illumination. The animals were housed in clear plastic cylinders about 10 inches in diameter and 20 inches in height. The cylinders rested on the rough surface of a piece of "Masonite."

For his initial test a male was introduced into the cylinder and allowed at least one-half hour of adaptation prior to the introduction of an estrous female. If he failed to initiate sexual behavior with a given female within 10 to 15 minutes, that female was removed and a second female was offered to the male. If the male again failed to initiate sexual behavior, the second female was removed and a third female was introduced into the cylinder. When the male failed to copulate with the third female, he was scored as "negative" for that day and returned to his home cage. The male was then given an identical test on the following day. If he again scored "negative," he was rested for a week before the 2-day testing was repeated.

Testing continued as described above until a male successfully initiated and completed a copulatory series. This virginal series was observed, but not scored, since it was felt that it might be affected by a number of uncontrolled variables. For example, experience has been shown to affect at least one of the measures taken in the present study, i.e., the number of head mounts per series decreases with increased sexual experience (McGill, 1962b). After the initial ejaculation, the male was rested for 2 weeks before another test. All matings following the initial mating were individually observed and scored.

Scoring was accomplished as follows: An Esterline-Angus Operations Recorder was activated prior to the introduction of the estrous female. As the female was introduced, the male's strain and number were coded on the moving paper tape by means of four ink-writing pens operated by four microswitches. When the male mounted the female, one of the microswitches was pressed down and held in a depressed position until the male dismounted. If the male gained intromission, a second microswitch was depressed for each thrust of intromission. In this case, when the male dismounted both microswitches were released. Thus, one line on the operations recorder tape showed mounts, while the second recorded the number of intromissions and the number of thrusts for each intromission. The other two pens were used to record "head mounts" and certain aspects of the female's behavior. As the male fell to his side during ejaculation, both the thrust and mount switches were held in a depressed position until genital disengagement occurred.

The operator also used pencil and paper to record the number of intromissions and the number of thrusts in each intromission and to score the female for receptivity.

The procedure following the virginal mating was similar to that prior to the virginal mating, i.e., if an experienced male failed to copulate with any of three females on a given day he was retested the next day. If he again failed to mate, he was rested for a week before the 2-day testing cycle was repeated. When a male successfully completed a series, he was rested for 2 weeks before he was given the next sex test. A 2-week rest period was used since a previous experiment (McGill, 1963) had shown that certain elements of sexual behavior may be affected by the length of the period of sexual inactivity separating copulatory series. However, some of our unpublished observations indicate that matings separated by 2 weeks do not differ significantly from those separated by 4 weeks. Two weeks of sexual inactivity between series, therefore, is probably sufficient to permit the dissipation of any effects of the previous mating.

Occasionally, a male would cease to copulate prior to ejaculation. Such matings were scored as "incomplete" if 40 minutes passed without an intromission. The male was then rested for 2 weeks and tested again.

Since the behavior of the female might affect certain measures of male behavior, a five-point receptivity scale was devised as shown in Table 6. Each female was given a score following each intromission. The mean of these scores constituted the female's average receptivity score for that series.

Testing continued as described above until a given male had had five copulatory series during which the female averaged 3.0 or better on the receptivity scale. If a male reached the age of 8 months without having had five series that met the above criterion, he was discarded from the sample. Males were discarded at 8 months of age, since our unpublished evidence indicates that certain behavioral measures begin to change at about this time.

In addition to the 14 measures listed in Table 3, the following five measures were taken for each male:

Pass-ups. As noted above, a male was offered as many as three females in succession during a sex test. If he "passed-up" all three he was scored as a "skip" (see below) for that test. If he mated with the third female, he was scored as having "passed-up" the first two. If he mated with the second female, he scored one pass-up. It was thus possible for a male to score a total of 10 pass-ups while meeting criterion.

Incompletes. The total number of "incomplete" copulatory series that occurred *after* the virginal series and *before* the male met criterion.

Skips Before Virginal. The number of negative sex tests before the male successfully ejaculated.

TABLE 6

Five Point Scale for Judgments of Female Receptivity[a]

Score	Description of female's behavior
5	High receptivity. No squeaking. No attempt to avoid mounts. Feet well braced and no attempt to move during intromission. Male determines when genital contact will be broken.
4	Good receptivity. Perhaps some squeaking and small hops during intromission. Feet generally well braced. Male determines when genital contact will be broken.
3	Average receptivity. Some squeaking and some movement during mount or intromission. Occasional rear to hind feet in attempt to avoid mount. Male gains intromission with a little persistence and determines when genital contact will be broken.
2	Poor receptivity. Faces male in defensive posture when approached. Much squeaking, moving during intromission. Frequent rearing to hind feet, or sitting, to avoid mounts and end intromissions. Female determines when genital contact will be broken.
1	Unreceptive. Active avoidance of male. Much squeaking at mounts and during intromissions. Intromission brief as female breaks away. Much rearing, sitting, kicking by female. Ejaculation can still occur, but number of intromissions is increased. Mating apt to be incomplete as male ceases to mount.

[a]McGill. 1962. *Behavior*, 19: 341–350.

Skips After Virginal. The number of negative sex tests between the virginal series and the final copulatory series.

Recovery. After each male had met criterion, he was rested for 2 weeks and then given a "recovery of sex-drive test" as described above.

RESULTS

As noted above, the complete results of this study including quantitative genetic analyses of the data, are to be presented elsewhere. Our purpose here is to present the answers to some preliminary questions, and then to give an example or two of biometric analyses of the data. Finally, we will attempt to show how the results of a behavior genetic experiment can be used to create testable hypotheses.

Reciprocal Crosses

The first preliminary question concerns the reciprocal crosses. Reciprocal crosses are used in genetic designs to test for (1) sex-linkage and/or (2) maternal effects. In a search for such effects, t tests comparing all possible reciprocal crosses shown in Table 5 were applied to the 14 measures listed in Table 3 and the five additional variables listed in the Methods section.

For the two F_1 groups, none of the 19 t tests revealed a significant difference (0.05 level). For the two F_2 groups (strictly speaking, these are not reciprocal crosses), two measures revealed significant differences at the 0.05 level. These were Intromission Latency and Number of Head Mounts. For the backcross comparisons, a few significant differences were found. However, these differences showed no consistent pattern. Also, the number found was not above what one would expect by chance (1 out of 20 when accepting the 0.05 significance level). Furthermore, Intromission Latency and Number of Head Mounts were not found to be significantly different in the backcross groups. The picture that emerged, then, was that of a random occurrence of statistically significant differences between reciprocal crosses. For this reason we felt justified in ruling out significant sex-linkage and maternal effects and in combining the various reciprocal crosses as shown in the final column in Table 5.

Search for Single Gene Effects

Recent experiments (Fuller, 1967; Henry and Schlesinger, 1967; Thiessen, 1966; Werboff et al., 1967) have indicated that coat-color genes may affect behavior. Since our F_2 group segregated into four different coat colors (black, dilute black, dilute brown, and brown) it was possible to test for such effects on the 19 measures of sexual behavior used in the present study. A test for homogeneity of variance revealed significant heterogeneity for 14 of the 19 variables. A square-root transformation produced homogeneity for five further variables. F ratios proved insignificant at the 0.05 level for the 10 variables on which the assumption of homogeneity of variance was satisfied. Ignoring inhomogeneity of variance, the other nine measures were tested for significance. F ratios were significant for two measures using both raw data and square-root transformations. These two measures were Ejaculation Duration and Percent Bite.

Pursuing the search for single gene effects for the measures Ejaculation Duration and Percent Bite, a two-way analysis of variance using both raw data and square-root transformations was performed. The main effects of this analysis were color quality (dilute versus full) and hue (black versus brown). For the measure Ejaculation Duration both the square-root transformation and the raw data indicated that males of the dilute genotype had significantly longer scores than full color males.

For the measure Percent Bite, both the square-root transformation and the raw data indicated that brown animals tended to bite the female significantly more frequently following the Ejaculation Duration than did black animals.

The second result is in accordance with the data for the inbred parent strains as presented in Tables 2 and 7 where the DBA (dilute brown) males bite the female significantly more frequently than the *C57BL* (black) males. The result of the analysis for Ejaculation Duration is the opposite of what would have been predicted on the basis of the inbred parent strain data from Tables 2 and 7.

We continued the search for "major" gene effects by plotting histograms for all six groups for each of the 19 variables. Examination of these histograms did not reveal any obvious Mendelian segregation patterns. This does not mean that major genes were not operating in determining some of these traits, for, as Falconer has pointed out (1960, pp. 105–106), a trait determined by relatively few genes may still result in a continuous distribution, or even in a normal curve, since errors of measurement and other nongenetic sources of variation tend to "blur the edges of the genetic discontinuity."

The results of our search for single-gene effects are at best only suggestive. The strongest possibility would seem to be that brown coat color may be associated with a tendency to bite the female following the Ejaculation Duration.

Biometric Analysis

Table 7 presents the raw-score means for the six genotypes on the 19 variables. Measures 1 through 9 result in one score per animal, per series. For example, there is only one Mount Latency during a series. The mean for each male on his five copulatory series was computed; the number in the table represents the grand mean of these individual means. Measures 10 through 14 occur several times during a series. A mean was taken for each animal for each series, then, from these five means, an overall mean was determined for each animal for each measure, and, finally, a grand mean was determined from these means. Measures 15 through 19 result in a single number for each animal for each variable.

An interesting observation that can be made from Table 7 is that for all variables except 2 and 6 the direction of the difference between the means of B_{X_1} and B_{X_2} reflects the direction of the difference between the means of P_1 and P_2. This relationship of backcross means to parental means is predicted by the polygenic model and supports the hypothesis that, for most variables, genotype is having an effect on sexual behavior.

Inspection of Table 7 also reveals that the mean for the F_1 animals fell between means of the parental groups for nine measures and outside the range set by the parental means for 10. Of these 10, the F_1 mean is closer to P_1 on five measures and closer to P_2 on five. These results support a point previously made (McGill and Blight, 1963b, and above) to the effect that different measures of sexual behavior may have different patterns of inheritance.

TABLE 7
Mean Values

Measure	C57 (P_1)	DBA (P_2)	F_1	F_2	B_{x1}	B_{x2}
1. No. of I	19.83	12.92	17.06	13.67	16.55	12.15
2. EL	1368.91	1977.27	1189.82	1204.73	1354.35	1316.94
3. ML	44.56	53.19	37.74	45.39	45.11	64.00
4. IL	151.91	171.02	115.40	123.48	127.87	136.03
5. Tot. of Th.	417.80	283.27	429.81	323.55	377.66	288.06
6. ED	23.19	19.92	22.51	20.63	19.04	19.36
7. No. of M	22.19	15.00	13.32	12.88	16.88	12.01
8. No. of HM	2.83	.40	3.33	2.79	3.70	2.13
9. % Bite	0.00	.65	.03	.04	.01	.15
10. Th. per I	26.17	30.44	32.02	33.03	30.20	32.65
11. T of I	20.63	24.27	21.99	24.35	22.55	23.90
12. III	71.87	196.27	81.53	115.73	95.38	138.99
13. T of M	2.94	6.15	3.50	3.83	3.38	4.03
14. PIMD	2.00	4.06	2.32	2.68	2.23	2.92
15. Pass-ups	2.52	2.10	1.31	1.86	2.20	2.18
16. Incompletes	.61	.45	.25	.35	.46	.31
17. Skips before Virginal	3.74	1.45	1.71	4.93	4.94	4.02
18. Skips after Virginal	3.32	2.43	.74	2.17	2.61	1.92
19. Recovery	88.52	1.52	2.24	35.93	40.43	19.13

Intromission Latency

The first measure that we wish to consider in some detail is Intromission Latency, defined as the number of seconds from the introduction of the female until the male first gains intromission. Fig. 6 shows the mean values for the six genotypes plotted after the manner described by Bruell (1962). Table 8 summarizes the results of two-tailed *t* tests that were applied to the data in a search for significant differences.

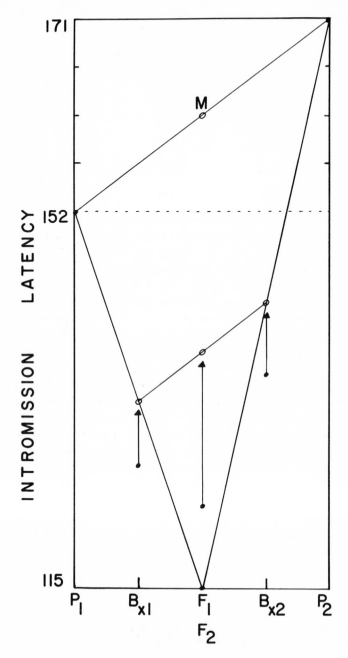

FIG. 6. Observed means (solid circles) and expected means (open circles, as predicted by formulas A, B, and C) for Intromission Latency. M = mid-parent.

TABLE 8

Significant Differences in Intromission Latency
(Two-tailed *t*-tests. NS = not significant)

	C57 (P_1)	*DBA* (P_2)	F_1	F_2	B_{x1}
DBA (P_2)	NS				
F_1	0.001	0.001			
F_2	0.001	0.001	NS		
B_{x1}	0.02	0.02	NS	NS	
B_{x2}	NS	0.001	0.01	NS	NS

As Table 8 shows, the difference between the means of P_1 and P_2 shown in Fig. 6 is not statistically significant. The F_1 group was significantly faster in gaining intromission than both inbred parent strains. These findings are similar to those for the much smaller sample shown in Table 2. The segregating populations are also significantly faster than the parents, except for the B_{x_2}-P_1 comparison. The pattern of inheritance is clearly that of overdominance in the direction of more rapid penetration of the estrous females.

Perhaps the most widely used model for the analysis of data such as these is that of Mather (1949) as summarized in Broadhurst and Jinks (1961). This model assumes that epistatic and genotype-environment interactions are absent. If such interactions are present, the model requires their removal by rescaling of the data. It is perhaps worth noting that, as Manosevitz (1967) has pointed out, genotype-treatment effects are worth studying in their own right. Furthermore, as Henderson (1967, p. 373) has noted, "It is especially important that investigators studying behaviour should consider the problems concerning scale transformations discussed by Falconer [1960], particularly with regard to cautions concerning transformations made without logical justification."

If nonallelic interactions are absent, the means of the nonsegregating populations are predictable from the means of the segregating populations in the following fashion:

$$\text{A.} \quad \overline{B}_1 = \frac{(\overline{P}_1 + \overline{F}_1)}{2}$$

$$\text{B.} \quad \overline{B}_2 = \frac{(\overline{P}_2 + \overline{F}_1)}{2}$$

$$\text{C.} \quad \overline{F}_2 = \frac{(\overline{P}_1 + 2\overline{F}_1 + \overline{P}_2)}{4}$$

The open circles in Fig. 2, 6, and 7 indicate these relationships. They may also be stated in the form of null hypotheses:

$$\text{A.} \quad 2\overline{B}_1 - \overline{P}_1 - \overline{F}_1 = 0$$

$$\text{B.} \quad 2\overline{B}_2 - \overline{P}_2 - \overline{F}_1 = 0$$

$$\text{C.} \quad 4\overline{F}_2 - 2\overline{F}_1 - \overline{P}_1 - \overline{P}_2 = 0$$

Use of appropriate standard errors (Bruell, 1962) allows tests of deviations from the predicted results.

Genotype-environment interactions are indicated by significant differences between the variances of the three segregating populations. Bartlett's test (McNemar, 1962, p. 249) may be used to test for significant inhomogeneity of variance among the three groups.

When these tests were applied to the raw data for Intromission Latency, test C proved significant, and Bartlett's test indicated significant inhomogeneity of variance of the nonsegregating populations. A logarithmic transformation was successful in removing both the genotype-environment interaction and the nonallelic interaction.

Once the scaling assumptions are met the variance of the F_2 population (V_P) can be partitioned into two major components, genotypic variance (V_G) and environmental variance (V_E). V_E may be estimated by a weighted mean of the variances of the three nonsegregating populations as follows:

$$V_E = \frac{n_{P_1} V_{P_1} + n_{P_2} V_{P_2} + n_{F_1} V_{F_1}}{n_{P_1} + n_{P_2} + n_{F_1}}$$

When this estimate of environmental variance is subtracted from the total variance of the F_2 generation the remainder is an estimate of the genotypic variance (V_G).

The total variance of the F_2 population (V_P) on the logarithmic scale was 0.052. $V_E = 0.045$. Therefore, $V_G = 0.007$.

These calculations permit us to determine "heritability in the broad sense," or what Falconer (1960) perhaps more appropriately called "degree of genetic determination."

$$^\circ GD = \frac{V_G}{V_P}$$

Degree of genetic determination for Intromission Latency on the logarithmic scale is 0.135. In other words, about 14 percent of the trait variance is due to genotypic variance.

V_G can be analyzed into an additive component, V_A, and a dominance component, V_D [this term also includes any interactions between V_D and V_A (Falconer, 1960, p. 139)].

Broadhurst and Jinks (1961) present formulas from which it is possible to derive the following:

$$V_A = 2V_{F_2} - V_{B_1} - V_{B_2}$$

$$V_D = V_{F_2} - V_A - V_E$$

These formulas are identical to those proposed by Wright (1952) and yield the following values when applied to the logarithmic transformations for Intromission Latency:

$$V_A = 0.008$$

$$V_D = -0.002$$

The occurrence of negative values when applying these and other bio-metric genetic formulas is not unusual (Wright, 1952, p. 27). Currently, however, there seems to be no adequate explanation for the occurrence of negative values. In this case, it may be safe to assume that most of the genotypic variance is due to V_A.

The calculation of V_A permits an estimate of "heritability in the narrow sense." This is defined as:

$$h^2 = \frac{V_A}{V_P}$$

Applying this formula to the current data results in an estimate of h^2 of 0.154.

We may conclude then that about 15 percent of the variance in the F_2 population is due to genetic differences and that this genotypic variance is composed primarily of additive genetic variance.

It is also possible to subdivide V_E, the environmental variance, into two components. These are called special environmental variance, V_{E_s} and general environmental variance, V_{E_g} (Falconer, 1960, p. 143). V_{E_s} refers to "within-individual variance arising from temporary or localized circumstances," according to Falconer, while V_{E_g} refers to "the environmental variance contributing to the between-individual component and arising from permanent or nonlocalized circumstances."

These components of environmental variance are obtained by deriving the between-individuals and within-individual components of variance for the F_2 generation. The between-individuals component will consist of V_G and V_{E_g}. When the between-individuals component is divided by the total variance of F_2, the result indicates the correlation between repeated measurements of the same animal. This intraclass correlation (r) is called the *repeatability* of the character. Therefore:

$$r = \frac{V_G + V_{E_g}}{V_P}$$

Calculating repeatability separates the component due to V_{E_s}, but leaves V_{E_g} confounded with genotypic variance. For Intromission Latency (raw data) this analysis resulted in the finding that 9 percent of the variance was due to the between-individuals component, while 91 percent was due to within-individual or special environmental variance, V_{E_s}. In other words, most of the variance in this trait can be accounted for by variation in the scores of an individual from test to test. This is perhaps not too surprising when we consider that the male's score on this particular measure depends in large part on the behavior of the female.

Repeatability also measures the gain in accuracy to be expected from repeated measurements. When repeatability is high, there is little gain in accuracy from repeated testing. In the present case, repeatability was quite low (0.16 for the raw score data) and the gain in accuracy from taking five measurements rather than one was about 65 percent (Falconer, 1960, p. 147). Taking five more measurements would only have increased accuracy by another 10 percent. Repeatability also sets the upper limits to h^2 and degree of genetic determination.

In summary, Intromission Latency is inherited in a heterotic fashion with genetic differences accounting for a relatively low proportion of trait variance [this finding is by no means unusual in quantitative genetics (Mather, 1949, p. 5)]. The genetic variance present seems to be due primarily to additive variance rather than to dominance variance. The large environmental component of variance is due primarily to special, nonlocalized variation that occurs from test to test within individual animals.

Ejaculation Latency

Our second example is that of Ejaculation Latency, defined as the number of seconds from the first thrust of intromission until the male begins to ejaculate. Fig. 7 shows the means for the six groups taken from Table 7 and rounded to the nearest 10 seconds. Note that this measure was one of two where the difference between the backcross means was not in the same direction as the difference between the parental means. Table 9 shows the significant differences resulting from two-tailed t tests.

As Table 9 shows, the parent generations differed significantly. The *C57BL* males required less time to attain ejaculation after mating had been initiated. The F_1 group was significantly faster than both of the parent strains. The direction of these differences is the same as that shown in Table 2, however, in the small-sample study summarized in that Table, none of the differences were significant. The F_2 group is also significantly faster than either inbred parent strain.

When Mather's scaling tests were applied to these data, tests B and C revealed significance, as did Bartlett's test for homogeneity of variance. Neither

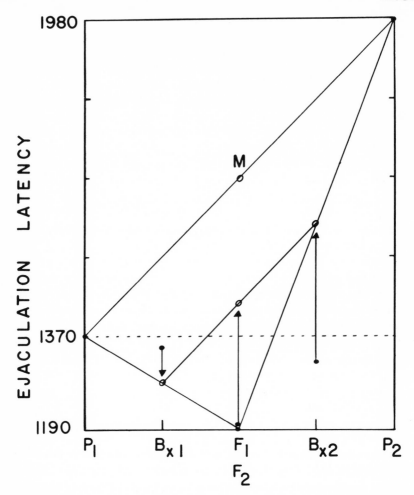

FIG. 7. Observed means (solid circles) and expected means (open circles, as predicted by formulas A, B, and C) for Ejaculation Latency. M = mid-parent.

square-root nor logarithmic transformations removed the significant differences. Therefore, the following analysis is based upon the raw data. It should be kept in mind that scaling requirements have not been met.

The analysis revealed the following:

$$°GD = 0.18$$

The h^2 calculation gave results considerably in excess of 0.18. However, repeatability, which, it will be recalled, sets an upper limit to degree of genetic determination and heritability, was 0.27. Since the formula for V_D produced

TABLE 9

Significant Differences in Ejaculation Latency
(Two-tailed *t*-tests. NS = not significant)

	C57 (P_1)	DBA (P_2)	F_1	F_2	B_{x_1}
DBA (P_2)	0.01				
F_1	0.05	0.001			
F_2	0.001	0.001	NS		
B_{x_1}	NS	0.001	0.02	NS	
B_{x_2}	NS	0.001	NS	NS	NS

a negative value, it is perhaps safe to estimate that both degree of genetic determination and heritability are about 15 to 25 percent.

Analysis of V_E showed that 73 percent of the environmental variance was due to V_{E_s} while 27 percent was due to $V_{E_g} + V_G$. Once again, then, we note that the major component of variance is that due to within-animal variation from test to test. As with Intromission Latency, it may be logical to attribute much of the variation in Ejaculation Latency to differences in the behavior of the females, in spite of the fact that an effort was made to control for this factor.

CONCLUSIONS

The first third of this chapter presented evidence supporting two limiting generalizations concerning the results of behavior-genetic experiments. These generalizations should be kept in mind in interpreting the results of any behavior-genetic experiment, particularly when one wishes to extrapolate from one genotype or population to another, or from one situation to another. However, the limitations do not preclude further experimentation into areas of research that may prove to be the most valuable fruits of the behavior-genetic approach. Within the confines set by the strains selected for study and the experimental situation under which the study was carried out, behavior-genetic analysis presents us with insights and testable hypotheses that may be derived in no other way. Testing these hypotheses may well lead to further experiments involving what many consider to be the basic goal of behavior genetics: the establishment of physiologic differences that underlie behavioral differences between genotypes.

The biometric analysis presented above has stressed the division of population variance into four major components: (1) additive genetic variance (V_A), (2) dominance variance (V_D), (3) special environmental variance

(V_{E_s}), and (4) general environmental variance (V_{E_g}). It is obvious from the formulas presented in this chapter that the estimates of these components are interdependent. For example, any environmental factor that the experimenter does not adequately control will decrease the estimate of genotypic variance and increase the estimate of environmental variance. Control of all possible sources of environmental variation is thus of extreme importance in behavior-genetic analysis.

Assuming that the experimenter recognizes the limitations under which he is working, and that he has controlled in every conceivable manner for the myriad sources of environmentally determined variance, what can be done to test his conclusions and perhaps to open the door for research concerned with the physiologic correlates of behavioral differences? The tests that will be selected depend upon the relative size of his estimates of the components of variance. If additive genetic variance (V_A) is relatively large, the character should respond fairly rapidly to the pressures of artificial selection. If non-additive variance (V_D) plus any interaction between V_D and V_A is relatively large, selection should result in rapid stabilization within the selected lines. If general environmental variance (V_{E_g}) is relatively large, the experimenter might be advised to search for prenatal and maternal effects by the transference of fertilized ova from one genotype to another and by cross-fostering of litters. When special environmental variance (V_{E_s}) is large, the experimenter should direct his attention toward possible sources of environmental variation that escaped his attention or appeared unimportant when the experiment was designed.

The results for the two measures analyzed above (Intromission Latency and Ejaculation Latency) indicate that genotypic variance accounts for a relatively low proportion of total variance. However, most or all of the genotypic variance seems to be additive. Thus we might hypothesize that a selection experiment for these two measures would be successful, but slow. The environmental variation present seems to be due to within-individual variation from test to test. The results of our analysis indicate that this is the major component of variance. In the present experiment, a number of uncontrolled factors might have contributed to V_{E_s}. Barometric pressure, recency of feeding or drinking, time of day of testing (or other biologic rhythms), nature of interaction with other males prior to the test, and behavior of the female, as discussed above, are a few of the possibilities. Tests of some of these variables are currently in progress in our laboratory. Successful tests of any of these hypotheses may lead to further hypotheses regarding physiologic differences underlying behavioral differences.

In summary, the behavior-genetic approach to the determinants of behavior has strong limitations and great advantages. Both must be recognized if the promise of behavior-genetic analysis is to be fulfilled.

REFERENCES

Broadhurst, P. L. 1960. Experiments in psychogenetics. *In* Eysenck, H. J., ed., Experiments in Personality, Psychogenetics and Psychopharmacology, New York, The Humanities Press, Vol. 1, pp. 3–102.

———— and J. L. Jinks. 1961. Biometrical genetics and behavior: Reanalysis of published data. Psychol. Bull., 58:337–362.

Bruell, J. H. 1962. Dominance and segregation in the inheritance of quantitative behavior in mice. *In* Bliss, E. L., ed., Roots of Behavior, New York, Harper and Row, Pubs., 48–67.

———— 1964. Inheritance of behavioral and physiological characters of mice and the problem of heterosis. Amer. Zool., 4:125–138.

Champlin, A. K., and T. E. McGill. 1967. A simple restraining device for injecting mice. Lab. Anim. Care, 17:600–601

Falconer, D. S. 1960. Quantitative Genetics, New York, The Ronald Press Co.

Fuller, J. L. 1964. Measurement of alcohol preference in genetic experiments. J. Comp. Physiol. Psychol., 57:85–88.

———— 1967. Effects of the albino gene upon behaviour of mice. Anim. Behav., 15:467–470.

———— and W. R. Thompson. 1960. Behavior Genetics, New York, John Wiley and Sons, Inc.

Goy, R. W., and J. S. Jakway. 1962. Role of inheritance in determination of sexual behavior patterns. *In* Bliss, E. L., ed., Roots of Behavior, New York, Harper and Row, Pubs, pp. 96–112.

Hafez, E. S., ed. 1962. The Behaviour of Domestic Animals. London Baillière, Tindall and Cox.

Henderson, N. D. 1967. Prior treatment effects on open field behaviour of mice: A genetic analysis. Anim. Behav., 15:364–376.

Henry, K. R., and K. Schlesinger. 1967. Effects of the albino and dilute loci on mouse behavior. J. Comp. Physiol. Psychol., 63:320–323.

Land, R. B., and T. E. McGill. 1967. The effects of the mating pattern of the mouse on the formation of functional corpora lutea. J. Reprod. Fert., 13:121–125.

Levine, L., G. E. Barsel, and C. A. Daikow. 1966. Mating behaviour in two inbred strains of mice. Anim. Behav., 14:1–6.

Manosevitz, M. 1967. The analysis of gene-environment interactions. *In* Fuller, J. L., Chm., The experimental analysis of genetics of behavior: The mouse as a model. Symposium presented at the American Psychological Association, Washington, D. C.

Mather, K. 1949. Biometrical Genetics, London, Methuen.

McGill, T. E. 1962a. Sexual behavior in three inbred strains of mice. Behaviour, 19:341–350.

———— 1962b. Reduction in "head-mounts" in the sexual behavior of the mouse as a function of experience. Psychol. Rep., 10:284.

———— 1963. Sexual behavior of the mouse after long-term and short-term postejaculatory recovery periods. J. Genet. Psychol., 103:53–57.

———— 1965. Studies of the sexual behavior of male laboratory mice: Effects of genotype, recovery of sex drive, and theory. *In* Beach, F. A., ed., Sex and Behavior, New York, John Wiley and Sons, Inc., pp. 76–88.

———— 1968. Sexual behavior in male mice. *In* Stokes, A. W., ed., Laboratory Manual of Animal Behavior, San Francisco, W. H. Freeman and Co., Pubs.

———— and W. C. Blight. 1963a. Effects of geneotype on the recovery of sex drive in the male mouse. J. Comp. Physiol. Psychol., 56:887–888.

———— and W. C. Blight. 1963b. The sexual behaviour of hybrid male mice compared with the sexual behaviour of males of the inbred parent strains. Anim. Behav., 11:480–483.

———— D. M. Corwin, and D. T. Harrison. 1968. Copulatory plug does not induce luteal activity in the mouse *Mus musculus*. J. Reprod. Fert., 15:149–151.

———— and T. W. Ransom. 1968. The effects of genotypic change on conclusions regarding modes of inheritance of elements of behavior. Anim. Behav., 16:88–91.

McNemar, Q. 1962. Psychological Statistics, 3rd ed., New York, Wiley and Sons.

Parsons, P. A. 1967. The Genetic Analysis of Behaviour, London, Methuen.

Thiessen, D. D. 1966. Pleiotropism in behavior genetics: The mouse as a research instrument. Percept. Motor Skills, 23:901–902.

Werboff, J., A. Anderson, and S. Ross. 1967. Mice of a four-way cross: Coat color associated with behavior and response to d–amphetamine. J. Psychol., 66:99–117.

Wright, S. 1952. The genetics of quantitative variability. *In* Reeve, E. C. R., and C. H. Waddington, eds., Quantitative Inheritance, London, H. M. Stationery Office, pp. 5–41.

SECTION II

Gene-Environmental Interplay

4

Genetic Variation and Hoarding*

MARTIN MANOSEVITZ AND GARDNER LINDZEY
Department of Psychology
The University of Texas at Austin

INTRODUCTION

Hoarding

The retrieval and storage of food, whether in the natural habitat or under controlled laboratory conditions, presents a fascinating and, in some respects, baffling empirical phenomenon. Within a few moments an animal may accumulate sufficient food to enable him to survive for a considerable period of time. Why does the animal hoard? What are the motivational, physiologic, genetic, and learned determinants of this complex, but well-integrated behavior pattern?

The general purpose of this chapter is to focus on genetic and experiential determinants of hoarding. Specifically we shall provide some historical observations, identify the interest behavioral scientists have displayed in hoarding, briefly summarize that portion of the literature which concerns genetic variation and hoarding, and present data from our own research on the determinants of hoarding. In our studies we have investigated strain variation in hoarding, estimated genetic parameters, manipulated hoarding through a variety of adult and early experiences, and studied genotype-treatment interactions.

Zoologists and psychologists have been interested in hoarding since early in the century. Zoologists typically have been concerned with seasonal variation, type and amount of food or other materials retrieved and stored, and the survival or adaptive value of this behavior. Early psychologists also have been interested in hoarding (James, 1950; McDougall, 1923), and have viewed hoarding or acquisitiveness as a significant component of behavior in both

* The research reported here was supported in part by grants from NIMH #MH-11276 and #MH-14076. The preparation of this chapter was facilitated by NSF grant #GU-1598.

lower animals and man. For example, McDougall considered hoarding to be an instinct displayed by almost all humans and by many animals.

The experimental study and quantitative measurement of hoarding began with Wolfe's (1939) study. From that time until the mid 1950's hoarding was actively studied by a number of investigators. These experiments focused on variables such as previous experience with food and deprivation, physiologic and cortical factors, learning, temperament, and testing conditions. A number of reviews of the hoarding literature have been presented (e.g., Bindra, 1959; Cofer and Appley, 1964; Ross et al., 1955; Thorpe, 1963).

Naturalistic observations indicate that many species hoard, i.e., hamsters, gophers, squirrels, beavers, chipmunks, mice, and rats. Although the rat has been used most extensively in laboratory studies of hoarding, studies using hamsters (e.g., Bevan and Grodsky, 1958; Koski, 1963; Smith et al., 1954; Smith and Ross, 1950) and mice (e.g., Ross et al., 1955; Smith and Powell, 1955) have been reported.

Why do many species engage in hoarding without direct instruction or training? The seasonal variation in the frequency of this behavior appears to suggest that the animal has some "foresight" into the availability of food in the future. This type of behavior has posed a challenging question which has interested learning (Marx, 1950; Mowrer, 1960), physiologic (Morgan, 1965), and motivational psychologists (Bindra, 1959). Attempts to ascertain the variables that control hoarding behavior have varied. It appears that the answer to the question posed several years ago by Bindra (1948), "What makes rats hoard?" has even at this time not been satisfactorily answered. Simple theories of hoarding which focus on single variables as *the* cause of hoarding, i.e., food deprivation, temperature or nutritional deficiencies appear clearly insufficient. There is ample evidence to demonstrate that this behavior pattern is complexly determined and that single variable formulations will not provide a satisfactory explanation.

The basic characteristic of all definitions of hoarding is that an animal carries or transports some object, almost always food, into the home cage, from a supply some distance from the home cage or territory. Although most studies of hoarding have been concerned with food, a few studies have focused on hoarding of nonfood objects such as wooden blocks (Miller and Viek, 1944; Ross et al., 1950), and water-soaked cotton dental pads (Bindra, 1947). Hoarding is usually measured by the number of food pellets retrieved during a specified test period. Depending on the particular study, the length of the test period may range from 15 to 60 minutes. Hoarding is considered to occur when the animal retrieves sufficient food to last him until the next opportunity to hoard. However, some investigators have chosen a specific number of pellets that an animal must retrieve in each trial before he reaches criterion and is considered to be a "hoarder." In these experiments, animals who do not meet the criterion are usually disregarded. In addition, the number of hoarding trials required before an animal reaches the criterion has been analyzed as an additional measure of hoarding. In almost all studies, hoarding is measured using multiple trials, but only on rare occasions has hoarding performance across trials received much attention.

Although there have been many studies which have attempted to assess determinants of hoarding, genetic variation has been largely neglected. The focus of this chapter will be upon this slighted source of variation.

EVOLUTIONARY CONSIDERATIONS

Retrieving and storing food has obvious adaptive value both at the individual and population level. Storing food during times of plenty, which often coincides with seasonal changes, may insure adequate supplies during lean years or dormant seasons. Animals who insure their food supply by hoarding will have a selective advantage over those members of the species who do not hoard. The fable of the ant and the grasshopper suggests that this insight is an established part of folk wisdom. Presumably the animals that hoard adequate supplies will contributė differentially to the gene pool for the next generation. Species that do not hoard or hoard very little may be at a selective disadvantage and may face extinction if periods of inadequate food prevail for any length of time.

Assuming that all rodents have been exposed to some periods of inadequate food supplies we would expect that only those who hoarded survived. Not all animals hoard equally, and it is possible that some strains have been living in environments where food supplies have not fluctuated. In such populations selective pressure for hoading behavior probably has been relaxed and the average animal may not hoard as much as animals chosen from environments where food supplies were not plentiful.

Even in environments in which an adequate food supply is continuously available, hoarding may confer selective advantage. For example, the animal which hoards a food supply in or close to his nesting area does not have to venture out as frequently to forage and find food. If predators live in the same environment, the more frequently an animal ventures from its nest the more exposed and prone it is to predators. This may be particularly important when the young are beginning to ingest solid food since the young animal may be particularly vulnerable to predatory attack. If the parents have provided an adequate food supply in or near the nesting site the potential hazard to the young is reduced. The evolutionary problems related to hoarding are of special interest to the behavior geneticist because of the obvious adaptive value *of hoarding!* of this character.

HOARDING AND HEREDITY

Over the years, a number of writers (McDougall, 1923; Morgan, 1947; Thorpe, 1963) have discussed the possibility that hoarding is instinctive. In Morgan's (1965) textbook, *Physiological Psychology,* hoarding is discussed in the chapter on instinctive behavior. The complex sequence of acts which comprise hoarding behavior appear to be universally present in several species and do not appear to be learned. When hoarding is considered as an instinct, the operation of genetic and biologic variables as factors in hoarding become of paramount interest. It should be recalled, however, that several psychologists (Holland, 1954; Marx, 1950; Mowrer, 1960) have argued that hoarding is a learned response.

Clearly it is not productive to present the issue in terms of instinct versus learning or nature versus nurture. The important question is whether, under specified environmental and biologic conditions, one can identify accurately the relative contributions of these two sources of variation. Hoarding, as all other behavior, is influenced by both hereditary and environmental determinants.

In spite of the interest in this question, up to the time of our publication in 1964 (Lindzey and Manosevitz, 1964) we were able to identify only four studies (Koski, 1963; Smith and Powell, 1955; Stamm, 1954, 1956) which investigated the influence of genetic variation upon hoarding. In several studies, genetic factors were mentioned and in some, explicit suggestions were made indicating the potential value such studies might have.

Koski (1963) used two strains of golden hamsters and reported no significant strain differences. However, he reported that the bimodal distribution of scores in his study supported a single gene theory of hoarding. Smith and Powell (1955) studying the relationship of emotionality and hoarding used mice from three inbred strains, and they observed significant strain differences. Stamm (1954) used three inbred strains of rats and also reported significant strain differences.

Only Stamm (1956) proceeded to investigate the nature of the genetic mechanisms involved. He mated animals from the high hoarding strain with animals from the low hoarding strain and produced an F_1 generation. He also produced one backcross generation by mating animals from the F_1 generation with animals from the low hoarding parental strain. This breeding design did not include the F_2 and the other backcross generation and thus precluded definitive conclusions on the mode of transmission of the gene(s) which influence hoarding. Stamm's results showed that subjects from the F_1 generation performed at a level similar to the high hoarding parental strain. The mean score of the backcross generation fell between the two parent generations, and the distribution of the backcross group was bimodal. Stamm suggests that his data are compatible with the operation of a single dominant gene. Presumably the high hoarding strains have either one or two of the dominant genes and the low hoarding strain has both recessive genes. These data are interpreted by Stamm as supporting a simple Mendelian model. However, he acknowledges that the data do not exclude alternative explanations, one of which could be polygenic.

Due to limitations in the breeding plan used by Stamm only a limited genetic analysis could be performed. It is impossible on the basis of his data to conclude what is the most probable mode of transmission for the gene(s) relevant to hoarding, or to estimate genetic parameters.

Our research program has had two major aims. First, to assess the extent to which genetic variation is a determinant of hoarding, and second, to discover how genetic factors interact with specified environmental variation.

GENETIC VARIATION AND HOARDING

Our early studies attempted to measure hoarding reliably in the laboratory and to demonstrate the effects of genetic variation upon hoarding using mice. In the initial studies we were unable to obtain stable hoarding at the levels reported by other workers (Lindzey and Manosevitz, 1964). However, we demonstrated that genetic variation was an important determinant of hoarding although the results were so unstable that an estimate of the magnitude of the genetic contribution was not possible. These studies also suggested the importance of genotype-treatment interactions.

An additional finding from our earlier studies, worthy of special note, was the shape of the curves for the 10 daily trials. We found in many of our studies a tendency for hoarding to decrease over trials. This was consistent with results reported by Smith and Ross, (1953) using *C3H* mice tested under deprivation conditions. In our subsequent studies we have not obtained curves showing decreasing trends. This may be due to the use of rather severe deprivation schedules in the early studies with the consequence that in successive trials animals became hungrier and therefore spent more time eating and less time hoarding. This interpretation fits with our later data (see below), where a nondeprivation feeding schedule was used.

Our initial studies revealed variable and infrequent hoarding which did not approach the levels reported by other workers. At this time our procedures and apparatus followed as closely as possible that used by Smith and Ross (1953). Several additional attempts were made to obtain stable hoarding at levels comparable to those reported by Smith and Ross. A variety of genotypes were used including animals from the same strain employed in the earlier study. Early experience, food deprivation, rearing conditions, and testing conditions were all manipulated with no success. On the basis of this pilot work changes in the apparatus and feeding schedule during testing were made.

The apparatus differed in many respects from that used by previous investigators studying hoarding in mice. To make the hoarding response prepotent, it was necessary to make the home cage relatively dark and sheltered, and the alley brightly illuminated. During the course of the pilot work, it was found that a large home cage and alley facilitated hoarding. Furthermore, we reduced the probability that other responses would be emitted during the hoarding trial by making it difficult for the Ss to climb up the alley sides, gnaw on the apparatus, or engage in similar activities. This was accomplished by making the alley sides smooth and all corners tight. Several feeding schedules were tried in the pilot studies. Eventually, we identified a feeding schedule that did not constitute deprivation, yet did make the Ss active enough to emerge from the home cage at the start of each trial. The apparatus and testing and feeding schedules used in our subsequent studies have been described previously (Manosevitz, 1965, 1967a).

After screening a large number of inbred mouse strains, we selected the two strains which were most divergent. One, the *JK* strain, hoarded better than any other strain in our laboratory. The *C57* strain was chosen from several strains which hoarded at low levels.

In our next studies (Manosevitz, 1965; Manosevitz and Lindzey, 1967) we attempted a more detailed inquiry into genetic factors. In these studies, starting with two inbred strains, the F_1, F_2, and reciprocal backcross generations were bred and tested. A biometric analysis was performed following the procedures outlined by Broadhurst and Jinks (1961).

The first part of the genetic analysis was performed using only the generation means. The expected midparent value was calculated using the observed means of the two parental generations. The observed F_1 and F_2 means did not deviate significantly from the midparent average (see Table 1). In addition, the F_1 and F_2 did not differ from each other. By computing the expected mean values for the two backcross generations, we can determine whether the observed means differ from the predicted means. As shown in Table 1, the observed means approximated the expected values quite closely. Generally, the nonsignificant differences between the expected and observed means suggest the operation of additive genetic factors. However, there is some suggestion that nonadditive factors may also be contributing to the observed means. The *C57* mean was significantly different ($p < 0.01$) from the F_1 mean. This indicates that the F_1 generation mean shifted toward the high hoarding *JK* strain average. Thus, there is some evidence that partial average dominance or potence may also be significant. Further information, regarding additive, dominance, and interaction components can be obtained by using Mather's model (1949) to estimate these components. However, before proceeding to this analysis, Mather's scaling tests must be

TABLE 1

**Observed and Expected Means for Each
Generation Using Raw Scores**

Generation	Observed	Expected
P_1 (*JK*)	114.8	—
P_2 (*C57BL/1*)	40.9	—
F_1	94.9	77.8[a]
F_2	78.6	77.8[a]
B_1 ($P_1 \times F_1$)	101.7	104.8[b]
B_2 ($P_2 \times F_2$)	65.5	67.9[c]

[a]Expected value = $(\bar{P}_1 + \bar{P}_2)/2$
[b]Expected value = $(\bar{P}_1 + \bar{F}_1)/2$
[c]Expected value = $(\bar{P}_2 + \bar{F}_1)/2$

applied to the data. Mather, among others (Broadhurst, 1960; Bruell, 1962), has emphasized that unless the scaling requirements are met only limited information can be obtained from further analysis.

It has been shown (Manosevitz and Lindzey, 1967) that Mather's scaling tests, A, B, and C, are nonsignificant on these data when computed using the raw scores. However, there is a significant genotype × environment interaction ($p < 0.01$) which can be removed by transforming each animal's hoarding score to its square root. In our discussion of the results, all analyses have been carried out using transformed scores. The means reported in Table 1 were recomputed using these transformed scores and the same results were observed.

After the scaling requirements are met, one may proceed with the estimation of heritable components. The only significant heritable component was additivity. Although there was some suggestion of dominance (F_1 mean significantly different from P_2 mean) the more extensive analysis using Mather's model indicates that this is not a significant source of variation. One possibility for this discrepancy may be the large variability found for dominance (h) using Mather's model. Further investigation using other strains and other genetic models, for example the diallel cross, should provide valuable information regarding these heritable components.

The biometric procedures suggested by Broadhurst and Jinks allows the estimation of the proportion of the variance due to various genetic factors using the variance or second degree statistics. By using a variety of estimation procedures several heritability estimates may be calculated.

Several writers (Lush, 1945; Broadhurst and Jinks, 1961) refer to the proportion of heritable variation in the population as heritability "in the broad sense." However, this component of variation is referred to by Falconer (1963) as the coefficient of genetic determination. Falconer reserves the use of the term heritability to designate heritability "in the narrow sense." Heritability defined in this way is the ratio of the additive variation to the sum of the additive and nonheritable variations. As in our previous publications, we shall employ Falconer's terminology.

The coefficient of genetic determination using data from the two parental generations and their F_1 and F_2 generations provides two estimates. In the F_2 generation 37 percent of the variation was estimated to be due to genetic factors. In the two parent lines the coefficient of genetic determination was estimated to be 49 percent. Thus, some general idea of the magnitude of the coefficient of genetic determination can be obtained from these two values.

Heritability was computed to be between 44 and 54 percent when the backcross data were not available. However, when the backcross data are used, it is possible to estimate the value of D and H directly. When this was done the heritability estimate was 29 percent. When the backcross data were missing, the heritability estimates we obtained were considerably higher. Therefore, before all six generations (P_1, P_2, F_1, F_2, B_1, and B_2) are tested estimates based on Mather's model should be made with caution.

In our next study (Manosevitz, 1967a) mice from three inbred strains were studied. Using a model developed by Bruell (unpublished) estimates of additive variation, dominant variation and heritability can be made using inbred strains only.

The results of this analysis indicate that when one assumes the absence of dominance, 31 percent of the total phenotypic variation is due to additive factors and the remainder is due to environmental factors. If one assumes the presence of dominance, additive variation is estimated at 27 percent, dominance variation at 13 percent, and the remainder is environmental variation. These results can be compared with those obtained by Manosevitz (1965) where two different inbred strains were used. In that analysis, 49 percent of the total phenotypic variance was estimated to be due to genetic factors. On the basis of data obtained from inbred strains, the general magnitude of genetic variation in hoarding behavior appears to be between 25 and 50 percent of the total genetic variation.

The results of our biometric analyses using several different strains, crosses, models, and methods of analysis can be summarized as follows. The analyses fit a polygenic model of hoarding. Therefore, any model of hoarding which suggests that only one or two genes are involved appears to be inadequate. The one component that appears to be significant and larger than any other component is additive or fixable genetic variation. There is some limited evidence that dominance is also important but, as shown above, its magnitude is smaller than the contribution of additive genetic variation. In general, genetic factors appear to be a significant proportion of the total phenotypic variation, the lower limit appears to be about one-third and the upper limit approximately one-half. Thus, in any study of hoarding it appears that genetic factors contribute a substantial portion of the variability and future investigators should be aware of this source of variability in designing their studies.

Of the total genetic variation about one-quarter to one-half appears to be heritable, i.e., due to additive genetic effects. However, comparison of heritability values between studies must be made with extreme caution.

As stated earlier, our research interests have two major aims: (1) assessing the extent and nature of genetic factors in hoarding and (2) determining the effects of environmental manipulations on hoarding and the interaction between environment and genotype. Let us now turn our attention to this latter set of concerns.

ENVIRONMENTAL VARIATION AND HOARDING

We have used a variety of environmental manipulations or "treatments" in our studies. Each treatment was selected because of a particular interest we had in that experience as a determinant of hoarding. The one common theme these treatments have is that all were designed in such a way that we expected them to increase hoarding.

In one of our early studies (Lindzey and Manosevitz, 1964), animals were reared in large cages with food pellets on the cage floor. It was expected that these animals would hoard at higher levels than previously observed because they had additional experience in carrying and manipulating food pellets. Within this large cage the animals indeed carried food from place to place and in some instances considerable amounts of food were stored in one area. However, the treatment did not markedly change the level of hoarding. In the next study in this series, half the animals from each genotype were exposed as adults to a severe food deprivation experience for 10 days, in groups of 12 animals each. The deprivation schedule was set to maintain the animals at 85 percent of their ad libitum weight. Active food competition was produced in this type of group setting. The remaining (control group) mice were maintained on the same deprivation schedule but they were housed individually. We expected increased hoarding responses in those animals who experienced group food competition. However, the treatment effect only approached significance ($p < 0.10$). In the analysis of the daily hoarding curves, there was a significant ($p < 0.001$) strain $\times$ treatment interaction. This interaction was mostly due to $C57$ animals who experienced the group competition feeding and failed to show the usual decrease in hoarding performance across trials.

At this point, the absence of illuminating findings suggested the examination of the procedures and apparatus we were using. On the basis of additional pilot work, the changes in the apparatus described above were instituted and we found that stable hoarding could be obtained with this apparatus and procedure. In all of the studies to be reported below this modified procedure and apparatus were used.

A further study (Manosevitz, 1965) was designed to see what effect an increase in fear would have upon hoarding. Several investigators have studied the association between emotion and hoarding, and shyness and hoarding. Smith and Powell (1955), using female mice from three inbred strains (*BALB/c, C3H,* and *C57BL/10*), studied the association between open-field defecation-urination, latency to leave open-field starting area, latency to reach the center of the open field, and hoarding. Latency to emerge from the hoarding cages on the initial hoarding trial was also recorded. Their study revealed that the *BALB/c* strain had the highest defecation-urination scores, the longest open-field and alley entrance latencies, and also hoarded the largest number of pellets. They concluded that their data supported the hypothesis that emotionality or timidity was positively related to hoarding.

Broadhurst (1958) in a study of strain differences in rats used some of the same strains which Stamm (1954) used and on the basis of his observations suggested that there may be an association between emotionality and hoarding. The order of high to low hoarding strains found in Stamm's study (black hooded II, brown hooded, and brown) is matched by Broadhurst's ordering of high to low emotionality. These data give some additional, although limited, support to the hypothesis that emotionality is positively associated with hoarding performance.

Bindra (1948) suggested that hoarding is to some extent dependent on the security offered by the home cage and the alley. He suggested that an animal's shyness was an important determinant of how secure the alley was, relative to the home cage. In his study, Bindra reported significant differences in hoarding performance between open and closed alleys. He hypothesized that there may be differences in hoarding as a function of the interaction between an animal's shyness or timidity and the nature of the alley. Bindra's hypothesis was tested by Hess (1953). He used shy and unshy rats and open, closed, and wire alleys. Hess found that shy rats hoard more in enclosed runways, whereas unshy rats hoard more in open alleys, thus supporting Bindra's hypothesis. Moreover, he demonstrated that shyness is an important determinant of hoarding under conditions which differ in the amount of security in the alley. It appears that the subject's shyness or timidity may serve as an important determinant of the strength of hoarding.

Collectively these studies provide some evidence that fear or emotionality may affect hoarding. However, they did not involve experimental manipulation of fear. The purpose of our experiment (Manosevitz, 1965) was to assess the effect of an aault fear arousing experience upon hoarding.

The treatment consisted of a 10-second immersion in room temperature water. This treatment was administered to half the animals 15 to 20 minutes before each daily trial. The results indicated that in the *JK* strain the males hoarded more than twice as many pellets as did the control group. In the *C57* strain the treated males hoarded about half as many pellets as the control group. Females from both strains were unaffected by the treatment. An analysis of variance showed that the treatment effect was significant ($p<0.025$) for inbred males. Presumably the treatment affected these animals by increasing level of fear. The significant strain × treatment interaction ($p<0.01$) reflects the fact that the treatment effect in the two strains was not uniform and in fact was bidirectional. If one subscribes to the notion that the relationship between fear and performance is best described by an inverted U then it is possible to consider the *JK* strain as representing a middle or facilitating level of fear while the *C57* strain may have experienced either extreme fearfulness or very little fearfulness. Based upon independent observations in our laboratory it appears probable that the *JK*'s represent a middle level of fear. However, further data is needed to provide firm warrant for this interpretation. At present there seems no plausible explanation of why the females from both strains were unaffected by the treatment.

In this study, data from both F_1 and F_2 animals were available. The F_1 and F_2 treated females hoarded less than their controls. The F_1 and F_2 males were unaffected by the treatment. The general magnitude of the treatment effect in the F_1 and F_2 generation was smaller than that found in the inbred generations. It is not surprising that the treatment was less effective with the crossbred animals. Several geneticists (e.g., Falconer, 1960; Lerner, 1958; Waddington, 1957) and psychologists (Lindzey, 1967; Winston, 1964) have argued that crossbred genotypes are better able to withstand environmental stress and they have provided strongly supporting experimental evidence. These writers assert that crossbred animals or heterozygous animals are better buffered, canalized, or have superior genetic homeostasis or balance than inbred animals. Therefore, heterozygous animals should react less to stress or changes in the environment because their genetic variability confers upon them greater stability in the face of changing environment.

The above studies involved largely adult treatments while the studies to be discussed next used treatments which exposed animals earlier in their developmental history to experimental manipulation. In the first of these experiments (Manosevitz, 1966), half the mice from a single inbred strain, *SWR/J*, were exposed on day of birth to a 300-R dose of X-irradiation. This study was concerned with the association between emotionality, activity, and hoarding. It has been suggested that a positive association exists between general activity and hoarding (Ross et al., 1955). Several workers have suggested that there is a positive association between emotionality or fear and hoarding (Bindra, 1948; Broadhurst, 1958; Hess, 1953; Smith and

Powell, 1955) and their findings were discussed above. In addition, our study (Manosevitz, 1965) briefly reported above also provides some limited support for this suggestion.

In the present study we predicted that irradiated animals would hoard more than nonirradiated animals. This prediction was based on several considerations. Previous findings in the literature indicated that hoarding was positively associated with general activity. In our earlier work (Manosevitz and Rostkowski, 1966) we demonstrated that irradiated animals are more active than nonirradiated animals. On the basis of previous findings (see above) there is some limited evidence to indicate that hoarding may be positively associated with fear or emotionality. We have shown (Manosevitz and Rostkowski, 1966) that irradiated animals eliminate more than nonirradiated animals in the open field.

Contrary to our expectation, animals that were exposed to the 300-R dose of radiation hoarded significantly less than the control animals. The correlations between hoarding and activity and hoarding and elimination were negative and therefore also contrary to our expectations. This reopens the question of just what the relation is between activity, elimination, and hoarding. One possibility is that irradiation affects mice differently than other species that have been studied rendering them more active and fearful but decreasing their hoarding performance.

In a pilot study we observed that mice exposed to an environment quite different from our standard laboratory environment, tended to hoard more than animals subjected to the usual laboratory conditions. The purpose of the next experiment (Manosevitz et al., 1968) was to follow up this observation and to control genetic variation so that genotype-treatment interaction could be assessed. We were interested in exposure to an enriched environment as an experimental manipulation because we wanted to identify those experiences that might increase hoarding. Approximately half the animals from two strains (*C3H/HeJ, JK*) were exposed to the enriched environment. They were placed at weaning in a large hardware cloth cage which was furnished with a platform, column, ramp, tunnel, and box. All animals were removed from the enriched environment at 102 days of age. Control animals were reared in standard cages. Because of breeding difficulties, animals for the *JK* female control group were not available. These missing data created some special problems in the data analysis and therefore three separate but interlocking analyses of variance were computed. The asymmetry in the design limited the conclusions we could draw from this experiment. A summary of the means and standard errors of the hoarding scores is presented in Table 2. The analyses of variance that are discussed below were performed using transformed scores.

The results of the first analysis of variance (strain, sex, and trials), performed on data from subjects exposed to the enriched environment only, indi-

TABLE 2

Means and Standard Errors of Transformed Hoarding Scores

Strain	Sex	Treatment	N	Daily hoarding scores transformed $\sqrt{X+1}$	
				$\overline{x}$	$s_{\overline{x}}$
C3H/HeJ	Male	C	16	1.403	0.037
C3H/HeJ	Male	E	15	1.346	0.034
C3H/HeJ	Female	C	15	1.319	0.034
C3H/HeJ	Female	E	20	1.501	0.035
JK	Male	C	10	1.475	0.059
JK	Male	E	11	1.746	0.067
JK	Female	E	8	1.781	0.068

cated that strain effects ($p<0.001$) were significant. The *JK* strain (males and females combined) hoarded approximately 24 percent more than the *C3H*'s. The second analysis of variance (strain, treatment, and trials), performed on data from males only, showed that strain was again significant ($p<0.001$).

The significant strain differences replicate our earlier findings (Lindzey and Manosevitz, 1964; Manosevitz, 1965, 1967a) and thus further highlight the importance of genetic variation in hoarding. The *JK* strain is the highest hoarding strain in this study as it was in an earlier study (Manosevitz, 1965) where it was compared to a different strain (*C57BL/1*).

The third analysis of variance performed on data from the *C3H* strain (sex, treatment, and trials) showed that there was a significant sex × treatment interaction ($p<0.005$). This indicated that in the *C3H* strain exposure to an enriched environment hardly affected hoarding in males (it actually *reduced* hoarding by 4 percent) and increased hoarding in females by 14 percent.

Strain × treatment interaction was significant ($p<0.01$) in the second analysis of variance. The postweaning enriched environment treatment increased hoarding in the *JK* male group but had almost no effect on *C3H* males. The general conclusion which can be drawn from this study is: the magnitude of the enriched environment effect and its direction are a function of both strain and sex.

The significant strain × treatment interaction obtained in this study is similar to the interaction observed in an earlier study (Manosevitz, 1965) in which a treatment quite different from the enriched environment treatment was used. In that study, the fear manipulation increased hoarding in the *JK* males but decreased hoarding in the *C57BL/1* males. Viewing the results of these studies together and focusing on *JK* males only, one may observe that hoarding behavior can be modified more easily in this strain than in other strains we have studied. It seems plausible that particular genotypes may be more plastic or responsive to environmental manipulations. If such genotypes can be reliably identified, they may be of considerable value in experiments on genotype-environment interaction (Manosevitz, 1967b).

In this experiment, hoarding performance across trials was analyzed in some detail. The daily performance of both strains, males and females combined, for the enriched environment treatment only are presented in Fig 1.

In the first analysis of variance the trial effect was significant ($p < 0.025$) and the strain × trial interaction approached significance ($p < 0.10$). As can be seen in Fig. 1, the curves for the two strains ascend in the first five trials and are similar in shape and elevation. In the remaining trials, the performance of the two strains diverge. The curve for the *JK* group continues to rise in a linear manner whereas the curve for the *C3H* group drops somewhat and levels off. The divergence of the two strains in the last five trials is reflected in the strain × trial interaction.

In the analysis of variance which included strain, treatment, and trials, for males only, the trials effect was significant ($p < 0.025$) and a three-factor interaction (strain × treatment × trials) approached significance ($p < 0.10$). The daily hoarding performance for both strains, control and enriched treatment, for males only is depicted in Fig. 2. There is a general tendency for males to hoard more in successive trials. However, the curve for the *JK* males who were exposed to the enriched environment treatment accelerates rapidly in the early trials and then flattens out during the last five trials. The three-factor interaction which approached significance reflects the fact that the shape of the *JK* experimental group curve is quite different from any other curves in Fig. 2.

The significant trials effect indicates a tendency for hoarding performance to increase as trials progress. The shape of these curves is quite dissimilar to those reported by Lindzey and Manosevitz (1964) and Smith and Ross (1953). This may be due to differences in feeding and deprivation schedules used in these various experiments discussed above.

The strain × trials interaction which approached significance indicated that the increase across trials may be different for the two strains (Fig. 1). It appears that the treatment facilitates the onset of high hoarding and this effect is stronger in the *JK* strain than in the *C3H* strain. The control animals seem to approximate the same level of hoarding by the end of the tenth trial. The data provide some evidence to indicate that postweaning exposure to an enriched environment may facilitate or release the response so that animals start to hoard earlier in the sequence of trials than the control animals.

There are several possible mechanisms which may be responsible for

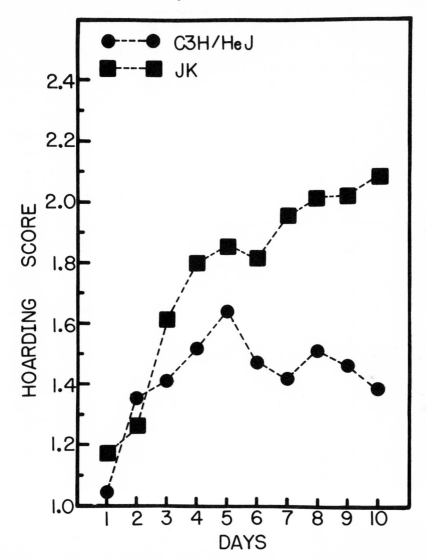

FIG. 1. Daily hoarding performance for two strains, males and females combined under enriched environment treatment only.

the effect produced by the enriched environment. The treatment effect may be mediated through increased activity. Several studies with rats have shown that environmental enrichment leads to increased activity (e.g., Duke and Seaman, 1964). If animals exposed to the enriched environment have higher general activity levels, we would expect them to hoard more and this effect would be most pronounced in the earlier trials. In addition, degree of fear-

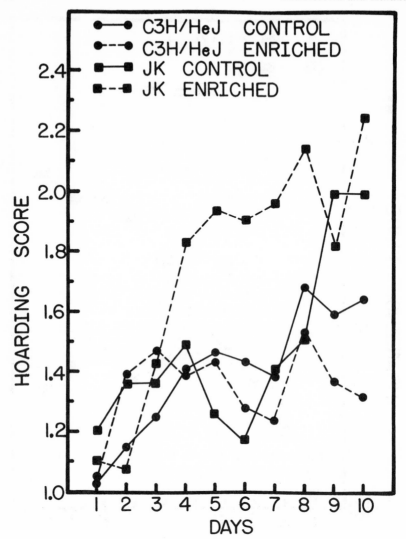

FIG. 2. Daily hoarding performance for two strains, control and enriched environment, males only.

fulness may interact with activity to influence hoarding. For example, at extreme levels of fearfulness an animal may freeze and remain inactive for most of the testing period, thus hoarding very little or not at all. However, an intermediate level of fear may actually arouse an animal to a fairly high activity level and potentiate hoarding. Nonfearful animals presumably will be relatively inactive during the hoarding test and may remain in the home cage or else eat during the testing period and thus hoard very little. Manipulating fear and activity by using different genotypes or treatments at several levels are needed to test this interpretation.

Another possibility is that control animals initially may engage in more exploratory behavior which in turn would interfere with the task of retrieving pellets to the home cage. This could be partly a function of the relative novelty which the hoarding alleys provide for the control animals. Previous research using rats has demonstrated that control or restricted animals explore more and this may interfere with other behaviors (e.g., Woods et al., 1961; Zimbardo and Montgomery, 1957). Future research in our laboratory will evaluate these possibilities.

Previous experience with food has been assumed to be one important determinant of hoarding behavior. Several investigators have manipulated food experience in young and adult animals to determine the importance of food experiences as factors affecting hoarding (see Ross et al., 1955).

An early experiment by Hunt (1941) provided some support for the hypothesis that early feeding frustration increases adult hoarding after adult deprivation. Several efforts to replicate this finding have been attempted with only limited success (Albino and Long, 1951; Hunt et al., 1947; McKelvey and Marx, 1951). Uncontrolled genetic variation, genotype-treatment interaction, variation in treatments and testing probably contributed to the limited success of these replications. As demonstrated above, genetic factors play an important role in hoarding and thus the next experiment (Manosevitz, 1970) was designed to systematically vary genetic factors. Moreover, since it was observed that genotype interacted with treatment in several of the earlier studies this study was also designed so genotype-treatment interaction could be assessed.

Mice from *AKR/J* and *C57BL/6J* strains were used in this experiment. Approximately half the animals in each strain were assigned to the aperiodic feeding group. Treatment started when the animals were 22 days old and continued until 102 days of age. The essential features of the design were: (1) prolonged aperiodic feeding treatment, (2) initial hoarding trials, (3) severe adult food deprivation experienced by all animals, and (4) a second set of five hoarding trials. The transformed hoarding scores for block 1 (five hoarding trials prior to adult severe deprivation) and block 2 (five hoarding trials postadult severe deprivation) are summarized in Table 3. An unweighted means analysis of variance with three between-subject factors (strain, sex, treatment) and two within-subject factors (blocks, trials) was performed on the transformed hoarding scores.

Only the significant results of the analysis of variance are summarized in Table 4. Once again, strain differences are significant and large. The *AKR* strain hoarded approximately 87 percent more than the *C57* strain. Strain interacted with sex; *AKR* males hoarded considerably more than females; however, among *C57* animals males and females hoarded at similar levels.

Manosevitz and Lindzey

TABLE 3
Summary of Subjects and Transformed Hoarding Scores

Strain	Sex	Treatment	N	Blocks	
				1	2
AKR/J	Male	A	13	3.298	3.378
AKR/J	Male	F	12	3.805	3.525
AKR/J	Female	A	12	2.252	3.055
AKR/J	Female	F	8	2.195	3.022
C57BL/6J	Male	A	18	1.397	1.551
C57BL/6J	Male	F	17	1.716	1.785
C57BL/6J	Female	A	21	1.427	1.791
C57BL/6J	Female	F	17	1.544	1.914

TABLE 4
Analysis of Variance of Transformed Hoarding Scores, Summary of Significant Effects

Source of Variation	MS	F
Strain[a] (St)	551.75	118.08[b]
Sex[a] (S)	44.95	9.62[c]
Blocks[a] (B)	24.18	29.18[b]
Trials (TR)	62.33	187.75[b]
St X S[a]	58.43	12.51[b]
S X B[a]	23.27	28.08[b]
St X Tr	17.65	52.16[b]
St X S X B[a]	7.36	8.88[c]
St X S X Tr	1.18	3.54[c]
Treatment (T) X B X Tr	0.82	2.46[d]
St X S X T X B X Tr	0.78	5.38[b]

[a]df = 1/110, all others df = 4/440
[b]$p < 0.001$
[c]$p < 0.01$
[d]$p < 0.05$

The prolonged aperiodic feeding experience effect was not statistically significant. This finding is similar to that reported by Hunt (1941). However, as
one may observe in Table 3, the treatment generally increased hoarding by a
small amount.

Average hoarding performance was higher in all groups after the adult
severe feeding deprivation experience. According to Hunt's (1941) earlier
results, we should have observed a significant interaction between aperiodic
feeding experience and adult deprivation (i.e., treatment × blocks). The
animals who experienced prolonged aperiodic feeding should have hoarded
significantly more than the control animals after the adult deprivation; however, this interaction failed to appear in our data.

The significant treatment × blocks × trials interaction indicated a pattern
opposite what one would predict on the basis of Hunt's results. In block 1, the
aperiodic feeding treatment group (F), across trials, hoarded more than the ad
libitum treatment group (A). However, in the second set of five trials (block 2)
the two treatment groups were quite similar. On the basis of Hunt's research one
would have expected the A and F treatment groups to be more dissimilar in the
second set of trials (after the severe adult deprivation treatment) than in the
first block of trials. Thus, the results of this experiment failed to confirm
Hunt's original experiment and conclusions.

Future Research

Our research on the determinants of hoarding has left several important
questions unanswered. Some of these questions are briefly listed here. Generally, the association between hoarding and other behavioral domains remains
to be investigated.

> 1. The effects of variation in general activity level upon hoard-
> ing have not been investigated systematically, although Ross et al.
> (1955) suggest that activity and hoarding should be positively
> associated. A number of studies with mice have shown strain varia-
> tion in general activity level (e.g., Bruell, 1967; McClearn and
> Meredith, 1966). Thus, strain variation in activity could be cor-
> related with strain variation in hoarding performance using a
> large number of strains. Another approach to this question would
> be to measure activity and hoarding on the same genetically hetero-
> geneous population of mice and correlate each animal's perform-
> ance on both measures.
> 2. The association between exploration and hoarding has not
> been investigated. Exploration in early trials would presumably
> interfere with hoarding performance. As suggested above, this may
> be why the control animals hoard less than the animals exposed
> to the enriched environment.
> 3. Although we have provided some limited evidence on the

role of fear or emotionality upon hoarding, the generality of these findings should be tested using other genotypes (inbred and cross-bred) and fear manipulations. Certainly our suggestion that the relationship between level of fear and hoarding performance is best described as an inverted U must be put to experimental test.

4. Our postweaning aperiodic feeding treatment only slightly affected adult hoarding. Perhaps a similar feeding treatment, experienced by animals earlier in the life cycle, would have a more pronounced effect upon adult hoarding. Although there has been considerable interest in feeding experience, the effects of pre-weaning food deprivation have rarely been studied and it would appear that this would be a fruitful approach in further analysis of the effects of feeding experience and hoarding.

SUMMARY AND IMPLICATIONS

In summary, the experiments described above have provided considerable data to aid our understanding of this complex response system. There appears to be no question that genetic variation is an important determinant of hoarding in mice. Some evidence of the magnitude of these genetic determinants, and their mode of operation, has been provided by the studies in which genetic parameters were estimated. In addition, our studies have shown that hoarding performance can be modified through a variety of environmental manipulations. However, these treatments do not affect all strains in a similar manner. Treatments interact with genotype to determine the magnitude and direction of treatment effects. Future studies on the determinants of hoarding should investigate not only genotype and treatments, but also genotype-treatment interactions.

In studies where genetic parameters are estimated, the environment to which Ss are exposed may influence the values obtained (Henderson, 1967). For example, if the environment is an "expressive" one, then heritability estimates may be increased. On the other hand, if the environment is "suppressive," then heritability estimates may be lowered (Fuller and Thompson, 1960, pp. 64–65). Data presented above indicate that postweaning exposure to an enriched environment may be more "expressive" for some genotypes than for others. In selection programs for high or low hoarding, establishing and maintaining a specific environment may influence the selection progress made and the magnitude of the selection gain.

The results of the prolonged food deprivation study indicate that animals exposed to an environment where the food supply is uncertain can respond, after a severe adult food deprivation experience, with appropriate adaptive behavior, i.e., increased adult hoarding. The implications of this behavior for the survival of the population are evident.

The results of our experiments clearly indicate that formulations of hoarding which are univariate are inadequate. The determinants of hoarding are complex and require consideration of the genotype, components of the environment, and their interaction.

REFERENCES

Albino, R. C., and M. Long. 1951. The effect of infant food-deprivation upon adult hoarding in the white rat. Brit. J. Psychol., 42:146–154.

Bevan, W., and M. A. Grodsky. 1958. Hoarding in hamsters with systematically controlled pretest experience. J. Comp. Physiol. Psychol., 51:342–345.

Bindra, D. 1947. Water hoarding in rats. J. Comp. Physiol. Psychol., 40:149–156.

———— 1948. What makes rats hoard? J. Comp. Physiol. Psychol., 41:397–402.

———— 1959. Motivation: A Systematic Reinterpretation, New York, The Ronald Press Co.

Broadhurst, P. L. 1958. Determinants of emotionality in the rat. III. Strain differences. J. Comp. Physiol. Psychol., 51:55–59.

———— 1960. Experiments in psychogenetics. *In* Eysenck, H. J., ed., Experiments in Personality, London, Routledge and Kegan Paul, Vol. 1, pp. 1–102.

———— and J. L. Jinks. 1961. Biometrical genetics and behavior: Reanalysis of published data. Psychol. Bull., 58:337–362.

Bruell, J. H. 1962. Dominance and segregation in the inheritance of quantitative behavior in mice. *In* Bliss, E. L., ed., Roots of Behavior, New York, Hoeber Medical Division, Harper and Row, Pubs., pp. 48–67.

———— 1967. Behavioral heterosis. *In* Hirsch, J., ed., Behavior-Genetic Analysis, New York, McGraw-Hill Book Co., pp. 270–286.

———— n.d. Variation in natural populations, inbred strain variance, and heritability. Unpublished manuscript, Western Reserve University.

Cofer, C. N., and M. H. Appley. 1964. Motivation: Theory and Research, New York, John Wiley and Sons, Inc.

Duke, J. D., and J. L. Seaman. 1964. Postweaning experiences and emotional responsiveness. Psychol. Rept., 14:543–546.

Falconer, D. S. 1960. Introduction to Quantitative Genetics, New York, The Ronald Press Co.

———— 1963. Quantitative inheritance. *In* Burdette, W. J., ed., Methodology in Mammalian Genetics, San Francisco, Holden-Day, Inc., pp. 196–216.

Fuller, J. L., and W. R. Thompson. 1960. Behavior Genetics, New York, John Wiley and Sons, Inc.

Henderson, N. D. 1967. Prior treatment effects on open field behaviour of mice—A genetic analysis. Anim. Behav., 15:364–376.

Hess, E. H. 1953. Shyness as a factor influencing hoarding in rats. J. Comp. Physiol. Psychol., 46:46–48.

Holland, J. G. 1954. The influence of previous experience and residual effects of deprivation on hoarding in the rat. J. Comp. Physiol. Psychol., 47:244–247.

Hunt, J. M. 1941. The effect of infant frustration upon adult hoarding in the albino rat. J. Abnorm. Soc. Psychol. 36:338–360.

———— H. Schlosberg, R. L. Solomon, and E. Stellar. 1947. Studies of the effects of infantile experience on adult behavior in rats. I. Effects of infantile feeding frustration on adult hoarding. J. Comp. Physiol. Psychol., 40:291–304.

James, W. 1950. The Principles of Psychology, New York, Dover Publications, Vol. 2.

Koski, J. 1963. The bimodal distribution of hoarding scores in the golden hamster. Proc. Penn. Acad. Sci., 37:291–294.

Lerner, I. M. 1958. The Genetic Basis of Selection, New York, John Wiley and Sons, Inc.

Lindzey, G. 1967. Some remarks concerning the incest taboo, and psychoanalytic theory. Amer. Psychol., 22:1051–1059.

———— and M. Manosevitz. 1964. Hoarding in the mouse. Psychon. Sci., 1:35–36.

Lush, J. L. 1945. Animal Breeding Plans, Ames, Iowa State College Press.

Manosevitz, M. 1965. Genotype, fear and hoarding. J. Comp. Physiol. Psychol., 60:412–416.

———— 1966. Hoarding and neonatal irradiation. Psychon. Sci., 5:183–184.

———— 1967a. Hoarding and inbred strains of mice. J. Comp. Physiol. Psychol., 63:148–150.

———— 1967b. The analysis of gene-environmental interactions. Paper presented at American Psychological Association Convention, September 2, Washington, D. C. Division 3 Symposium, The Experimental Analysis of Genetics of Behavior: The Mouse as a Model.

———— 1970. Prolonged aperiodic feeding and adult hoarding in mice. J. Comp. Physiol. Psychol., in press.

———— R. B. Campenot, and C. F. Swencionis. 1968. The effects of an enriched environment upon hoarding. J. Comp. Physiol. Psychol., 66:319–324.

———— and G. Lindzey. 1967. Genetics of hoarding: A biometrical analysis. J. Comp. Physiol. Psychol., 63:142–144.

———— and J. R. Rostkowski. 1966. The effects of neonatal irradiation on postnatal activity and elimination. Radiat. Res., 28:701–707.

Marx, M. H. 1950. A stimulus-response analysis of the hoarding habit in the rat. Psychol. Rev., 57:80–94.

Mather, K. 1949. Biometrical Genetics, London, Methuen.

McClearn, G. E., and W. Meredith. 1966. Behavioral genetics. Ann. Rev. Psychol., 17:515–550.

McDougall, W. 1923. Introduction to Social Psychology, 16th ed., Washington, D. C., Robert B. Luce, Inc.

McKelvey, R. K., and M. H. Marx. 1951. Effects of infantile food and water deprivation on adult hoarding in the rat. J. Comp. Physiol. Psychol., 44:423–430.

Miller, G. A., and P. Viek. 1944. An analysis of the rat's response to unfamiliar aspects of the hoarding situation. J. Comp. Physiol. Psychol., 37:221–231.

Morgan, C. T. 1947. The hoarding instinct. Psychol. Rev., 54:335–341.

———— 1965. Physiological Psychology, New York, McGraw-Hill Book Co.

Mowrer, O. H. 1960. Learning Theory and Behavior, New York, John Wiley and Sons, Inc.

Ross, S., W. I. Smith, and C. W. Nienstedt, Jr. 1950. The hoarding of nonrelevant material by the white rat. J. Comp. Physiol. Psychol., 43:217–225.

———— W. I. Smith, and B. L. Woessner. 1955. Hoarding: An analysis of experiments and trends. J. Gen. Psychol., 52:307–326.

Smith, W. I., A. J. Krawczun, N. J. Wisehaupt, and S. Ross. 1954. Hoarding behavior of adrenalectomized hamsters. J. Comp. Physiol. Psychol., 47:154–156.

———— and E. K. Powell. 1955. The role of emotionality in hoarding. Behaviour, 8:57–62.

———— and S. Ross. 1950. The effect of continued food deprivation on hoarding in the albino rat. J. Genet. Psychol., 77:117–121.

———— and S. Ross. 1953. The hoarding behavior of the mouse. I. The role of previous feeding experience. J. Genet. Psychol., 82:279–297.

Stamm, J. S. 1954. Genetics of hoarding: I. Hoarding differences between homozygous strains of rats. J. Comp. Physiol. Psychol., 47:157–161.

———— 1956. Genetics of hoarding. II. Hoarding behavior of hybrid and backcrossed strains of rats. J. Comp. Physiol. Psychol., 49:349–352.

Thorpe, W. H. 1963. Learning and Instinct in Animals, 2nd ed., London, Methuen.

Waddington, C. H. 1957. The Strategy of Genes, London, George Allen and Unwin.

Winston, H. D. 1964. Heterosis and learning in the mouse. J. Comp. Physiol. Psychol., 57:279–283.

Wolfe, J. B. 1939. An exploratory study of food-storing in rats. J. Comp. Physiol. Psychol., 28:97–108.

Woods, P. J., A. S. Fiske, and S. I. Ruckelshaus. 1961. The effects of drives conflicting with exploration on the problem-solving behavior of rats reared in free and restricted environments. J. Comp. Physiol. Psychol., 54:167–169.

Zimbardo, P. G., and K. C. Montgomery. 1957. Effects of "free environment" rearing upon exploratory behavior. Psychol. Rept., 3:589–594.

5

The Sound of One Paw Clapping: An Inquiry into the Origin of Left-Handedness*

ROBERT L. COLLINS
*The Jackson Laboratory,
Bar Harbor, Maine*

INTRODUCTION

Right- and left-handedness are alternative expressions of what may be the oldest recorded example of human behavioral variation. Stone Age man left behavioral fossils of his laterality captured in the artifacts he fashioned (Yakovlev, 1964). There is scarcely an aspect of human culture that has not been permeated by a legacy of superstition concerning the left-handed individual. The right-left duality is expressed in religion, politics, history, art, and the everyday commerce of an individual with his environment. Often the first question raised when considering this fundamental behavioral asymmetry is, "What are its origins?" A fascinating scientific literature and popular folklore surround this question (Uhrbrock, 1963). While laterality has been a favorite Gordian knot of genetic studies, a surprising diversity of opinion concerning its possible antecedents has emerged. It has been proposed that hand preference is regulated according to a Mendelian pair or alleles (Rife, 1950; Trankell, 1955; Annett, 1964), according to multiple genetic factors (Kaplan, 1960), and according to genetic-environment interactions (Falek, 1959). By contrast, strong environmentalist positions have been advanced by nongeneticists (Hildreth, 1949; Penfield and Roberts, 1959).

* This research was supported in part by PHS research grants MH 11327 and MH 1775 from the National Institute of Mental Health and by training grants NSF-GY-826 and NSF-GW-682 from the National Science Foundation. I am grateful to Drs. Earl L. Green, John L. Fuller, and James F. Crow for their advice and valuable suggestions. I thank Dr. Earl L. Green for suggesting the algebraic form of equation (9). Mr. Geddes Simpson, Miss Helen Giles, and Mr. Thomas Reisser provided able technical assistance.

Perhaps a clearcut assessment of the genetic hypothesis has remained elusive due to the entanglement of environmental effects with the presumed genetic influences. It is commonly believed that hand preference is modifiable through formal and informal training during early human development. If this is the case, it may be possible to gain insight into the origin of lateral asymmetries by studying them in laboratory animals whose genetic constitution has been controlled and whose social environment presumably does not overzealously perpetuate right-handedness. This chapter is directed towards a resolution of the question, "Is hand preference regulated genetically either in whole or in part?" Experimental studies of laterality in genetically standardized mice are first presented. Later, published data on hand preference in human pairs of known gentic relationship are reexamined. The results will show that the relationship of hand preference to genetic antecedents is remarkably similar in both species.

LATERALITY IN MICE

If mice are indeed handed, one paw should be used more often in a situation in which the subject is given free choice to use either paw; this lateral bias should not be task specific; the asymmetry should be enduring; and the preferred paw should also be the superior paw in certain forced-use tasks.

Mice from highly inbred strains obtained from the Jackson Laboratory production department were tested for paw preference. Detailed observations of mice from *C57BL/6J* and *DBA/2J* strains are reported. Subjects were initially tested between 8 and 12 weeks of age. They were deprived of food for 24 hours and then placed into plastic testing chambers on which a 9 mm glass feeding tube was attached. The tube was equally accessible from the right or left. Maple flavored rolled wheat (Maypo) was placed in the tube so that a subject could withdraw food by using a single paw. Fifty reaches were observed for each subject and his score was defined as the number of right hand reaches in 50 observations (RPE's). Fig. 1 illustrates a portion of the behavioral sequence. Mice were classified as dextral or sinistral by considering them to be coins whose "toss" could land either R or L. R was assigned to a subject scoring above 25 RPE's, and L, to one scoring below 25 RPE's.

Fig. 2 illustrates the frequency distribution of RPE scores for 320 *C57BL/6J* mice. This strain has been maintained by brother-sister matings for beyond 90 generations. The distribution of scores is clearly bimodal with maxima at 0 to 2 and 48 to 50 RPE's, and minima at 27 to 29 RPE's. The proportion of right paw reaches was 0.492, and left, 0.508. Fig. 3 presents the same information for 120 *DBA/2J* mice. This strain has also been

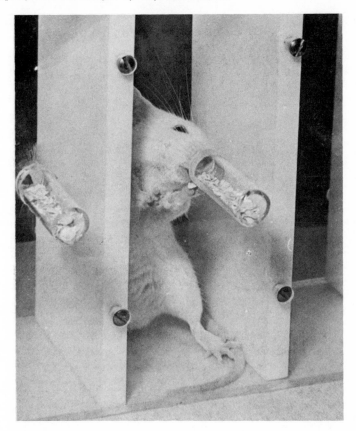

FIG. 1. A mouse retrieving a food flake using the left paw. The photograph was taken near the end of the behavioral sequence. The feeding tube is equidistant from the cubicle walls.

maintained by close inbreeding for beyond 90 generations. Maxima occurred at 3 to 5 and 41 to 44 RPE's, and minima at 21 to 23 RPE's. The proportion of right paw entries was 0.501, and left, 0.499.

According to the procedure for assigning classifications, the proportions of right paw preferent mice in *C57BL/6J* and *DBA/2J* strains were 0.51 and 0.52 respectively. In both cases, mice used one paw more frequently than the other to retrieve food even through the likelihood of observing a right or left reach was equal across subjects. Since both right and left paw preferent mice were observed in each genetically uniform population, it would appear that genetic differences are not required to explain the alternative forms of lateral preference.

Fig. 4 illustrates the distribution of paw preference scores for 961 mice from nine genetically uniform populations. Data for the additional groups have

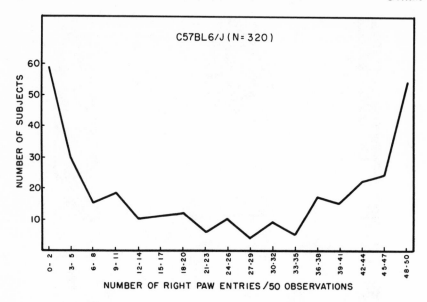

FIG. 2. Frequency distribution of lateral preference scores for 320 *C57BL/ 6J* mice. The abscissa is divided into 17 blocks of 3 right paw entries.

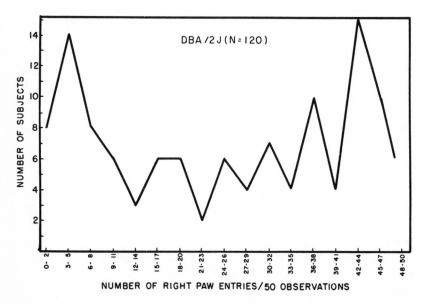

FIG. 3. Frequency distribution of lateral preference scores for 120 *DBA/2J* mice.

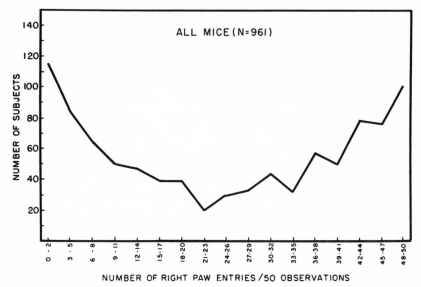

FIG. 4. Frequency distribution of lateral preference scores for 961 mice from genetically uniform populations.

been summarized previously (Collins, 1968). The distribution is bimodal with maxima at 0 to 2 and 48 to 50 RPE's, and minima at 21 to 23 RPE's. The proportions of right paw reaches was 0.497, and left, 0.503. In all, 50.6 percent of the mice were classified as dextral, and 49.4 percent as sinistral.

To what degree is the lateral preference observed in the cubicle task specific? To answer this question, David Honig, working in this laboratory, compared the RPE scores for 19 mice tested in the cubicle and in a circular arena of diameter 17.4 cm. Food was placed into a watchglass underneath a 6 mm aperture and 50 food reaches were observed. The correlation of RPE scores between the two tests was 0.96. This finding supports the notion that the behavioral asymmetry measured in the cubicle is representative of a characteristic which is transitive across other free-use tasks demanding alternative response topographies.

Consistency of Preference

The consistency of paw preference scores for individual subjects was determined in three ways. Within the initial testing session, data from 100 randomly chosen subjects were divided into two blocks of 25 reaches and compared statistically. To evaluate short-term consistency, 15 mice were tested four times at 4-day intervals. An additional 30 *C57BL/6J* and 30

DBA/2J mice were tested four times at monthly intervals to evaluate long-term consistency.

Table 1 summarizes the Pearson product moment correlations for the RPE performance and tetrachoric correlations for laterality classifications for all groups. Within the initial test session 20 percent of the mice maintained identical RPE scores in the two blocks of 25 observations; 35 percent increased the use of the preferred paw slightly, while 45 percent decreased its use. All 100 mice maintained the same laterality assignments during both halves of the test. During short-term retesting, the RPE scores for some mice showed small variations in number. All 16 mice maintained their initial classification during short-term retesting. Under long-term retesting, 4 *C57BL/6J* and 8 *DBA/2J* mice exhibited a single fluctuation in their laterality assignments. The classifications for 80 percent of the mice were invariant through 4 monthly retests. In view of the numerous sources of uncontrolled variation that frequently plague behavioral measurement, the consistency of the paw preference performance in mice is remarkable. All 26 correlations listed in Table 1 were statistically significant ($p < 0.05$ that ρ or $\rho_t = 0$).

TABLE 1

Intercorrelation Matrix of Paw Preference Scores and Laterality Classifications for Mice Retested at Intervals of Four Days (n = 16) and One Month (n = 60)

Four Day Retest

	1	2	3	4
1	0.92[a]	0.92[b]	0.92	0.93
	1.00[a]	1.00[c]	1.00	1.00
2	0.88	—	0.93	0.95
	0.98	—	1.00	1.00
3	0.82	0.93	—	0.97
	0.96	1.00	—	1.00
4	0.75	0.91	0.94	—
	0.85	0.97	0.97	—

Monthly Retest

[a]Within session split-block correlation (n = 100)
[b]Product moment correlation of the number of right paw entries in 50 observations.
[c]Tetrachoric correlation of laterality classifications.

Lateral Superiority

To determine whether the paw preferred in food reaching was also the superior paw on a forced-use task, 29 *A/J* and 32 *A/HeJ* males were tested for grip strength. Tape was first placed over one paw. Mice were picked up by the tail and permitted to grasp a wire mesh screen connected to a force displacement transducer and recorded (Fig. 5). Mice were pulled upward until they released the wire screen, and the force displacement was measured in grams of grip strength. Mice were tested 20 to 30 times on the first day. On the second day, they were tested with the opposite paw. The experimental design was balanced with respect to the order of testing preferred and non-preferred sides, and all procedures were performed under double-blind conditions. The mean of the five highest readings for each mouse were compared between preferred and nonpreferred sides within and between subjects.

Average grip strength for mice tested on the preferred side was 64.9 g, while the mean for mice tested on the nonpreferred side was 61.2 g. The 3.7 g difference in favor of the preferred side was statistically significant using a two-tailed test ($p < 0.025$, $t = 2.33$, $df = 60$). In the between subjects comparison, mice tested with the preferred paw on the initial day pulled 68.5 g,

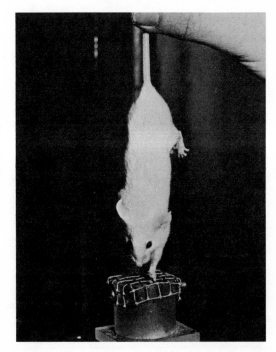

FIG. 5. A suspended mouse grasping the force displacement apparatus with the left paw. The right paw was covered with a tape glove. On being pulled upward the mouse releases the wire mesh and the maximum force displacement is recorded.

while mice tested on the nonpreferred side pulled 62.7 g. This 5.8 g difference did not exceed a conventional level of statistical significance ($p \simeq 0.08$, $F = 5.17$, $df = 1/57$). These preliminary findings are consistent with the view that the paw preferred in free-use situations possesses an advantage over the nonpreferred side in certain forced-use tasks.

Evidence for functional laterality in mice was provided by the experimental data. Mice used one paw more frequently than the other in a free-use task; the lateral preference was transitive across a different free-use task; lateral preference was enduring; and lateral superiority in grip-strength was positively associated with the preferred paw.

Since both right- and left-handed mice were observed in each genetically uniform population, it would appear that genetic differences are not responsible for the variation in paw preference. In fact, near maximal phenotypic variance was observed in the two intensively studied strains. This is somewhat surprising considering that these strains possess minimal genetic variance. Accordingly, it is concluded that the degree of genetic determination for the alternative forms of laterality in these mice was essentially zero (Collins, 1968, 1969).

If right- and left-handedness in mice are not attributable to genetic influences, are these forms also nongenetic in humans? It is imperative that data from human studies be reexamined to determine whether the genetic hypothesis is indeed tenable.

HUMAN HANDEDNESS

If the degree of genetic resemblance between individuals is known or can be inferred with reasonable accuracy, the genetic hypothesis for the regulation of hand preference can be supported by demonstrating that human pairs more closely related genetically are more alike phenotypically. Let us turn to a reanalysis of published data on the concordance of hand preference in populations of human pairs of known genetic resemblance. The populations and their expected genetic relatedness are monozygous twins (1.00), dizygous twins (0.50), paired siblings (0.50), and parental pairs (0.00).

Consider an urn filled with a large number of thoroughly mixed cubes labeled "*R*" or "*L*" in the proportions *P* and *Q*, respectively. An experimenter repeatedly withdraws two cubes and observes if both labels are similar, *R-R, L-L,* or dissimilar, *R-L,* order being unimportant. If the label of each cube is independent of that of its pair, and the proportions do not change throughout the experiment, the expected proportions of the three kinds of paired outcomes will be given by the binomial distribution as follows: $E(R\text{-}R) = P^2$, $E(R\text{-}L) = 2PQ$, and $E(L\text{-}L) = Q^2$. Conversely, if such an experiment results in a binomial distribution of paired outcomes, the knowledge of the label of one cube provides no additional information concerning the label of the other, and the label of each cube is statistically independent of that of its pair.

Consider "i" populations of paired individuals with known genetic resemblance consisting of those being right-handed, R, and those being left-handed, L, in the proportions P_i and Q_i, respectively. If the proportions of paired outcomes for each population follow the binomial proportions $(P_i + Q_i)^2$, it must be concluded that the hand preference of each individual is statistically unassociated with that of its pair. If phenotypic similarity is found to be uncorrelated to genotypic similarity, the experiments have not provided evidence to suggest that human lateral preference is regulated by genetic components.

Reanalyses of intrapair similarities and differences in hand preference were performed on twin data compiled by Newman et al. (1937), Wilson and Jones (1932), and Rife (1940, 1950); these data are conveniently summarized by Rife (1940, 1950). For each study the proportion of right-handers, P, and left-handers, Q, was computed as follows: $P = [2N(R\text{-}R) + N(R\text{-}L)]/2N$, where N is the total number of pairs and $(R\text{-}R)$ means the proportion of right-handed pairs. If the distribution of paired outcomes is random, the number of $R\text{-}R$, $R\text{-}L$, and $L\text{-}L$ pairs should be given by the respective terms: $N(P^2, 2PQ, Q^2)$. The deviation from the binomial is here defined as $\triangle = (R\text{-}R)^{1/2} + (L\text{-}L)^{1/2} - 1$. $E(\triangle) = 0$ if the proportions correspond to the binomial form.

Fourfold point correlations, Φ, were computed for the hand preferences of all pairs (Ferguson, 1959). The number of discordant pairs was divided equally between $R\text{-}L$ and $L\text{-}R$ cells. The observed correlation was compared to the expected genetic correlation for each population. The phenotypic and genetic correlations should be associated if handedness is influenced by genetic differences.

Table 2 summarizes the observed and expected frequencies for the intrapair concordance of hand preference for monozygous (MZ) twins. Since MZ twins possess an identical genotype, it is expected that these data should be most sensitive to the effects of possible genetic antecedents. An examination of Table 2 indicates the numbers of MZ pairs that are similar and dissimilar in hand preference are in remarkable agreement with the frequencies expected on the basis of chance. While the expected genetic correlation of MZ pairs is 1.00, the observed phenotypic correlation was essentially zero ($\Phi = -0.026$). Thus, these data indicate that the hand preference for one MZ twin is statistically independent of that of the other, and the knowledge of one twin's preference provides no additional information concerning the hand preference of the co-twin.

TABLE 2

Observed and Predicted Concordance of Handedness
in Monozygous Twins

Number of pairs

Investigator	R-R	R-L	L-L	N	Q	ϕ
Wilson and Jones (1932)	56	13	1	70	0.107	0.029
Expected by chance	55.8	13.4	0.8	70		
Newman (1937)	30	17	3	50	0.230	0.028
Expected by chance	29.6	17.7	2.6	49.9		
Rife (1940)[a]	176	41	6	223	0.119	0.122
Expected by chance	173.2	46.7	3.1	223		
Rife (1950)	261	76	6	343	0.128	0.009
Expected by chance	260.6	76.7	5.6	342.9		
Total	347	106	10	463	0.136	0.026
Expected by chance	345.6	108.9	8.6	463.1		
Proportions observed	0.749	0.229	0.022			

[a]Not included in total. Deviation of total from the binomial form: $\Delta = 0.013$

TABLE 3

Observed and Predicted Concordance of Handedness
in Dizygous Twins

Number of pairs

Investigator	R-R	R-L	L-L	N	Q	ϕ
Wilson and Jones (1932)	97	24	2	123	0.114	0.032
Expected by chance	96.6	24.8	1.6	123		
Newman (1937)	39	11	0	50	0.11	−0.124
Expected by chance	39.6	9.8	0.6	50		
Rife (1940)[a]	104	39	3	146	0.154	−0.025
Expected by chance	104.5	38.1	3.5	146.1		
Rife (1950)	164	45	2	211	0.116	−0.039
Expected by chance	164.8	43.3	2.8	210.9		
Total	300	80	4	384	0.115	−0.027
Expected by chance	301.1	77.9	5.0	384		
Proportions observed	0.781	0.208	0.010			

[a]Not included in total. Deviation of total from the binomial form: $\Delta = -0.014$

Table 3 summarizes the observed and predicted frequencies of lateral concordance for dizygous (DZ) twins. The results indicate that the chance expectations accurately predict the observed frequencies for each study and for all studies combined. While the average genetic correlation within DZ pairs is 0.50, the observed phenotypic correlation was essentially zero ($\Phi = -0.027$).

Table 4 presents the same information for paired siblings and parental pairs. Paired sibs share the same expected genetic resemblance as DZ twins, although they have different pre- and postnatal environments. The chance predictions of paired outcomes agree favorably with the observed frequencies. The observed phenotypic correlation was essentially zero ($\Phi = 0.075$).

Parental pairs share no or very little genetic resemblance by descent. It is seen in Table 4 that the binomial expectations accurately reconstruct the observed frequencies in parental pairs. The expected genetic correlation of parental pairs is 0.00. The observed phenotypic correlation concurred ($\Phi = 0.041$).

The chance expectations for the concordance of lateral preference in populations of human pairs differing in their genetic resemblance are remarkably close to the observed frequencies for each population, each study, and all studies combined. Since these data do not permit a rejection of the null hypothesis that the observed frequencies resulted from chance pairing, the experimental data do not support a view that lateral preference is determined by genetic factors.

It is of interest to note that the frequencies of left-handedness in the four sets of data comprise two nonoverlapping statistical groupings. The 95 percent confidence intervals for values of Q in MZ and DZ twins are 0.114 to 0.158 and 0.092 to 0.138, respectively. The ranges for paired sibs and parental pairs are 0.072 to 0.084 and 0.076 to 0.106, respectively. Accordingly, these find-

TABLE 4

**Observed and Predicted Concordance of Handedness
in Paired Siblings and Parental Pairs**

Investigator	Number of pairs			N	Q	ϕ
	R-R	R-L	L-L			
Paired Siblings						
Rife (1940)	3067	475	41	3583	0.078	0.075
Expected by chance	3047.7	513.4	21.5	3582.6		
Proportions observed	0.856	0.133	0.011			
Parental Pairs						
Rife (1940)	620	62	5	687	0.045	0.041
Expected by chance	626.6	59.2	1.4	687		
Proportions observed	0.902	0.090	0.007			

ings pose an additional burden upon any genetic interpretation of the concordance data.

A General Probability Model for Concordance

Two important genetic theories have been advanced to explain the inheritance of human handedness. Rife (1950) hypothesized that hand preference is controlled by an allelic gene pair in which there is partial penetrance in heterozygotes. Trankell (1955) presented an alternative gene pair model in which there is partial penetrance in the recessive homozygote. These two models and any of the remaining six possible autosomal gene pair models may be evaluated for their ability to predict the observed twin concordances by methods developed in this section. A general probability gene pair model for the concordance of alternative characters in twins is presented. The general model may be reduced to any special allelic pair model by the appropriate specification of what are termed parameters of noncongruence.

The decision to accept any gene pair model will be made if it meets three criteria: (1) Since the observed frequencies of concordance listed in Tables 1 and 2 are indistinguishable from the binomial expectations, the prediction equations according to a genetic hypothesis must also be indistinguishable from the binomial. If and only if a set of paired outcomes is in the binomial form, the following relationship must hold: $[Pr(R\text{-}R)]^{1/2} + [Pr(L\text{-}L)]^{1/2} = 1 + \triangle$, where $E(\triangle) = 0$. To give any genetic hypothesis a little leeway, the magnitude of the discrepancy from the binomial form should be no larger than that observed between the observed and binomial; i.e., for MZ twins, $\triangle = 0.013$, and for DZ twins, $\triangle = -0.014$. (2) An acceptable model should generate equations whose solutions under an arbitrary set of parameters produce frequencies of lateral concordance in twin pairs which are statistically indistinguishable from the summary values observed in Tables 1 and 2. These values for MZ twins are $(R\text{-}R) = 0.75$ and $(L\text{-}L) = 0.02$, and for DZ twins, $(R\text{-}R) = 0.78$ and $(L\text{-}L) = 0.01$. It follows from the first consideration that we need only be concerned with the frequencies of twins that are similar in hand preference. (3) If any genetic model meets the above criteria, the required parametric values should inspire confidence. For example, if a model fits only when the frequency of one allele is essentially zero, the model will be judged unreasonable or implausible.

Let us provisionally assert that a gene pair exists in which allele A predisposes to right-handedness, and allele a predisposes to left-handedness. The allelic frequencies are p and q, respectively, where $p + q = 1$. We now define parameters of genotype-phenotype noncongruence (s, t, u). These specify the proportion of individuals that will exhibit a phenotype contrary to that provisionally asserted to be determined by a particular genotype. These are expressed in the following manner where Pr signifies probability:

$$Pr(AA = R) = 1 - s \qquad\qquad Pr(AA = L) = s$$
$$Pr(Aa = R) = 1 - t \qquad\qquad Pr(Aa = L) = t$$
$$Pr(aa = R) = u \qquad\qquad Pr(aa = L) = 1 - u$$

where $0 \leq (s, t, u) \leq 1$.

Under the tested assumption that parental combinations are formed without regard to the lateral preference of their members (Table 3), a gene frequency analysis for the concordance of alternative characters in MZ and DZ twins is developed. Tables 5 and 6 summarize the parental mating combinations, gene frequencies, and the probabilities for the concordance of handedness in twins. The tables were constructed in the following manner: Consider $AA \times AA$ matings. These produce AA offspring only. The probability for both members of a twin pair to be similar in handedness is given by $Pr(AA = R, AA = R) = (1 - s)^2$, $Pr(AA = L, AA = L) = s^2$, and to be dissimilar by $Pr(AA = R, AA = L) = 2s(1-s)$. Probabilities for concordance in the remaining mating combinations were computed, multiplied by the probabilities of the mating combinations, and the reduced equations for concordance are summarized on the bottom lines of Tables 5 and 6. We may now examine the proposed genetic models for the inheritance of human hand preference.

MODEL 1. $s = u = 0$, $t = 0$ or 1.—This completely deterministic genetic model specifies that heterozygous individuals are invariably either right-handed or left-handed. This model can be dismissed immediately since it does not permit monozygous twin pairs to exhibit any discordance in lateral preference. Therefore, it cannot satisfy criteria 1 and 2.

MODEL 2. $s = u = 0$, $0 \leq t \leq 1$.—Superficially, this model of incomplete penetrance of heterozygotes resembles the hypothesis advanced by Rife (1950). However, it differs in an important respect. Rife assumed that one-half of Aa individuals are right-handers, and one-half, left-handers. He further assumed that Aa twins are never concordant. This assumption requires a special type of environmental determination of handedness in twin pairs which should not be accepted in the absence of conclusive evidence. Rife's model has been recast in a form which considers the probability of right-handedness of each Aa twin to be independent of the handedness of its co-twin and dependent solely upon the noncongruence parameter, t. For example, if $t = 0.5$, then $Pr(Aa = R, Aa = R) = Pr(Aa = L, Aa = L) = 0.25$, and $Pr(Aa = R, Aa = L) = 0.50$.

Substituting $s = u = 0$ into the reduced equations in Tables 5 and 6 produces the following probabilities for concordance:

TABLE 5

Gene Frequency Analysis of the Probability Model for Concordance of Handedness in Monozygous Twins

Mating	Probability	Phenotypic Probabilities of Pairs of Offspring		
		R-R	R-L	L-L
$AA \times AA$	p^4	$(1-s)^2 p^4$	$2s(1-s)p^4$	$s^2 p^4$
$AA \times Aa$	$4p^3 q$	$2[(1-s)^2 + (1-t)^2]p^3 q$	$4[s(1-s) + t(1-t)]p^3 q$	$2(s^2 + t^2)p^3 q$
$Aa \times Aa$	$4p^2 q^2$	$[(1-s)^2 + 2(1-t)^2 + u^2]p^2 q^2$	$2[s(1-s) + 2t(1-t) + u(1-u)]p^2 q^2$	$[s^2 + 2t^2 + (1-u)^2]p^2 q^2$
$AA \times aa$	$2p^2 q^2$	$2(1-t)^2 p^2 q^2$	$4t(1-t)p^2 q^2$	$2t^2 p^2 q^2$
$Aa \times aa$	$4pq^3$	$2[(1-t)^2 + u^2]pq^3$	$4[t(1-t) + u(1-u)]pq^3$	$2[t^2 + (1-u)^2]pq^3$
$aa \times aa$	q^4	$u^2 q^4$	$2u(1-u)q^4$	$(1-u)^2 q^4$

$Pr(R-R) = (1-s)^2 p^2 + 2(1-t)^2 pq + u^2 q^2$

$Pr(R-L) = 2s(1-s)p^2 + 4t(1-t)pq + 2u(1-u)q^2$

$Pr(L-L) = s^2 p^2 + 2t^2 pq + (1-u)^2 q^2$

$Pr(AA = R) = 1 - s, Pr(AA = L) = s; Pr(Aa = R) = 1 - t, Pr(Aa = L) = t; Pr(aa = R) = u, Pr(aa = L) = 1 - u.$

TABLE 6

Gene Frequency Analysis of the Probability Model for Concordance of Handedness in Dizygous Twins

Phenotypic Probabilities of Pairs of Offspring

Mating	Probability	R-R	R-L	L-L
AA X AA	p^4	$(1-s)^2 p^4$	$2s(1-s)p^4$	$s^2 p^4$
AA X Aa	$4p^3 q$	$[(1-s)+(1-t)]^2 p^3 q$	$2(s+t)\,[(1-s)+(1-t)]\,p^3 q$	$(s+t)^2 p^3 q$
Aa X Aa	$4p^2 q^2$	$1/4[(1-s)+2(1-t)+u]^2 p^2 q^2$	$1/2[s+2t+(1-u)]$ $[(1-s)+2(1-t)+u]\,p^2 q^2$	$1/4[s+2t+(1-u)]^2 p^2 q^2$
AA X aa	$2p^2 q^2$	$2(1-t)^2 p^2 q^2$	$4t(1-t)p^2 q^2$	$2t^2 p^2 q^2$
Aa X aa	$4pq^3$	$[(1-t)+u]^2 pq^3$	$2(t+u)\,[(1-t)+(1-u)]\,pq^3$	$[t+(1-u)]^2 pq^3$
aa X aa	q^4	$u^2 q^4$	$2u(1-u)q^4$	$(1-u)^2 q^4$

$Pr(R\text{-}R) = (1-s)^2 p^2 + \{[(1-s)+(1-t)]^2 - 2(1-s)^2\}pq + \{2[(1-s)^2 - u^2] - [(1-s)+(1-t)]^2\} + \{[(1-s)/2 + u/2]^2 + (1-t)[(1-t) - (1-s) - u]\}p^2 q^2 + u^2 q^2$

$Pr(R\text{-}L) = 1 - [Pr(R\text{-}R) + Pr(L\text{-}L)]$

$Pr(L\text{-}L) = s^2 p^2 + [(s+t)^2 - 2s^2]pq + \{2[s^2 - (1-u)^2] - (s+t)^2 + [t+(1-u)]^2\}pq^2 + \{[s/2 + (1-u)/2]^2 + t[t - s - (1-u)]\}p^2 q^2 + (1-u)^2 q^2$

$Pr(AA = R) = 1 - s, Pr(AA = L) = s; Pr(Aa = R) = 1 - t, Pr(Aa = L) = t; Pr(aa = R) = u, Pr(aa = L) = 1 - u.$

129

Monozygous Twins:

1. $$Pr(R\text{-}R) = p^2 + 2(1-t)^2pq$$
2. $$Pr(L\text{-}L) = 2t^2pq + q^2$$

Dizygous Twins:

3. $$Pr(R\text{-}R) = p^2 + (t^2 - 4t + 2)pq + 4(1-t)pq^2 + (t^2 - t + \tfrac{1}{4})p^2q^2$$
4. $$Pr(L\text{-}L) = t^2pq + q^2 + 2tpq^2 + (t^2 - t + \tfrac{1}{4})p^2q^2$$

It is clear that equations (1) and (2) are not algebraically equivalent to equations (3) and (4). In addition, $[Pr(R\text{-}R)]^{1/2} + [Pr(L\text{-}L)]^{1/2}$ must be greater than 1.0 for all practical values of q in the MZ equations since $p^2 + 2(1-t)^2pq > p^2$ and $2t^2pq + q^2 > q^2$, where $p + q = 1$. In addition, $\triangle$ is a minimum for fixed values of q when $t = q$, since

$$\frac{\mathrm{d}\triangle}{\mathrm{d}t} = \frac{4tpq}{2\sqrt{2t^2pq + q^2}} - \frac{4(1-t)pq}{2\sqrt{p^2 + 2(1-t)^2pq}}$$

Set $\dfrac{\mathrm{d}\triangle}{\mathrm{d}t} = 0$ and let $t = q$. The minimum is now satisfied:

$$\frac{1}{\sqrt{2pq + 1}} - \frac{1}{\sqrt{1 + 2pq}} = 0.$$

It follows then that if $\triangle \leq 0.013$, $q_{max} \leq 0.013$. However, if q is very small and p is very large, then $Pr(R\text{-}R) \geq 0.97$. Since this value is substantially larger than the observed 0.75, it is readily seen that there are no possible values for the parameters q and t such that the model of incomplete penetrance of heterozygotes would be admissible as a tenable hypothesis for the inheritance of human lateral preference.

MODEL 3. $s = t = 0$, $0 \leq u \leq 1$.—According to Trankell's (1955) hypothesis, a portion of the individuals with aa genotypes will become right-handers, and a portion will become left-handers, while AA and Aa genotypes will invariably produce right-handers. Substituting $s = t = 0$ into the general equations in Tables 5 and 6 produces the following probabilities for concordance:

Monozygous Twins:

5. $$Pr(R\text{-}R) = p^2 + 2pq + u^2q^2$$
6. $$Pr(L\text{-}L) = (1-u)^2q^2$$

Dizygous Twins:

7. $$Pr(R\text{-}R) = p^2 + 2pq + u^2q^2 + (2u - u^2 - 1)pq^2 + \tfrac{1}{4}(u^2 - u + 1)p^2q^2$$
8. $$Pr(L\text{-}L) = (1-u)^2q^2 + (1-u)^2pq^2 + \tfrac{1}{4}(1-u)^2p^2q^2$$

Trankell's model is evaluated easily by numerical methods. Values of q and u may be computed such that equation (5) is 0.75 and equation (6) is approximately 0.02. The following provisional values were obtained: $p = 0.05$, $q = 0.95$, and $u = 0.85$. For MZ twins these values produced $Pr(R\text{-}R) = 0.75$, $Pr(L\text{-}L) = 0.02$, and $\triangle = 0.008$. For DZ twins these values produced $Pr(R\text{-}R) = 0.75$, $Pr(L\text{-}L) = 0.02$, and $\triangle = 0.012$. Thus, Trankell's model is able to generate a parametric fit to the observed data and meet criteria 1 and 2. To admit this model as a reasonable genetic explanation for human hand preference, one must carefully consider its implications. For example, (1) the frequency for the allele influencing left-handedness must be at least 20 times larger than the frequency of the allele influencing right-handedness; (2) more than 90 percent of the population of twins are "genotypic" left-handers; (3) at least 85 percent of these native left-handers become right-handed due to environmental influences, and (4) the effect of environmental variables must produce at least seven times as many right-handers as does the influence of genetic variation.

There is another objection to Trankell's hypothesis. If one would admit it as a tenable explanation for the determination of human hand preference, then one could not deny that the converse genetic model is also admissible as in the following:

MODEL 4. $(1 - t) = u = 0$, $0 \leq s \leq 1$.—In this model a portion of the individuals with AA genotypes become right-handers, and a portion, left-handers, while individuals with Aa and aa genotypes are invariably left-handers. The equations for this model are analogous to those of Model 3 and are not shown. According to numerical methods, the following parametric values would be required to meet criteria 1 and 2: $p = 0.99$, $q = 0.01$, and $s = 0.125$. The implications of Model 4 are that: (1) the gene for left-handedness would be rare, (2) one-eighth of the genotypic right-handers would become left handed due to environmental influences, and (3) environmental influences would produce at least six times as many left-handers as would genetic variables.

Thus, while Trankell's hypothesis of incomplete penetrance of recessive homozygotes can provide a parametric fit to the observed data, it generates a set of implausible implications. In addition, it cannot be distinguished from its mirror image genetic hypothesis, Model 4.

A Nongenetic Model for the Inheritance of Handedness

Circumstantial data of two sorts seem to perpetuate a view that human hand preference is determined genetically. First, it has been observed repeatedly that a higher proportion of left-handed children are from families in which one or both parents are left-handed. Table 7 summarizes the incidence

TABLE 7

Conditional Probabilities for Left-Handed Offspring According to Parental Mating Types

Investigator

	R-R	R-L	L-L	N
Ramaley (1913)	0.1205	0.3235	0.8571	1140
Chamberlain (1928)	0.0425	0.1142	0.2800	7714
Rife (1940)	0.0758	0.1954	0.5455	2178
Weighted Average	0.0564	0.1757	0.4419	11,032

of left-handedness of progeny according to the phenotypes of the parents. An obvious parent-offspring correlation exists. It is emphasized that while such a correlation may be used to support a genetic hypothesis, it is not sufficient evidence to establish a genetic relationship. Such an association may arise from environmental influences confounded within family mating combinations (Falconer, 1960). Second, more than 90 percent of the tested human population is right-handed. This seems to suggest that the phenotypic frequencies are maintained by differences in allelic frequencies. This section will show how a nongenetic model for the regulation of hand preference can be formulated which leads to a stable incidence of left-handedness in the general population. It will be assumed that offspring resemble their parents in handedness according to an unspecified environmental mechanism.

Consider an initial generation G_0 of a large random breeding population in which the probability of right-handed adults is P, and left-handed, Q. If parental combinations are formed randomly and the progeny subject to the conditional probabilities for left-handedness summarized in Table 7, there will be a different proportion of left-handers in G_1. If G_1 parental matings are formed without regard to lateral preferences and their progeny are subject to the same conditional probabilities, G_2 individuals will comprise a different population of left-handers. This may be carried to the "$n + 1$th" generation to obtain an equilibrium of phenotypic frequencies as in the following.

Let $Q_n = Pr_n(L)$ in G_n and let the conditional probability for left-handedness given the mating type be $m_i = Pr(L/i)$, where

$$i = 0, \text{ for } L\text{-}L \text{ matings}$$
$$i = 1, \text{ for } R\text{-}L \text{ matings}$$
$$i = 2, \text{ for } R\text{-}R \text{ matings}$$

According to Table 7

$$m_0 = 0.4419$$
$$m_1 = 0.1757$$
$$m_2 = 0.0564$$

We may then write an equation relating $Q_{n+1} = f(Q_n)$ as follows:

9. $Q_{n+1} = (m_0 - 2m_1 + m_2)Q_n^2 + [2(m_1 - m_2) - 1]Q_n + m_2.$

Now set $Q_{n+1} = Q_n$ under equilibrium, substitute values of m_i, and solve for Q_n. This produces:

$$Q_n = \frac{[0.7614 \pm (0.7614)^2 - 4(0.1469)\ (0.0564)]^{\frac{1}{2}}}{2(0.1469)}$$

$$Q_n = 0.0752 \text{ and } (1 - Q_n) \text{ or } P_n = 0.9248.$$

These values agree favorably with the following frequencies of handedness in the general population as summarized by Rife (1950) from the results of four independent investigations: $Q = 0.073$ and $P = 0.927$.

Perhaps it is not so surprising that a stable equilibrium develops or that the asymptotic frequency for left-handedness is close to the value for Q in the general population; these results might have been expected. Of greater interest is the effectiveness of the model for maintaining a relatively stable incidence of sinistrality. To visualize the approach to equilibrium an IBM 1620 computer was programmed to calculate the probabilities for sinistrality according to equation (9) for eight generations of nonassortative mating in which the initial population was composed of Q left-handers, where $0 \leq Q \leq 1$, in steps of 0.1.

Fig. 6 illustrates the proportions of left-handers in each generation of random mating. By the fourth generation, the proportion of left-handers approaches the equilibrium value of 0.075. This rapid approach to equilibrium indicates that the regulation of hand preference by nongenetic mechanisms would be efficient and effective in maintaining a small but stable incidence of sinistrality in the general population.

DISCUSSION

Findings from the reanalyses of concordance in twins and the gene frequency analysis of the proposed genetic models present a consistent picture of the relationship of human hand preference to genetic variation. Since not one of the extensive investigations reanalyzed herein provides evidence which leads to a rejection of the hypothesis that the concordance of hand preference in MZ and DZ twins results from random pairing, none provides evidence to support a view that hand preference is determined by genetic differences either in whole or in part. It is remarkable that this chance relationship has not received prior emphasis. Merrell (1957) correctly observed it; however, he supported Rife's genetic explanations although admitting that they "required specialized and arbitrary assumptions."

Four of the eight possible autosomal gene pair models have been examined

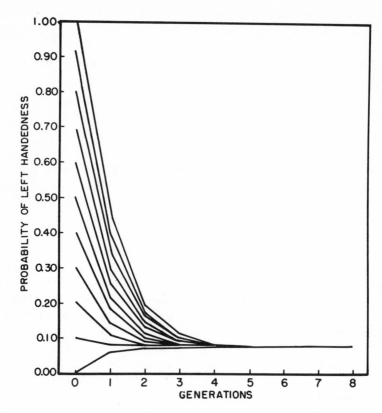

FIG. 6. Probabilities of left-handedness in the general population according to the environmental model for eight generations of nonassortative mating.

for their ability to reconstruct the observed frequencies of lateral concordance in MZ and DZ twins. The model of strict genetic determinism (1) and the revision of Rife's hypothesis (2) are incapable of generating predictions which are in accord with the observed data. Only Trankell's model (3) and the alternative mirror image equivalent (4) could be fitted to the data. However, since both models required frequencies for one allele to be near zero, and neither model could be distinguished from the other, neither model is tenable. Certainly, if all three hypothesized genotypes are subject to noncongruence (s, t, and u all variable), it should be possible to generate other genetic models which fit the observed data. However, it is emphasized that the search for alternative genetic explanations using the noncongruence methods or using any multi-locus models is rather futile. For any genetic model so derived would be placed in the unenviable position of predicting frequencies of concordance which are identical to those frequencies assigned to the null hypothesis. Since statistical support for a research hypothesis usually requires a rejection of a null hypothesis, and since both hypotheses produce identical predictions, the research hypothesis could never be supported nor could the null hypothesis ever be rejected.

The well-defined association between the phenotypes of offspring and their parents indicates that human hand preference is clearly heritable. Since prior considerations do not support a view that mechanisms of transmission genetics are responsible for this association, the crucial question remaining is whether these correlations result from cultural inheritance or from an unknown extrachromosomal biologic predisposition to asymmetry. This question could be readily tested. J. L. Fuller (personal communication) suggested that parent-offspring relationships for the hand preferences of fostered and nonfostered children might provide a sensitive test of the relative importance of prenatal and postnatal factors.

In view of the host of characters admittedly regulated by genetic influences, one wonders why there seems to be a failure of genetic control over laterality. If maximal phenotypic variation can be maintained by nongenetic influences, then a genetic regulation of hand preference would convey little advantage to a species. Phenotypic variation arising from environmental sources would serve admirably to preserve the species against environments demanding uniformity of lateral specialization. Nongenetic influences could, in fact, convey more plasticity for adaptation than could genetic influences. Perhaps genetic variation may simply provide the ability for differentiated hand preference without specifying the alternative phenotypic form, just as genetic variation might favor acquisition of language without specifying the vocabulary and grammar.

Yakovlev (1964) considered that ". . . preferential, either right or left, handedness in skilled acts of human dexterity is one of the most recent products of evolution, indissolubly bound with the evolution of the human consciousness of the environment and of the language. . . ." In light of the studies with mice, this seems not to be the case. Rather it would appear that when neolithic man first picked up a charcoal stick to draw pictures on cave walls, he was following an old phylogenetic tradition of asymmetrical specialization shared by others, if not most, bilaterally organized species. Indeed, if asymmetrical specialization is shared by diverse phylogenetic forms, then functional laterality is a far more primitive behavioral polymorphism than is commonly believed.

Mus musculus has provided a useful analog, if not homolog, to human hand preference. If the merit of an animal model is proportional to its derived conceptual generalizations and testable extrapolations, then the inbred mouse has contributed significantly to understanding human hand preference by providing a phylogenetic perspective and a demonstration of how the knot binding laterality to genetics may be cut more neatly than untied.

REFERENCES

Annett, M. 1964. A model for the inheritance of handedness and cerebral dominance. Nature, 204:59–60.

Chamberlain, H. D. 1928. The inheritance of left handedness. J. Hered., 19:557–559.

Collins, R. L. 1968. On the inheritance of handedness: I. Laterality in inbred mice. J. Hered., 59:9–12.

———— 1969. On the inheritance of handedness: II. Selection for sinistrality in mice. J. Hered., 60:117–119.

Falconer, D. S. 1960. Introduction to Quantitative Genetics, New York, The Ronald Press Co.

Falek, A. 1959. Handedness, a family study. Amer. J. Hum. Genet., 11:52–62.

Ferguson, G. A. 1959. Statistical Analysis in Psychology and Education, New York, McGraw-Hill Book Co.

Hildreth, G. 1949. The development and training of hand dominance. III. Origin of handedness and lateral dominance. J. Genet. Psychol., 75:255–275.

Kaplan, A. R. 1960. A theory on the biology of lateral dominance. Acta Genet. Med., 9:318–324.

———— 1965. On the problem of lateral neuromuscular dominance. Psychiat. Neurol., 149:369–374.

Merrell, D. J. 1957. Dominance of eye and hand. Hum. Biol., 23:314–328.

Newman, H. H., F. N. Freeman, and K. J. Holzinger. 1937. Twins, a Study of Heredity and Environment, Chicago, University of Chicago Press.

Penfield, W., and L. Roberts. 1959. Speech and Brain Mechanisms, Princeton, Princeton University Press.

Ramaley, F. 1913. Inheritance of left-handedness. Amer. Nat., 47:730–738.

Rife, D. C. 1940. Handedness, with special reference to twins. Genetics, 25:178–186.

———— 1950. Application of gene frequency analysis to the interpretation of data from twins. Hum. Biol., 22:136–145.

Trankell, A. 1955. Aspects of genetics in psychology. Amer. J. Hum. Genet., 7:264–276.

Uhrbrock, R. S. 1963. Laterality of Function Bibliography, Washington, D. C., American Documentation Institute.

Wilson, P. T., and H. E. Jones. 1932. Left-handedness in twins. Genetics, 17:560–571.

Yakovlev, P. I. 1964. Telokinesis and handedness (an empirical generalization). In Wortis, J., ed., Recent Advances in Biological Psychiatry, New York, Plenum Press, Vol VI; pp. 21–30.

SECTION III

Single Gene Modification of Behavior

6

Single Gene Substitutions and Behavior[*]

JAMES D. HAWKINS
Department of Psychology,
San Jose State College,
San Jose, California

INTRODUCTION

This chapter is concerned with an area within behavior-genetics which has seen relatively little work but, in which, considerable interest has recently been shown. The theme is that, while the number of investigators interested in behavior-genetic phenomena is increasing, the field is at a crucial stage of development such that a sharper focusing of fundamental theoretical problems and research strategies is necessary to its further substantive growth. Single locus mouse research can contribute to this process. This chapter will attempt to provide a conceptual framework for the general evaluation of problems and methods of behavioral research with mutant-bearing stocks of *Mus musculus*. It will include consideration of the conceptual background for single gene research, some reflections on the goals of behavior-genetics and their current transitional status, the role of the mouse as a prototype, and a position on how single gene experimentation may contribute to the further growth of the field. Special reference is given to the quest for "natural units" of analysis, to the ecology of the laboratory as an inbreeding environment, and to questions of behavior development and evolution within the laboratory.

PREFATORY CONSIDERATIONS

In considering genic effects and the evolution of insect behavior, Aubrey Manning observes: "The speculation-to-fact ratio in this discussion has in-

[*] This paper was written at the University of Texas and was supported by postdoctoral fellowship HD 16,374 G & D of the National Institutes of Health (USPHS). I am grateful to my sponsor Dr. Delbert D. Thiessen for his help and encouragement.

evitably been rather high" (Manning, 1967). The ratio is even higher with regard to *Mus* because there is less data available. The need of a rational framework for theory construction is apparent, nonetheless, regardless of its ultimate usefulness or longevity. In the past few years there have been several papers encouraging increased research activity with existing stocks of mutant mice. Fuller (1967), is concerned with both the potentialities and limitations of mutants as behavior subjects. He also defines criteria for research which, he suggests, may integrate the genetic, physiologic, and behavioral levels of analysis. Earlier, Merrell (1965) had proposed two somewhat less specific approaches to evaluation of single gene effects on behavior. This, in the context of criticizing existing polygenic work. Thiessen, in the meantime, had reported on the first continuing program involving the systematic use of numerous mutants and multiple behavioral tests (Thiessen, 1966, 1967). All of these approaches look toward the analysis of the physiologic pathways between genes and behavior. None of them precludes any of the others. Each, however, differs in its emphasis as to what is likely to be the most fruitful approach to behavior study utilizing mutant animals.

Two investigators have suggested the use of mutant stocks in the analysis of complex behavior sequences (van Abeelen, 1965, 1966; Hawkins, 1964, 1966a). This work, while recognizing the importance of tracing gene-behavior pathways, is of greater relevance with regard to evolutionary considerations, and will be discussed in a later section. As a final primary antecedent to this discussion, there is the programmatic prediction of 1960 by Fuller and Thompson:

> ... we have argued for a lack of correspondence between genes and behavior traits. Single genes characteristically affect many forms of behavior; single psychological traits have variance ascribable to many genes. Between the genetic and behavioral levels of description is a network of physiological processes. We propose that behavior is insensitive to a substantial portion of the metabolic shunting which occurs in this network, for a neuron may not be concerned with the source of its energy provided it gets enough. ...
>
> Confirmation of the non-congruence model requires manipulation of genotypes instead of phenotypes. ... Manipulating genes, chromosomes, and gametes rather than selecting for phenotype will be the next important phase of behavior genetics. The conceptual framework of these experiments will be the use of genes as treatments affecting behavior, rather than the determination of the heritability of traits. (pp. 343–345)

Thus, these authors suggest the method by which their model might be confirmed. Nonetheless, the ultimate lack of direct testability was recognized even then:

> If the contribution of a gene to a behavioral trait varies with circumstances in a highly irregular fashion, suitable scale trans-

formations may not be possible. The relationships between genotype and phenotype must still be lawful, but each specific
combination of genotype and life history may have its own rules
of developing response patterns. (p. 344)

At a more molar level, Fuller and Thompson expected that heritability testing
would utilize the concepts of population genetics rather than attempting to
specify particular genes which control particular kinds of behavior.

Most readers would probably concur that in spite of nearly a decade of
subsequent behavior genetic research, the volume of which has grown enor-
mously, we can add little more than detail to the general Fuller and Thompson
position today. Despite its lack of absolute testability, their view is probably
widely held. Experiments with mice, which fulfill the prophecies, have been
reported but they have been relatively few in number. Implementation of such
research is a current dilemma of the field.

GOALS AND STRATEGIES IN BEHAVIOR-GENETICS

We can be reasonably specific about the genetic properties of the *Mus*
resources available for behavior study (e.g., see the relevant chapters on
systems of mating, mutant genes and linkage testing systems, summaries on
genetic strains and stocks, and the development and use of random-bred stocks
in: Burdette, 1963; E. L. Green, 1966; McClearn, 1967). Mutant mice are, of
course, just one specialized kind of mouse resource. We can also be reasonably
specific about the experimental procedures and controls to be employed. This
methodology has generally been drawn from experimental psychology and is
similar whether one is considering effects due to different allelic combinations
at a given locus or differences between strains.

Our problems and goals, however, can be stated presently in only the
most ambiguous terms. The initial thrust of behavior-genetics was the effort
to counter a pervasive environmentalist bias in psychology and to establish
clearly that heredity does contribute to behavioral variation. It is undoubtedly
correct that this reason for the existence of the field "is now so well established that it no longer provides a satisfactory integrating influence or sense of
direction for this rapidly growing discipline" (Rodgers, 1967). Nature versus
nurture need never have become an issue in psychology. Further, the problems
it raises are not germane to methodologic behaviorism (e.g., see the comments
of Beach, 1951).

In psychology, revamped statements of the nature-nurture issue continue
to exist side by side with statements of their nonutility. For example, Diamond
and Chow (1962) state: "the classical problem concerns the relative role of
hereditary and environmental *variation*—not heredity and environment." The
human geneticists David and Snyder (1962), however, go beyond what most
psychologists are yet willing to affirm:

> Genetics, since its inception, has been shackled with the dichotomy of "heredity" and "environment"—both of which were concept names before genetics was born—and it shares these shackles with psychology. . . . we have been asking the wrong questions when we have tried to decide *which* is responsible—heredity or environment—for such-and-such observed variability, or when we have inquired *how much* of the variability is attributable to hereditary factors and how much to environmental. Instead, we should be inquiring *how,* that is, investigating the developmental mechanisms through which the variability arises. (p. 48)

If behavior genetics represents a point of attack on the nature-nurture issue in psychology; and if that issue is not, of itself, sufficient to serve as an integrating problem for the subarea; the question, "Where does behavior genetics go from here?" is clearly pertinent. One general question upon which geneticists such as David and Snyder and some psychologists (e.g., Anastasi, 1958) have long agreed is: "How?" That is, how does "the" genetic material and "the" environment affect behavior? Another pertinent question, which again involves both heredity and environment, is: "What?" That is, what behaviors are likely candidates for genetic study and under what conditions? The identification of what constitutes a useful behavioral phenotype is a problem long recognized and ever open to new methodology.

Within psychology, mouse research usually is categorized as falling within the physiologic and/or comparative subareas. In zoology, Professor John King (1967) has suggested that behavior-genetics relates most directly with ecologic and developmental genetics. While it is an oversimplification, the question "How?" seems most closely tied to the physiologic, and developmental mechanisms which lie between genes and behavior. The question "What?" is, conceptually, an essentially evolutionary one and ties most readily with comparative and ecologic considerations. The boundary between the two approaches is not rigidly drawn. It is both possible and desirable for the individual investigator to experiment in both domains.

THE MOUSE AS A PROTOTYPE

Because of the title of this volume, it is clear that contributors to it either affirm or tacitly accept the proposition that laboratory mice can serve as a prototype in behavior-genetic investigation. It may not be immediately appreciated just what *Mus* is prototypic of. Indeed, many readers may question whether it is prototypic of other life forms at all. Beach (1960), has noted that, for most people, laymen and psychologists alike, psychology is a science of *human* behavior. He notes: ". . . I have come to question the logical defensibility of *any* concept of a comparative psychology." Beach (1960) has not given up his comparative orientation, of course, and suggests the following point of view:

> I am sure that I do not have to defend the proposition that the study of any kind of behavior in any species of animal is a scientifically respectable enterprise. . . . if the goal is to build a *comparative science of behavior,* two desiderata seem obvious.
> First the behavior selected for examination should be, so to speak, "natural" to the species. Insofar as possible it should be species-specific, . . . Secondly, the kinds of behavior chosen for analysis should be as widely distributed as possible, phylogenetically speaking. This would increase the opportunity for interspecific comparison and improve the probability of arriving at valid and broad general principles or laws.
> Let me repeat that what I am proposing differs from current practice in animal experimentation in that animals are not to be used as substitutes for people, and the kinds of problems investigated are not to be exclusively or even principally derived from human behavior or psychology. (p. 2)

W. R. Thompson, disagrees with (at least) the last part of this position (Thompson, 1968, p. 80). In the same book Ginsburg (1968) expresses his marked disagreement with the Thompson view, while Fuller (1968) concurs with it:

> First, I shall record agreement with Dr. Thompson's view that the primary *raison d'être* of behavioral genetics is its value to the behavioral sciences. Thus, its phenotypes ordinarily will be chosen with respect to their relevance to problems of behavior, not for convenience of genetic analysis (p. 111). . . . If, as Dr. Thompson states, behavior genetics exists primarily to serve the behavioral sciences, it should select its problems with reference to important issues in psychology and sociology. (p. 116)

Now, Beach and Ginsburg are certainly interested in general principles of behavior and include man in their considerations. Thompson and Fuller, on the other hand, are known to be broadly comparative in both their research and their outlook. Nonetheless, there seems sufficient difference of opinion to warrant the idea that these workers differ in their basic view of what their animal subject's role is, as a prototype. In the one case a genome-dependent homology is assumed for some aspect of behavior common to mammals phylogenetically as advanced as mice and higher. Questions of interest then hinge on what the relevant parameters are, within the genetic-environmental interaction, that will affect a specific behavior of psychologic importance. Such an approach has a long medical and pharmacologic history to draw on (Fuller, 1965). In this usage, inbred mice are simply a convenient, standardized tool.

In the second case, the mouse is prototypic in a more general sense, in that it is viewed as a unique species peculiarly suited to separate or simultaneous manipulation of its behavior, genetic make-up, physiology, and ecology. There is probably no other mammal for which as much is known in these areas taken collectively. This knowledge allows any of the four to be a dependent variable or, alternatively, to be systematically varied with reference to a given experimental problem. To the extent that a science of comparative behavior is possible, laboratory mice seem well suited as "hypothesis generators." Experiments with other species, in which less is known of behavior, genetics, physiology, and ecology, will often have more meaningful conclusions where mouse research is available as a comparative standard. The intent of such research is the study of *Mus musculus,* not the testing of theoretical models using mice as the data source. Kuhn (1961) has noted that physical scientists had to study the thermometer as an object, before it could be used as an instrument for the investigation of thermal phenomena. Mice, as tools in behavioral research, seem to be in much the same position today.

Mutant gene experimentation, at present, relates most directly with the latter of these two approaches. Even so, such research is not, of itself, prototypic. Rather, it may eventually lead to a more empirically based use of *Mus musculus* as a behavioral research prototype. The tenuousness with which single gene observations can now be related to human behavior is clearly illustrated by Lindzey (1967). Lindzey's thesis is that human morphology (broadly defined to include attributes such as hirsuteness, symmetry, color, and esthetic attractiveness as well as physical and structural aspects of the person) is "significantly intertwined" with function. After citing studies which indicate that genetic variation is clearly a contributor to human morphologic variation, he states:

> Given this observation, it seems altogether reasonable to anticipate that the chromosomal loci that influence variation in morphology may have multiple or pleiotropic effects, some of which are behavioral. To use an illustration made somewhat remote by phylogenetic regression, my colleague Harvey Winston and I (Winston and Lindzey, 1964), working with mice, have produced evidence suggesting that albinism (which is determined by a recessive gene *cc* in linkage group 1) is associated with a relative deficiency in escape learning. Whether this particular finding is sustained or not, there is every reason to expect that comparable pleiotropic effects linking physical and behavioral components will be found in a variety of behavioral and physical areas. Even at the human level we may expect comparable, although perhaps more complex, examples. (pp. 231–232)

A group of studies which indicate an activity decrement in yellow $(A^y)C57BL/6J$ mice supports the Lindzey thesis. In mice yellow coat color may be produced by either of two mutant alleles, A^{vy} (viable yellow) or A^y (yellow), occurring at the agouti locus on chromosome V (M. Green, 1966, p. 87). A^y induces enlarged body size, increased fat deposition, and linear size increase of the skeleton among other pleiotropic effects (Grüneberg, 1943; Wolff, 1963). Hawkins (1965), found that body-weight differences between A^ya (yellow) and aa (nonagouti, black) 120-day-old *C57BL/6J* mice accounted for about one-half of the activity score variance in a 165-minute activity test using a 16 compartment test arena. Later, using a smaller open-field arena and a 15-minute test period, he again found a highly significant activity decrement associated with A^y in 120-day-old (but not in 35- or 80-day-old) subjects (Hawkins, 1966b). Professor Jan Bruell (personal communication) has also tested A^ya and aa *C57BL/6J* animals both in activity wheels and in an exploration apparatus. He also found a highly significant depression of activity in the yellow animals on both tests. Black and tan (a^t), another agouti locus mutation, does not affect body weight in this strain. Neither does it affect behavior scores in a four-unit test arena or in activity wheels (Bruell, 1964a, 1964b).

The foregoing results are congruent with Lindzey's position that morphology pleiotropically affects behavior. On the other hand, it should be noted, in the Hawkins (1965) experiment, 52 percent of the variance was not weight dependent. It *was* associated with the mutant allele, however. The degree of such a pleiotropic effect would be expected to differ with other allelic substitutions which affect morphology: *ob/ob* obesity could present a very different picture. Its effects in this regard apparently have not been tested. Single gene work of this kind is narrowly specific and would hardly warrant experimental designs based directly upon the broad view of Lindzey, in which mice would be given prototypic status. Tinbergen (1968), has illustrated the way in which animal research may be of help in developing generalizations with regard to human problems of crucial significance. Whether or not useful (i.e., genetically intelligible) behavioral phenotypes are likely to emerge from the central issues of contemporary psychology is still highly problematical (e.g., see Fuller, 1965, 1968). Behavior-genetics is too young for one to be prescriptive about which approaches may or may not have utility. It is suggested here that the study of known mutants and their alleles can now contribute considerably to choices among phenotypes for study in mouse research and to an improved understanding of gene-behavior pathways.

Gene profiles, within the species, would seem to be the most tenable goal for this subarea at present. This provides a situation where neither the alleles tested, the strains or stocks used, nor the carrier species is a prototype. Rather, such intraspecies testing would lead to a comparative psychology of laboratory mouse behavior. Conclusions drawn within the species could be expected to set limits upon cross-species generalizations. This view is based upon four assumptions which are consistent with current behavior-genetic and evolutionary theory (Mayr, 1963, pp. 263–296): (1) Any gene may act on a plurality of traits; (2) No gene acts independently; each interacts with all the other genes which constitute its background; (3) Differences found between different phenotypes (under experimental conditions constant for them all) may be considered to result from the genetic differences between those phenotypes; (4) For groups of subjects differing genetically only with respect to a single pair of allelic genes, behavioral differences found between them (under constant experimental conditions) may be considered to result from the action of that pair of genes. With these proscriptions, it seems clear that mutant alleles must be evaluated against various genetic backgrounds before generalizations about their effects can be made. Given genetically suitable stocks for work to begin on a given problem, single locus findings are likely to eventually encompass the use of inbred, hybrid, backcross, random bred, and possibly even "wild type" genetic backgrounds, the testing of multiple alleles and other effective loci, and changed experimental, rearing, and even prenatal environments, before any given problem or effect will be considered thoroughly explored.

BEHAVIORAL MECHANISMS

Lockard (1968), has discussed the accident and irony of the overextensive use of laboratory rats in the history of experimental psychology. He suggests that at least as often as not, they are inappropriate subjects in the uses to which they are put. Many of his criticisms apply equally well to mice. Rodgers (1967) also notes this well documented propensity for animal psychologists to overgeneralize from rat research. He suggests, however, that the behavior-genetic subarea is on an opposite course—that of overparticularization. With regard to single gene experimentation, Rodgers correctly warns that it may duplicate an earlier cul de sac of behavior-genetics involving strain differences. He and others (e.g., Fuller, 1960; McClearn and Meredith, 1966) have pointed out that strain differences, with regard to behavior traits, are now an expectation. Reports of strain difference in a given behavior, usually contribute little to the advancement of the field anymore though they contributed greatly to the formulation of the Fuller and Thompson noncongruence model cited earlier (Fuller and Thompson, 1960, pp. 343–346). Similarly, a pyramid of single gene difference findings will not be likely to help us advance much from that general position. To counter the trend toward the "... study of populations defined by their genetic characteristics in such a way that generalizations will be restricted to defined genotypes ...," Rodgers suggests "... the study of 'mechanism specific' behaviors instead of behaviors uniquely associated with given genotypes" (Rodgers, 1967).

In his approach to behavioral single gene research, Fuller (1965) is clearly in agreement with the Rodgers thesis. "The results of the single locus studies, including some unpublished ones of our own, are relatively meager. Eventually, as we obtain more biochemical mutations with which to work, the picture may change" (Fuller, 1965, p. 251). Later (Fuller, 1967), working with the point mutation to *c* which had been found in inbred *C57BL/6J* animals, he confirmed the Winston and Lindzey (1964) finding of deficient water escape performance in albino mice. This eliminated the question (Meier and Foshee, 1964) of whether genes closely linked with *c* were producing the effect. In this context, Fuller (1967) suggests the guiding principle that:

> ... gene substitution should be used as a treatment in behavioural experiments when it produces a physiological change which could be accomplished by other means only with difficulty or not at all. It follows as an ideal that mutant genes in behavioural research should actually be methods of producing biochemical and anatomical effects which are related to important behavioural systems. (p. 470)

Clearly, his work does not meet this criterion but he does note that "Albino mice are deficient in tyrosinase and may therefore differ from pigmented mice in the concentration of brain catecholamines." He suggests the possibility that catecholamine imbalance in the CNS, rather than peripheral effects of the gene could be the effective mechanism. Thus, "... the albino gene is promising material for additional psychophysiological investigation" (Fuller, 1967, p. 470).

Thiessen also shares the Rodgers "mechanism specific behavior" position (Thiessen and Rodgers, 1967). The emphasis of his single gene screening program has been placed on finding substitutions which effectively modify sensory-motor systems of clear a priori behavioral importance (Thiessen, 1966, 1967). The latest details of this project are reported in another chapter of this volume.

Tracing gene-behavior pathways is of obvious importance. It is likely the only way to avoid an unintegrated catalog of single gene difference findings. However, much more is known concerning most physiologic-behavioral mechanisms than is known about their genetic bases. Pathways between known single locus substitutions, their structural effects, and complex behaviors are currently speculative at best. Given the present state of knowledge of single locus effects, many problems exist in integrating behavioral gene effects with known physiology. For example, well specified mutants may have diffuse (i.e., pleiotropic) and nonspecific (i.e., qualitative or quantitative differences in trait expression) effects. This, even on controlled genetic backgrounds and with either physiology or behavior. For example, A^{vy} produces four coat-color phenotypes: clear yellow, lightly spotted, heavily spotted, and agouti (Wolff,

1965). The clear and spotted varieties are heavier than agouti (with more weight variability among females than males) due to increased fat and water content in carcass and liver. "Variability of expression of the $A^{vy}a$ phenotype and the correlation of coat-color with body and liver composition suggest that A^y and A^{vy} alleles have similar effects on the synthesis of hair pigment as well as on fat metabolism. However, the characteristics of the 'agouti' $A^{vy}a$ phenotype indicate that these effects of the $A^{vy}a$ genotype are more easily modified than those of the A^ya genotype" (Wolff, 1965). As another example, Thiessen (1967), found A^{vy} produced a lower count than A^y in an activity wheel test and the lowest square crossing count of 15 mutants tested in an open-field arena. Numerous examples of substitutions which produce both physiologic and behavioral pleiotropy and variations in expressivity can be found in M. Green (1966) and Wolfe and Coleman (1966).

There are further problems concerned with the diffuse and nonspecific effects which may be found with mutants. A given behavior may depend primarily on either central or peripheral processes, on a particular genome for expression, or on environmental conditions of either rearing or testing. As noted earlier, Fuller (1967) suggests that the cc constitution may affect some behaviors via a central biochemical process. Albinism plainly has peripheral effects for visually mediated behaviors (DeFries et al., 1966). Also noted earlier, Hawkins (1965) found that one-half of the variance in activity scores attributable to A^y was assignable to a peripheral factor (body-weight) while the other one-half was unaccounted for and could represent a central mechanism or mechanisms.

The effects of the total genome and of environmental factors, in the lack of specificity of single gene effects, can also be illustrated with results from A^y experimentation. Hawkins (1967), tested five genotypes [$C57BL/6J$-(aa), $C57BL/6J$-A^ya, $DBA/2J$-(aa), $B6D2F_1$-aa, and $B6D2F_1$-A^ya] in a mating situation. Ten isolate-reared males of each genotype were tested five times each. All 10 $C57BL/6J$-A^ya males made vicious attacks on one or more of the females they were paired with: two of them attacked in all five tests. One $C57BL/6J$-aa animal also attacked all five of his females. Neither of the hybrid genotypes nor the $DBA/2J$ males showed the behavior. Clearly, A^y was important to the aberrant pattern but was effective on only one genetic background. Since neither the $DBA/2J$ nor either of the hybrids showed the attack pattern, while one $C57BL/6J$-aa male did, the inbred genome as a whole was also implicated. Finally, because of earlier experimental results, the method of rearing was thought to play a role in the expression of the trait. In work done later at the University of Texas (Hawkins, unpublished data), 20 group-reared males carrying different mutant combinations on the $C57BL/6J$ background have been similarly tested. These animals were: CCA^ya, C-aa, and cc aa. Only two of the 20 yellow males attacked in their mating tests and none of the black animals did. Further, these attacks were much less severe. This result tends to confirm the importance of rearing condition as a factor in the expression of this trait. Surprisingly, two of the albino males were also attackers. This suggests that an incipient strain trait, rather than a particular gene effect, is involved. *C57BL/6J* animals are usually thought of as a less aggressive strain than the *DBA/2J* (Calhoun, 1956; van Abeelen, 1966). Certain nonspecific single gene substitutions may bring out an aggressiveness not normally found with this inbred genome.

Given that individual mutant alleles often are not static in either their physiologic or behavioral expression, a general picture of behavioral single gene effects seems unlikely, in the near future at least. It is plain that the manipulation of mutants represents a method of making physiologic changes of potential behavioral importance in intact animals. However, the existence of pleiotropy and variations in expressivity make mutant bearing animals something less than precision tools in many behavioral tests. The systematic development of gene-behavior profiles both within and across strains is an obvious present need.

EVOLUTIONARY RESEARCH

Earlier it was noted that the ideas developed within this chapter are not compatible with defining problems by primary recourse to questions of psychologic importance. In like manner, it is suggested that in using laboratory mice, we should carefully avoid problems having primary reference to that which is adaptive in nature. Most writings on evolutionary theory are concerned with free-living populations and questions ultimately devolve to the theory of natural selection. The interpretative transition to laboratory populations is often ambiguous at best. At the cellular level, the genetic machinery is presumed to be the same in both wild and laboratory populations. Environmentally contingent differences are pervasive, however, at both the genetic and developmental levels. The on-again-off-again quandary of the laboratory evolutionist is well illustrated by Lane-Petter (1963).

> All laboratory species have been derived from wild ancestors. In the course of breeding in captivity they have been altered, often profoundly, but their derivation is still completely self-evident.... It is reasonable to postulate that their innate pattern of behavior may also be to a very large extent archetypal.
> There are, of course, obvious differences, both in the physical appearance and in behavior, between the wild and the laboratory bred versions. Colonies of albino rats, or of mice with a high incidence of leukemia, for example, do not occur in nature. Similarly, the laboratory versions are normally tame and easily handled, while the wild ones can be guaranteed to bite if cornered. If it

were possible to feed laboratory rats on exactly the same diet as wild rats subsist on in nature, it is doubtful whether the former would thrive.

In the course of laboratory breeding the group behaviour of rats and mice has undergone profound modification of the archetypal pattern. On the other hand, laboratory rats and mice which are normally docile still retain at least some of their capacity for aggressive behavior and can be provoked to attack each other or bite their handlers in certain circumstances. . . .

These considerations are more valid when the number of generations separating a laboratory colony from a wild ancestor is small. The point has been beautifully illustrated by McLaren and Michie (personal communication) in their attempts to breed in captivity a group of Gough Island mice (*Mus musculus*). They found that even in a large cage no breeding took place until, following the earlier experience of Professor L. C. Dunn in the United States with wild caught mice, they provided a large exercise drum, which the mice used with great enthusiasm. Breeding ensued. The wild mice roam freely over the uninhabited Gough Island and it would seem that a great deal of exercise is a necessary prerequisite for breeding in this animal.

. . . A reasonable first assumption is that the laboratory rat or mouse, being ultimately derived from wild ancestors, possesses ancestral characteristics; the onus of breaking down this assumption, piecemeal, rests on the laboratory worker, who must satisfy himself about the evidence for believing that what is undesirable in the wild archetype has been eliminated from the laboratory counterpart. (p. 1–19)

Among the more prominent behavioral-evolutionary theorists working with mice is Jan Bruell. His research concerned with behavioral heterosis dates to the early 1960's and has influenced numerous other experimenters. In a recent summary of this work (Bruell, 1967) he discusses his theoretical position with regard to genetic and evolutionary mechanisms and suggests the course that future research should take.

We pointed out that it does not make much sense to speak of the gene pool of a species. Thinking and experimentation must center around gene pools of local interbreeding populations for reasons discussed extensively by Mayr (1963, Chap. 7). The inbred strains of most laboratory animals were not created with this in mind. Thus one cannot hope that strains chosen at random from those available through dealers would carry a representative sample of genes from any gene pool shaped by the forces of a common environment. The wild base populations from which the ancestors of our laboratory strains were drawn are shrouded by the mist of incomplete records and thus, in most cases, we do not know anything about the natural environmental conditions under which the founders of our strains evolved. This reduces inferences from breeding experiments about "the natural population" or "the natural environment" of our laboratory strains to the status of speculations. . . .

The remedy here seems to lie in a departure from current practice. Inbred strains have their important place in behavior-genetic research; they provide the best available estimates of environmental variance (R. C. Roberts, 1967, Chapter 11, page 225; Falconer, 1960, p. 130); because of their uniformity they are ideally suited for pilot studies designed to develop new tests and measuring techniques; and, as illustrated in this chapter, they may help one to develop tentative hypotheses about heterotic and intermediate inheritance of certain traits. But to study behavioral inbreeding depression and heterosis in a biologically meaningful way, it seems one should avoid inbred strains. (p. 284)

Bruell's programmatic position is well founded and can be expected to produce useful evolutionary insights. In the next few pages another almost completely programmatic position is developed which seeks to supplement, rather than supplant the earlier inbred-based, quantitative genetics approach to questions of behavioral evolution. It suggests that inbred mice, particularly those which are mutant-bearing, may still make a contribution. This, *if* their peculiarities in both breeding structure and habitat are kept constantly in mind. Specifically acceded to is Bruell's (1967, p. 285) position that: "In the future, before embarking on genetic research, we should do our best to become well acquainted with the environment and the life as it is lived by the animal we study. Only such intimate knowledge will ensure that we select for detailed genetic analysis behavior whose adaptive significance, or lack of it, we understand." There has hardly been a start to this kind of research with inbred animals. "In the past we often tended to study only those forms of behavior that were conveniently measured; our approach was test-oriented" (Bruell, 1967, p. 285).

The key notion advanced here is that highly inbred strains can profitably be considered as being not only essentially isogenic but also as having some degree of coadaptation in their genomes based on their specialized ecology. Neither of these ideas seems to be currently in vogue and either or both of them may eventually be demonstrated as untrue. They are testable. Margaret E. Wallace has asked the question: "How homozygous are our inbred lines and closed colony stocks?" In reviewing the literature, she has come to an equivocal conclusion (Wallace, 1965a, 1965b).

. . . Perhaps my gravest handicap is that I haven't any axe to grind —I'm not convinced that the theory of inbreeding is entirely fulfilled in practice, nor that the by-products of current research have done much to damage the theory: my only conviction is that, so long as there is doubt, it is worth designing experiments specifically to test the predictions of theory, and little has been done in this field so far. (p. 165)

I think it may be said that in general, practice has confirmed the predictions of theory; sib-mated strains are the most uniform, more so than closed-colony systems; inbred strains seem to have satisfied their users, though this is really subjective evidence. Perhaps the best objective evidence lies in the fact that the highly inbred strains appear to be homozygous for all the loci controlling immunological reaction—and blood constituents generally; a fairly large number of loci are known to be concerned and a great many people have used the strains.

In addition two or three attempts have been made—in Drosophila rather than mice because of the great numbers of progeny needed—to select within inbred lines. This process can be successful only if some loci have remained heterozygous; these attempts have failed. It is in the individual strains and substrains of mice intensively studied that genetic variation occasionally appears, and gives rise to a measure of doubt. (Wallace, 1965a, pp. 169–170)

Mrs. Wallace has developed a rationale for assessing homozygosity which involves the use of a tester stock carrying modifier "polygenes" and highly inbred strains bearing dominant mutants of variable expression (Wallace; 1965a, 1965b). Personal correspondence with Mrs. Wallace and another worker in the United States indicates that little is being done to resolve the problem as she has raised it.

Ginsburg has discussed the behavioral plasticity of the *C57BL* strains generally and has presented data which indicate that the variability found is not genotype dependent (Ginsburg, 1967). Ginsburg (1965) has also noted that this behavioral variability is associated with morphologic variability.

The *C57BL/10* strain, which is most affected by these handling differences, is particularly labile with respect to both behavior and morphology (Ginsburg, 1958; Grüneberg, 1947). We have attempted to test this strain for genetical variability in behavior differences by breeding from the extremes of the distribution. However, selection for seven generations has not produced any divergence in progeny. Comparable results have been reported for the morphological variability. It appears that this strain is poised on a threshold with respect to many environmental factors, both behaviorally and anatomically, and one result is that minor environmental changes against which other genotypes are well buffered may make a tremendous difference in mice of the *C57BL/10* type. (p. 72)

In selection, the global phenotype of the organism, not discrete attributes, is acted upon. Therefore, while genes are the unit of inheritance, entire genotypes are the unit of the selection process (Fuller and Thompson, 1960; Mayr, 1963). Genes which complement, modify, or mask the expression of other genes may become a part of the gene pool of an inbred strain. Srb et al. (1965, pp. 427, 465, 478) note that the rate of homozygosity increase with brother × sister inbreeding is sufficiently slow for selection to be effective. Further, the general conditions and controls of the inbreeding environment can be specified and clearly constitute a very special, if stabilized, niche for this highly adaptable animal to fill. Among the ecologic conditions which inbred animals face are the following. They are produced in the numbers needed to meet experimenter demand. They experience little if any seasonal variation. Temperature and humidity are controlled. They are free of disease and parasites. Man is their only major predator. The inbred pedigreed foundation stocks are caged under constant but confined conditions which limit the expression of some behaviors (e.g., a given animal can only retreat from another; it cannot escape). Adaptation to these and other environmental differences from animals in nature (diet, sanitation, etc.), permit inbred animals to reproduce more offspring in the laboratory than they would in any conceivable wild environment. Wild animals, we have seen, may have no genetic (reproductive) fitness under standard laboratory conditions (Lane-Petter, 1963).

> ... Obviously any drastic improvement under selection must seriously deplete the store of genetic variability. An open, natural population has to cope with numerous conditions to which the closed and sheltered laboratory population is not exposed. Protected against the effects of immigration and an adverse and variable environment, such a population can afford to develop genotypes that would be of inferior viability in nature. A closed population can respond to special selection pressures in the new uniform environment and on the standardized genetic background of a fixed and limited number of genes (Mayr, 1963, pp. 289–290).

Most of what Lane-Petter said, in the example cited earlier, is acceptable. Given the ideas developed subsequently, however, his interpretive assumption is taken to be incorrect. A more reasonable assumption is that laboratory animals are better adapted to their inbreeding environment than to any other; that they possess laboratory characteristics (Hawkins, 1964). The onus for breaking this assumption rests with those who would use inbred animals as a mammalian model which is independent of its breeding structure. If we are to have generalizations in behavior-genetics, a theoretical base is necessary. The position taken here is that, for lower animals at least, the theory of natural selection is sufficient. The study of laboratory *Mus musculus* (in its multitudinous forms), in this context, is a special case. If the Gough Island mice of the Lane-Petter quotation were studied or a "wild" *Mus musculus* population living in a Central Texas barn, these two would be special (and different) cases. In each of these populations the ecology, selection pressures, degree of adaptation, and the gene pool can safely be assumed to be different. For convenience, we speak of artificial selection when man is involved in choosing which organisms shall mate and which shall not. However, whether we use the word "natural" or "artificial" the biologic process which is operative in wild, farm, pet, laboratory, or other animals is the same. A complex of inherent biologic attributes in interaction with a complex of life situations, affects both their developmental adaptation and reproductive potential. Hence, it is held that: (1) The general theory of natural selection is appropriate to interbreeding populations in nature; (2) The genetic theory which has come to underlie natural selection theory is appropriate to both wild and inbred populations; (3) The testing of naturalistic hypotheses in a laboratory context has usually been counter to the acceptance of these principles since the ecology and breeding structure of laboratory *Mus* is very different from any found in nature. Point three stands or falls with the theory of natural selection since it is simply a logical extension of that theory.

> The adaptive nature of behavior is almost a truism. In order to survive, organisms must respond to stimuli in a way which results on the average in the satisfaction of tissue needs and the execution of reproductive functions. The accepted explanation for the correspondence between needs and behavior is the evolution of behavior mechanisms through natural selection.
> Briefly, the natural-selection theory of behavioral evolution postulates three related processes. First, random genetic variation occurs within a population. Second, this results in variable behavior, some forms of which are better adapted than others to the environmental challenges which are encountered. Third, the better-adapted individuals are more successful in reproduction, and the genes which are necessary for superior adaptation increase. The process has no definite end point, and evolution is a contemporary process as well as a historical one. (Fuller and Thompson, 1960 p. 319)

Such then, is the rationale for the position that each highly inbred strain represents a single coadapted genotype which is the same for all members of a strain (within sexes), except for the mutations which may arise in production. If this is true, behavioral tests which are *appropriate to the inbreeding environment* give utility to the use of inbred animals in testing evolutionary hypotheses. Some in behavior genetics will question this general position. Adoption of these ideas by other workers should lead to new kinds of experimentation by proponents and opponents alike, however, and thus it is hoped, to the growth and improvement of the field. The evolutionary implications of single gene experiments have their most direct interpretation when given this context since major substitutions can be expected to have selective effects on gene complexes which have become coadapted in their absence. Two examples may help to make this last point more clear. Both involve A^y; one is at the biochemical "mechanism specific" level, the other is more traditionally psychologic in nature.

Marsden and Bronson (1964) found that male urine, when topically applied to the external nares of previously grouped female *C57BL/6J* mice, produced estrous synchrony. More recently, it has been found (using *SJL/J* females) that the estrous inducing pheromone of male mice is volatile and can be transported by air movement (Whitten et al., 1968). Bartke and Wolff (1966), using *YS/ChWf* animals, found that A^ya males do not synchronize estrous while *aa* males do so regularly. In a later test, Bartke found that A^ya and *aa* males do not differ, however, in their ability to block pregnancy:

> The comparison of data from my earlier work on estrus synchrony with the present results has to be made rather carefully as the mice were moved to a different environment. Nevertheless these observations may suggest that the Whitten effect and the Bruce effect are mediated by different pheromones or at least that there is a difference in the threshold, i.e., both yellow and non-yellow males produce enough pheromone to block pregnancy but only *aa* males produce enough odor to synchronize the estrus. This, however, is a speculation rather than the conclusion. (Bartke, 1968. Mouse News Letter, 38:26 and personal communication. Both are cited with Dr. Bartke's permission).

A recent test for aggressiveness (Hawkins, unpublished) was carried out using *C57BL/6J* males of genotype *CCA^{y}a, C- aa,* and *cc aa.* The subjects were tested in their home cages using essentially the procedure of Bauer (1956). Tester animals were pure strain *C57BL/6J* males. So little agonistic behavior was engendered that there were essentially no results. As noted previously, there was some aggression exhibited by four of these subjects toward females in an earlier breeding test. Also noted earlier, under certain conditions A^ya males are prone to attack other mice of either sex (Hawkins, 1967). Paradoxically, from these results and the vicious attacks made against females (Hawkins, 1966a), A^y-bearing *C57BL/6J* males appear more likely to attack estrus females than strange males in the home-cage test situation.

It is clear that the A^y substitution has marked effect on the biochemistry, physical structure, and behavior of mice. It is not at all clear how these various levels of effect are related since various inbred strains have provided the initial findings. It seems reasonable that A^y may act as a powerful modifier at several developmental levels. If so, the magnitude of its effect can be expected to vary systematically on different inbred genomes.

In the first chapter ever devoted exclusively to behavior genetic analysis, Hall (1951) observed "... psychogenetics is as yet more a promise than an actuality." By 1960, behavior genetic problems were being ever more actively pursued in several disciplines. Speaking to psychologists, Fuller said: "It is neither necessary nor desirable that there be a mass conversion of psychologists from the investigation of environmental variables to the study of genetic variables" (Fuller, 1960). Behavior genetics has continued its growth. There is considerable evidence in the literature that the time is now ripe for a more active interest in mammalian single gene experimentation. To temper the writer's enthusiastic speculations, a double paraphrase is clearly in order. As a special interest area within behavior genetics, single gene experimentation with mice is as yet more a promise than an actuality. Further, it is plainly neither necessary nor desirable that behavior genetic workers make a mass conversion to single gene techniques.

REFERENCES

Anastasi, A. 1958. Heredity, environment and the question "how"? Psychol. Rev., 65:197–208.

Bartke, A. 1968. Research news. Mouse News Letter, 38:26.

——— and G. L. Wolff. 1966. Influence of the lethal yellow (A^y) gene on estrous synchrony in mice. Science, 153:79–80.

Bauer, F. J. 1956. Genetic and experiential factors affecting social relations in male mice. J. Comp. Physiol. Psychol., 49:359–364.

Beach, F. A. 1951. Instinctive behavior: Reproductive activities. *In* Stevens, S. S. ed., Handbook of Experimental Psychology, New York, John Wiley and Sons, Inc., pp. 387–434.

——— 1960. Experimental investigations of species-specific behavior. Amer. Psychol., 15:1–18.

——— 1965. Experimental studies of mating behavior in animals. *In* Money, J., ed., Sex Research: New Developments, New York, Holt, Rinehart, and Winston, Inc., pp. 113–134.

Bruell, J. H. 1964a. Inheritance of behavioral and physiological characters of mice and the problem of heterosis. Amer. Zool., 4:125–138.

——— 1964b. Heterotic inheritance of wheelrunning in mice. J. Comp. Physiol. Psychol., 58:159–163.

——— 1967. Behavioral heterosis. *In Hirsch,* J., ed., *Behavior-Genetic Analysis,* New York, McGraw-Hill Book Co., pp. 270–286.

Burdette, W. J., ed. 1963. Methodology in Mammalian Genetics, San Francisco, Holden-Day, Inc.

Calhoun, J. B. 1956. A comparative study of the social behavior of two inbred strains of house mice. Ecol. Monogr., 26:81–103.

David, P. R., and L. H. Snyder. 1962. Some interrelations between psychology and genetics. *In* Koch, S., ed., Psychology: A Study of a Science, Biologically Oriented Fields: Their Place in Psychology and in Biological Sciences, New York, McGraw-Hill Book Co., Vol. 4, pp. 1–50.

DeFries, J. C., J. P. Hegmann, and M. W. Weir. 1966. Open field behavior in mice: Evidence for a major gene effect mediated by the visual system. Science, 154:1577–1579.

Diamond, I. T., and K. L. Chow. 1962. Biological psychology. *In* Koch, S., ed., Psychology: A Study of a Science, Biologically Oriented Fields: Their Place in Psychology and in Biological Sciences, New York, McGraw-Hill Book Co, Vol. 4, pp. 158–241.

Falconer, D. S. 1960. Introduction to Quantitative Genetics, New York, The Ronald Press Co.

Fuller, J. L. 1960. Behavior genetics. *In* Farnsworth, P. R., and Q. Mc-Nemar, eds., Annual Review of Psychology, Palo Alto, Annual Reviews, Inc., Vol. 11, pp. 41–70.

———— 1965. Suggestions from animal studies for human behavior genetics. *In* Vandenberg, S. G., ed., Methods and Goals in Human Behavior Genetics, New York, Academic Press, Inc., pp. 245–253.

———— 1967. Effects of the albino gene upon behaviour of mice. Anim. Behav., 15:467–470.

———— 1968. Genotype and social behavior. *In* Glass, D. C., ed. Genetics, New York, Rockefeller University Press and Russell Sage Foundation, 111–117.

———— and W. R. Thompson. 1960. Behavior Genetics, New York, John Wiley and Sons, Inc.

Ginsburg, B. E. 1958. Genetics as a tool in the study of behavior. Perspect. Biol. Med., 1:397–424.

———— 1965. Coaction of genetical and nongenetical factors influencing sexual behavior. *In* Beach, F. A., ed., Sex and Behavior, New York, John Wiley and Sons, Inc., pp. 53–75.

———— 1967. Genetic parameters in behavioral research. *In* Hirsch, J., ed., Behavior-Genetic Analysis, New York, McGraw-Hill Book Co., pp. 135–153.

———— 1968. Breeding structure and social behavior of mammals: A servomechanism for the avoidance of panmixia. *In* Glass, D. C., ed., Genetics, New York, Rockefeller University Press and Russell Sage Foundation, pp. 117–128.

Green, E. L., ed. 1966. Biology of the Laboratory Mouse, 2nd ed., New York, McGraw-Hill Book Co.

Green, M. 1966. Mutant genes and linkages. *In* Green, E. L., ed., Biology of the Laboratory Mouse, 2nd ed., New York, McGraw-Hill Book Co., pp. 87–150.

Grüneberg, H. 1943. The Genetics of the Mouse, Cambridge, Cambridge University Press.

———— 1947. Animal Genetics and Medicine, New York, Hoeber.

Hall, E. S. 1951. The genetics of behavior. *In* Stevens, S. S. ed., Handbook of Experimental Psychology, New York, John Wiley and Sons, Inc., pp. 304–329.

Hawkins, J. D. 1964. Wild and domestic animals as subjects in behavior experiments. Science, 145:1460–1461.

———— 1965. Effects of the A^y substitution on mouse activity level: Spurious pleiotropy. Psychon. Sci., 3:269–270.

———— 1966a. The effect of genotype on scores of sexual responsiveness in female Mus musculus. Doctoral Dissertation, Claremont Graduate School, Claremont, California.

——— 1966b. Developmental effects of the A^y substitution on mouse activity level. Psychon. Sci., 4:105–106.

——— 1967. Abnormal sexual behaviors in male A^y-bearing mice. Psychon. Sci., 9:31–32.

King, J. A. 1967. Behavioral modification of the gene pool. *In* Hirsch, J., ed., Behavior-Genetic Analysis, New York, McGraw-Hill Book Co., pp. 22–43.

Kuhn, T. S. 1961. The function of measurement in modern physical science. *In* Wolff, H., ed., Quantification, New York, Bobbs-Merrill Co., pp. 31–63.

Lane-Petter, W. 1963. The physical environment of rats and mice. *In* Lane-Petter, W., ed., Animals for Research, New York, Academic Press, Inc., pp. 1–19.

Lindzey, G. 1967. Behavior and morphological variation. *In* Spuhler, J. N., ed., Genetic Diversity and Human Behavior, Chicago, Aldine Publishing Co., pp. 227–240.

Lockard, R. B. 1968. The albino rat: A defensible choice or a bad habit? Paper presented at Western Psychological Association Convention, San Diego, March 28.

Manning, A. 1967. Genes and the evolution of insect behavior. *In* Hirsch, J., ed., Behavior-Genetic Analysis, New York, McGraw-Hill Book Co., pp. 44–60.

Marsden, H. M., and F. H. Bronson. 1964. Estrous synchrony in mice: Alteration by exposure to male urine. Science, 144:1469.

Mayr, E. 1963. Animal Species and Evolution, Cambridge, Harvard University Press.

McClearn, G. E. 1967. Genes, generality, and behavior research. *In* Hirsch, J., ed., Behavior-Genetic Analysis, New York, McGraw-Hill Book Co., pp. 307–321.

——— and W. Meredith. 1966. Behavioral genetics. *In* Farnsworth, P. R., and Q. McNemar, eds., Annual Review of Psychology, Palo Alto, Annual Reviews, Inc., Vol. 17, pp. 515–550.

Meier, G. W., and D. P. Foshee. 1964. Albinism and water escape performance in mice. Science, 147:307–308.

Merrell, D. J. 1965. Methodology in behavior genetics. J. Hered., 56:263–266.

Roberts, R. C. 1967. Some concepts and methods in quantitative genetics. *In* Hirsch, J., ed., Behavior-Genetic Analysis, New York, McGraw-Hill Book Co., pp. 214–257.

Rodgers, D. A. 1967. Behavior genetics and overparticularization: An historical perspective. *In* Spuhler, J. N., ed., Genetic Diversity and Human Behavior, Chicago, Aldine Publishing Co., pp. 47–59.

Srb, A. M., R. D. Owen, and R. S. Edgar. 1965. General Genetics, 2nd ed., San Francisco, W. H. Freeman and Co., Publishers.

Thiessen, D. D. 1966. Pleiotropism in behavior genetics: The mouse as a research instrument. Percept. Motor Skills, 23:901–902.

——— 1967. Chromosome mapping of mutations in the mouse. Paper presented at American Psychological Association Convention, Washington, D.C., September 2.

———— and D. A. Rodgers. 1967. Behavior genetics as the study of mechanism-specific behavior. *In* Spuhler, J. N., ed., Genetic Diversity and Human Behavior, Chicago, Aldine Publishing Co., pp. 61–71.

Thompson, W. R. 1968. Genetics and social behavior. *In* Glass, D. C., ed., Genetics, 79–101, New York, Rockefeller University Press and Russell Sage Foundation.

Tinbergen, N. 1968. On war and peace in animals and men. Science, 160: 1411–1418.

van Abeelen, J. H. F. 1965. An ethological investigation of single-gene differences in mice. Doctoral Dissertation, University of Nijmegen, The Netherlands.

———— 1966. Effects of genotype on mouse behaviour. Anim. Behav., 14: 218–225.

Wallace, M. E. 1965a. How homozygous are our inbred strains and closed colony stocks? Food Cosmet. Toxic., 3:165–175.

———— 1965b. The relative homozygosity of inbred lines and closed colonies. J. Theoret. Biol., 9:93–116.

Whitten, W. K., F. H. Bronson, and J. A. Greenstein. 1968. Estrus-inducing pheromone of male mice: Transport by movement of air. Science, 161: 584–585.

Winston, H. D., and G. Lindzey. 1964. Albinism and water escape performance in the mouse. Science, 144:189–191.

Wolfe, H. G., and D. L. Coleman. 1966. Pigmentation. *In* Green, E. L., ed., Biology of the Laboratory Mouse, 2nd ed., New York, McGraw-Hill Book Co., pp. 405–425.

Wolff, G. L. 1963. Growth of inbred yellow (A^ya) and non-yellow (aa) mice in parabiosis. Genetics, 48:1041–1058.

———— 1965. Body composition and coat color correlation in different phenotypes of "Viable yellow" mice. Science, 147:1145–1147.

7

Chromosome Mapping of Behavioral Activities[*]

D. D. THIESSEN, KEITH OWEN, AND MAL WHITSETT
Department of Psychology,
University of Texas

INTRODUCTION

Without question, single gene activities extend into behavioral phenotypes, and their effects are often exceedingly distinct (Merrell, 1965; Thiessen, 1966, 1967). In fact, it is difficult to conceive of a neutral gene with respect to behavior. A behavior is an organism's answer to constant selection for an optimal adjustment between the available genetic material and the impositions of the environment. Each gene specifies a chemical reaction that in some way must fit into a coordinated system of activities. If the fit is good, the organism gains a reproductive advantage; if it is not, a loss of fitness results —biochemical neutrality is inefficient, surely a genetic burden and a phenotypic improbability.

Genes with well-defined influences on behavior can theoretically be mapped on chromosomes as readily as genes that demarcate biochemical pathways or morphological traits. Mapping has already been accomplished, or is well underway, for genes with extraordinary deviant phenotypes like the "waltzer" or "shaker" series in the mouse (Sidman et al., 1965). Neurologic mutants such as these often segregate as good Mendelian genes and, accordingly, are easy to locate on linkage groups relative to each other and to other marker genes. Genes with less dramatic effects on behavior should also yield to this mapping technique, although the problems of defining the nature and extent of behavioral involvement are seriously magnified. Despite the added difficulties, we consider it extremely important to work with genes that modify behavior within the normal range of variation.

[*] This research was supported by NIMH Grant MH 14076-01 and NIMH Research Development Award MH-11, 174-01 to Delbert D. Thiessen.

Our laboratory is embarked on a long-term project directed toward devising a chromosome map of gene influence on basic response patterns. Our attention is focused on known loci and alleles of *Mus musculus,* but in principle, the technique is applicable to any species. The genes are initially identified by observable morphologic characteristics, usually variations in coat color, and only secondarily related to particular behaviors. Briefly, the experimental procedure is to test animals differing by only a single gene over a battery of behavioral tasks. When a behavioral difference is found, it can be ascribed to the allelic difference at that locus. A chromosomal pattern of behavioral influence will gradually unfold with the screening of many alleles over several representative tasks.

This chapter is intended to serve several functions. First, the rationale and advantages of single gene analyses are outlined. Second, a review of single gene effects is presented with special attention to the mouse. Third, the importance of coat color markers for behavioral investigations is stressed. And fourth, results from our initial testing program are described. To date, we have surveyed the albino allele over 11 problems and 14 additional coat-color alleles over four problems. The observations end support to our belief that individual genes can be characterized by their behavioral influence as well as their primary metabolic effect.

THE STRATEGY OF SINGLE GENE RESEARCH

Mechanisms of Gene Action on Behavior

Single genes have been known to alter traits ever since 1865 in the time of Gregor Mendel. His report to the Brünn Natural History Society on pea characteristics remained unnoticed until hereditary principles were rediscovered in 1900 by the three botanists, Hugo de Vries, Carl Correns, and Eric Von Tschermak. Detailed work on behavior came much later. It is nevertheless highly interesting to note that G. von Guaita (see Grüneberg, 1952) uncovered systematic laws of inheritance two years prior to the general recognition of Mendel's work using the behavioral characteristics of "waltzing" in the mouse. The remarkable spinning behavior of waltzing mice was one of the earliest examples of a recessive gene influence. Investigators now have located the relevant gene on Linkage Group X and associated this and similar mutants with a variable amount of vestibular damage and deafness. Apparently the gene acts on the inner ear and related pathways.

Single genes' influence on behavior is probably more pervasive than first realized. While it is true that many behaviors of psychologic interest are controlled by several genes interacting in complex ways, often a single gene positioned in a critical pathway of development significantly alters a response.

The gene need not necessarily be as dramatic in its biologic influence as the waltzing gene. There is a strong suggestion that the most easily identified gene-behavior correlations are often indirectly associated with the gene action usually considered primary. An example is phenylketonuria, where human intellectual impairment appears distant from the principal metabolic block. The reduced activity of the enzyme phenylalanine hydroxylase in the liver, caused by a homozygous recessive condition, prevents phenylalanine from being converted to tyrosine. Several metabolic and behavioral consequences follow, including severe mental retardation. Many other single genetic units have indirect effects on behavior, for example, several neurologic lesions and some coat color variants found in the mouse.

A gene necessarily works through an intricate network of developmental channels. Fig. 1 gives a general picture of interlocking steps from the initial gene action to the ultimate behavioral phenotype. The first step involves the production of a protein, usually an enzyme. The enzyme initiates the reaction for histogenesis (tissue formation), later morphogenesis (organ formation), and the organization of organ systems. At every level of increasing complexity the interactions become more involved. Initially, only the intracellular metabolic features appear, which eventually organize themselves

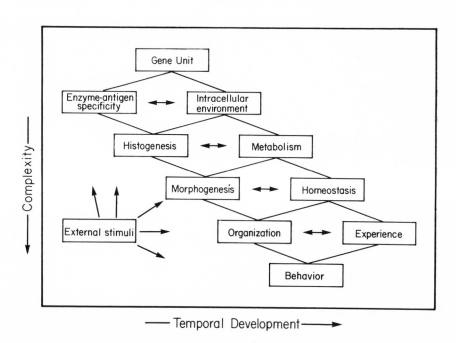

FIG. 1. Development of gene action on behavior. Organization on left depicts structural changes and organization on right depicts functional changes. Interactions occur at all levels of development.

into homeostatic feedback systems. At every point in time, the development is canalized by specific experiences and quite general environmental pressures. Little wonder that the correlation between the first gene reaction and the final behavior is inexact. Consider, too, that all active genes interact at all levels of development and manifestation; hence, the description of development depicted in Fig. 1 is quite simplified.

The lengthy and often tortuous path from gene specificity to metabolic synchrony explains why behavior must be considered as a pleiotropic reflection of physiologic processes. Gene influence on behavior is always indirect. Grüneberg (1952) refers to secondary action of this sort as "spurious pleiotropism" and Hadorn (1961) calls it "allophenic action." Whatever the term used to define secondary consequences of DNA-RNA activity, it is clear that behavior often acquires its characteristics by virtue of intermediate metabolites, chemical shunts, and metabolic blocks. Both the intermediate activities and the range of behaviors affected may be extensive. The clarity of gene-behavior relations appears to correlate directly with the extent of metabolic involvement.

This is to say that there are no genes for behavior per se. There are genes affecting enzyme activity, hormone level, tissue sensitivity, membrane permeability, and other functions and structures; but there are no genes that are directly translated into behavior. The influence of heredity is always indirect. Thus, finding a gene substitution which affects behavior is not the same as discovering *the* genetic key for the behavior. All genes in one way or another contribute to the entire organismic phenotype, even though some may contribute substantially more than others. Nevertheless, by studying many individual gene variations we may be able to specify the relative importance of various pathways intervening between gene action and behavior. Using this technique, it is possible to map behavioral phenotypes on chromosomes. Mapped loci can then be used to trace the genetic strategy of behavioral development. In some instances, the primary and secondary biologic consequences of the gene are known, such as a change in enzyme activity or hormone secretion. When a behavioral effect for this same gene is found, the biologic explanation may be very near. One might argue, therefore, that one of the major tasks of the behavior geneticist is to find out what behaviors are affected by the many genes which are available for experimentation.

The Extent of Single Gene Influences on Behavior

Certainly enough is now known about point mutations (single allelic changes) in several species to warrant a concerted search for behavioral effects. At least 479 loci have been identified in *Drosophila* (Altman and Dittmer, 1964). The extensive knowledge of the constitution of its genome should be particularly useful for the research strategy suggested in this paper. However, the amount of behavioral research on single gene effects is thus far disappointingly small. The available studies are summarized in Table 1. Recent reviews by Manning or Manning and his collaborators (1965, 1967; Ewing and Manning, 1967) cover the details of several of these studies. Most involve the effects of various mutations on mating success. With the exception of one gene which appears to be neutral with respect to this measure (Merrell, 1949), all alleles surveyed show a depression of mating success. Wing-beat frequency and phototaxis have been investigated to even a lesser degree. Mutations either elevate or depress the frequency of wing beat, and change the direction or intensity of the phototaxic response.

The paucity of studies prevents any generalization at all. The lack of interest in *Drosophila* (with the notable exception of Hirsch and his associates, e.g., 1967) reflects the relative difficulty of studying behavioral and physiologic processes in small organisms. It is believed, as well, that *Drosophila* have a limited behavioral repertoire. Behaviors of traditional interest to psychology, such as emotion, motivation, and learning, are probably best pursued with more complex organisms; however, there are many basic behaviors of living that are common to most organisms and open to experimentation. Examples are feeding, drinking, reproduction, and reactions to stimulus gradients. Recent evidence even suggests that *Drosophila* are capable of performing instrumental learning tasks (Murphy, 1967). It is no doubt the case that refined instrumentation and a broader biologic orientation will allow investigators to take greater advantage of the genetic polymorphism existing in this species.

In humans, the store of alleles of potential significance for behavior is even greater than for *Drosophila*. McKusick (1966) catalogues 68 identifiable loci on the sex chromosome and indicates that more than 600 loci exist on the nonsex chromosomes, the autosomes. Very few have been tested for behavioral effects, although sex-linked and autosomal alleles are known to influence color vision (Kalmus, 1965) and ABO blood factors may relate to psychoneuroses and other personality characteristics (Cattell et al., 1964). The number of human diseases now known to be inherited as monogenic units is well over 100 (Crow, 1967; United Nations, 1958), accounting for more than 1 percent of all birth defects. More specific to psychologic processes, the last three decades of research have led to the recognition of at least 38 gene-controlled biochemical errors in low-grade mental deficiency (Erlenmeyer-Kimling and Paradowski, 1966). In one study by Reed and Reed (1965) fully 29 percent of 289 mental retardates with IQ's of 70 or below was found to be due to a strict simple genetic effect. Since only about 16 percent of the cases appeared to be influenced by a polygenic system and the remainder was of environmental or unknown origin, the most substantial portion of the variance was obviously of simple genetic origin. Lastly, the presence and absence of certain muscles and their tendons, important for basic abilities of

TABLE 1

Observed Influences on Behavior:
Single Gene Substitutions in *Drosophila*

Gene Name	Gene Symbol	Linkage Group	Biologic Phenotype	Behavioral Phenotype	Reference
Bar	*B*	X	Eyes narrow in homozygote; kidney-shaped in heterozygote	Mating success reduced	Petit, 1959
Bar	*B*	X	Eyes narrow in homozygote; kidney-shaped in heterozygote	Phototaxis reduced	Brown & Hall, 1936; Medioni, 1959
Bar	*B*	X	Eyes narrow in homozygote; kidney-shaped in heterozygote	Activity of male reduced	Elens, 1965a
Crossveinless	*cv*	X	Crossveins absent or nearly so	Wing beat frequency reduced	Williams & Reed, 1944
Cut	*ct*	X	Wings cut to points, scalloped	Wing beat frequency increased	Williams & Reed, 1944
Cut	*ct*	X	Wings cut to points, scalloped	Mating success reduced	Merrell, 1949
Forked	*f*	X	Bristles short, gnarled	Neutral for mating success	Merrell, 1949
Raspberry	*ras*	X	Eyes dark ruby	Mating success reduced	Merrell, 1949
Scute	*sc*	X	Scutellar bristles missing, others missing or reduced	Wing beat frequency reduced	Williams & Reed, 1944
Vermilion	*v*	X	Eyes bright vermilion, ocelli colorless	Phototaxis reduced	Fingerman, 1952

TABLE 1 continued

Gene Name	Gene Symbol	Linkage Group	Biologic Phenotype	Behavioral Phenotype	Reference
White	w	X	White eyes, ocelli, testes, malpighian tubes	Mating success reduced	Reed & Reed, 1950; Geer & Green, 1962; Petit, 1959; Elens, 1965a
White	w	X	White eyes, ocelli, testes, malpighian tubes	Phototaxis reduced	Brown & Hall, 1936; Scott, 1943; Fingerman, 1952
White	w	X	White eyes, ocelli, testes, malpighian tubes	Wing beat frequency increased	Williams & Reed, 1944
White	w	X	White eyes, ocelli, testes, malpighian tubes	Activity of male reduced	Elens, 1965a
Yellow	y	X	Body yellow, bristles and hair yellow or brown in different alleles	Activity reduced	Sturtevant, 1915
Yellow	y	X	Body yellow, bristles and hair yellow or brown in different alleles	Mating success reduced	Merrell, 1949; Bastock, 1956
Yellow	y	X	Body yellow, bristles and hair yellow or brown in different alleles	Phototaxis increased	Medioni, 1959
Miniature	m	X	Wings small, dark	Wing beat frequency increased	Williams & Reed, 1944
Black	b	II	Body, legs, veins black	Mating success reduced	Elens, 1965b
Black	b	II	Body, legs, veins black	Activity of male reduced	Elens, 1965b

TABLE 1 continued

Gene Name	Gene Symbol	Linkage Group	Biologic Phenotype	Behavioral Phenotype	Reference
Brown	*bw*	II	Eye color brownish to garnet	Phototaxis reduced	Fingerman, 1952
Cinnabar	*cn*	II	Eye color bright scarlet, ocelli colorless	Wing beat frequency increased	Williams & Reed, 1944
Lobe	*L*	II	Eyes small, nicked at anterior edge	Wing beat frequency reduced	Williams & Reed, 1944
Ebony	*e*	III	Body color black	Mating success reduced	Rendel, 1951; Crossley, 1963; Elens, 1965b
Glass	*gl*	III	Eye color dilute, facets fused	Wing beat frequency increased	Williams & Reed, 1944
Sepia	*se*	III	Eye color brownish red, darkening to black	Phototaxis increased	Fingerman, 1952

168

locomotion and sensory-motor coordination, have been shown to be under fairly simple gene control in human populations (Spuhler, 1951). Clearly, the study of simple genetic mechanisms is an attractive and efficient means of investigating behavioral processes.

The extensive research for single gene mutations in humans now underway gives hope that many traits of a quantitative nature will soon yield to particulate divisions. Several characters once thought to be regulated by so many genes that only statistical analyses were possible are now known to be influenced by a composite of a few discrete genes. Mental deficiency is a case in point. If only the entire variance is considered, then the distribution of traits is certainly polygenic. However, from this distribution at least 38 separate entities have been diagnosed and more will undoubtedly be found. Similarly, adrenocortical errors and thyroid defects, obviously of relevance to behavior, appear to be regulated by numerous genes inherited in a polygenic fashion. Hamburg (1967) at Stanford University has shown that nearly 100 percent of the variance of such endocrine disorders can be accounted for by the thorough description of about 10 separate biochemical blocks. The point is that statistical representation of traits may exaggerate the role of quantitative variation and actually hide distinct genetic contributions. Of course, the power of behavioral and physiologic analysis is greatly magnified where simple genetic units are discernible. Robertson (1967), a quantitative geneticist, expressed it well when he said:

> ....in the last three or four years the fog of quantitative variation, which I had assumed to be the result of segregation at a large number of loci, has gradually begun to clear as I have recognized individual segregations almost as personal friends (p. 267).

In another mammalian species, the mouse, well over 300 mutant genes occupying more than 250 loci are known (Green, 1966). Of these at least 55 are nonlethal and easily identifiable coat-color genes, ideal for behavioral study. In addition, 81 neurologic mutants are recognized in the mouse, affecting almost every conceivable biologic function (Sidman et al., 1965). Representative neurologic mutants that in our opinion should be thoroughly investigated with regard to behaviors are listed in Table 2. They range in effect from relatively minor peripheral deviations to severe central defects.

Despite the availability of experimental material on the mouse, few single gene substitutions with "normal" functions have been studied with an eye toward behavior. Most genes tested have been shown to affect some behavior. These genes are listed in Table 3. There are also several reports of failures to observe behavioral effects due to single genes. These include: *short-ear* (Bundy, 1950), *short-ear* and *maltese dilution* (Ashman, 1959), *furless, yellow, short-ear, maltese dilution,* and *black-and-tan* (Les, 1959),

TABLE 2

**Examples of Neurologic Mutants Affecting
Various Targets in the Mouse**

Gene Name	Gene Symbol	Linkage Group	Biologic Phenotype	Behavioral Phenotype
Absent Corpus Callosum	*ac*	–	Partial to complete absence of corpus callosum	None now known
Cerebral Degeneration	*cb*	–	Degeneration of cerebral hemispheres and olfactory lobes	Diffuse
Dancer	*Dc*	–	Absence of macula of utriculus	Circular; failure to swim
Deafness	*dn*	–	Degeneration of Deiter's cell in organ of Corti	Accentuation of sniffling
Dilute-lethal	*d*	II	Myelin degeneration; lowered phenylalanine hydroxylase activity	Convulsions
Eyeless	*ey*	–	Absence of eye and optic tract	Visually incapable
Muted	*mu*	–	Absence of otoliths in one or both ears	Postural defects
Pironette	*pi*	XVII	Organ of Corti degenerate	Circular
Wabbler-lethal	*wl*	III	Myelin degeneration; elevated succinic dehydrogenase	Locomotor difficulties

t-allele (Martin and Andrewartha, 1962), *short-ear* and *pale-ear* (Denenberg et al., 1963), *yellow* (van Abeelen, 1963) and *maltese dilution* (van Abeelen, 1965). The limited nature of the observations for these alleles does not exclude the possibility of behavioral influence and, in fact, some genes listed by investigators as having no effect on behavior have been found by others to influence some performance. Aside from our own project, there has been only one systematic attempt to explore single gene effects on behavior (van Abeelen, 1963, 1964–65), and that was done with only a few genes and without careful standardization of tasks or observations.

Perhaps the most interesting gene thus far investigated in the mouse is the one for albinism. The double recessive allele, *cc,* at the *C* locus on Linkage Group I blocks the enzymatic activity of tyrosinase. This enzyme is necessary for the conversion of tyrosine to dopa, and hence finally to melanin. The simplified chain of events from the substrate to pigment formation is as follows:

$$
\begin{array}{l}
\text{Tyrosine} \\
\quad\downarrow \quad \text{(tyrosinase)} \\
\text{Dopa} \\
\quad\downarrow \quad \text{(dopa-oxidase)} \\
\text{Dopa-quinone} \\
\quad\downarrow \\
\text{Melanin}
\end{array}
$$

The two pertinent enzymes are indicated in parentheses. When tyrosinase is missing, as in the recessive condition, *cc,* tyrosine is not oxidized to dopa and the reactive sequence necessary for pigment formation is broken. The extensive behavioral deviations that follow (see Table 3) are obviously pleiotropic to this major enzymatic deficiency. However, the exact pathway to the behavior is unknown; although there is some indication that photophobic reactions, related to a loss of eye pigment, mediate the extreme hesitancy of the animal (DeFries et al., 1966).

Very few other species have been studied for allelic substitutions on behavior. Keeler and King (1942) several years ago attempted to relate coat color in the rat to emotionality, and more recently correlated variations in coat color of foxes to timidity (Keeler et al., 1968). Rothenbuhler (1967) believes he has isolated two genes controlling nest cleaning behavior in bees, and Crozier and Pincus (1936) suggest that three genes regulate negative geotaxis in the rat. For the most part, the genetic hypotheses are unconfirmed.

Here, then, is the genetic raw material for study, especially abundant in *Drosophila,* man, and mouse. Single genes are numerous and easily related to behavior. The beauty of single gene analyses lies in the possibility of tracing the physiologic interactions from the initial gene alteration to the behavior in question. The alleles can be easily identified, manipulated as independent variables, and, in several cases, related to known metabolic pathways. It is as if only one letter of a word is allowed to vary at a time in order to study its special influence. The procedure of screening single gene substitutions on a standardized battery of tasks as explained in a later section, will almost certainly uncover important behavioral relations. Already the literature reveals that one or two behavioral observations are sufficient to disclose single gene effects for *Drosophila* and mice in about 50 percent of the attempts. The genes involved have primary effects within the usual range of variation and cannot be considered abnormal. One might proceed, of course, with the faith that all gene substitutions modify behavior.

The Classification of Gene Effects

We can anticipate extensive behavioral effects from nearly every gene tested. However, success may ring the death knell for such studies if only

TABLE 3

Observed Influences on Behavior:
Single Gene Substitutions in the Mouse

Gene Name	Gene Symbol	Linkage Group	Biologic Phenotype	Behavioral Phenotype	Reference
Albino	c	I	Absence of pigment in hair and eyes	Decreased nibbling	Les, 1959
Albino	c	I	Absence of pigment in hair and eyes	Delayed audiogenic seizure	Gutherz & Thiessen, 1966
Albino	c	I	Absence of pigment in hair and eyes	Decreased water escape performance	Winston & Lindzey, 1964
Albino	c	I	Absence of pigment in hair and eyes	Decreased active avoidance; increased passive avoidance	Winston et al., 1967
Albino	c	I	Absence of pigment in hair and eyes	Decreased activity	DeFries et al., 1966
Albino	c	I	Absence of pigment in hair and eyes	Competitive advantage for sex partners	Levine & Krupa, 1966
Albino	c	I	Absence of pigment in hair and eyes	Decreased activity and alcohol preference	Henry & Schlesinger, 1967
Albino	c	I	Absence of pigment in hair and eyes	Decreased water escape performance and open field activity; poor black-white discrimination	Fuller, 1967

TABLE 3 continued

Gene Name	Gene Symbol	Linkage Group	Biologic Phenotype	Behavioral Phenotype	Reference
Pink-eyed dilution	*p*	I	Pink eyes, reduced or brown pigment	Decreased staring at E, less paw lifting, more grooming and shaking	van Abeelen, 1963, 1965
Dilute	*d*	II	Blue-gray color of hair	Decreased activity	Henry & Schlesinger, 1967
Short-ear	*se*	II	Reduced cartilaginous skeleton	Impaired avoidance learning to sound	Bundy, 1950
Yellow	*A^y*	V	Yellow or orange fur and black eyes	Decreased short and long term activity	Hawkins, 1965, 1966
Yellow	*A^y*	V	Yellow or orange fur and black eyes	Failure of male to synchronize mating in grouped females	Bartke & Wolff, 1966
Brown	*b*	VIII	Brown instead of black pigment	Increased grooming	van Abeelen, 1963, 1965
Misty	*m*	VIII	Dilute coat color, tail and belly spots	Decreased nibbling	Les, 1959
Pintail	*Pt*	VIII	Short tail	Fast avoidance extinction	Denenberg et al., 1963

173

individual differences can be disclosed. Simply pointing to individual differences has become a trite aphorism. Individual differences take on meaning only to the extent that they outline general laws of behavior. We wish to know, among other things, general principles of behavioral evolution, mechanisms underlying behavioral variation, and conditions of gene-environment interaction. We are interested only in passing with the *C57* mouse or a specific gene for coat color. Chromosome mapping is only the beginning.

The behavioral variations may be specific to a particular sensory or motor system or have general or patterned effects. All gene substitutions will eventually reduce to a structural change in the DNA molecule. Until this is feasible, behavioral alterations can be classified within a coordinated space of known biologic parameters such as that seen in Fig. 2. According to Muller's (1932) broad classification, alleles can be *amorphs* (inactive), *hypomorphs*

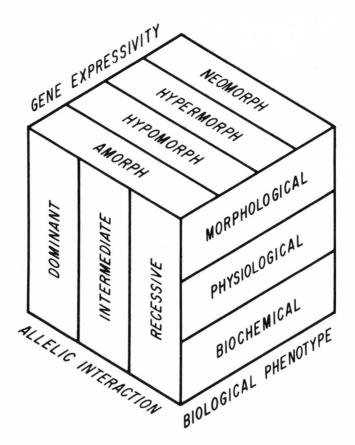

FIG. 2. Biological cube, showing dimensions of gene action for classifying behaviors according to "biological phenotype," "allelic interaction," and "gene expression."

(reduced function), *hypermorphs* (increased function), and *neomorphs* (changed function.) Alleles also interact to give dominant, intermediate, or recessive effects, and the primary phenotype can be described in terms of morphology, physiology, or biochemistry. Thus a three-dimensional analysis involving gene expressivity, allelic interaction, and biologic phenotype provides 36 observational categories in which to anticipate behavioral effects. For example, a genetic influence on general activity might be found to be associated with a dominant gene, acting on morphologic structures and displayed hypomorphically.

Other classification devices are of course possible. Functionally related genes tend to cluster on particular chromosomes (Elston and Glassman, 1967; Lewis, 1967), and this clustering may extend to behavioral phenotypes, thus providing natural groupings. Perhaps behaviors are decipherable into general classes of gene action according to mode of inheritance or primary phenotype. And perhaps alternative alleles at a single locus will provide gradated effects on the same behavior.

The simplest behavior-genetic analysis is the one described above, where a single unit is deliberately isolated from others and studied in a variety of circumstances. This procedure is not always feasible because genes can combine in various ways to produce an effect and complex behaviors are usually divisible into several genetic units. The literature in behavior genetics nevertheless suggests that behaviors of interest are determined for the most part by three or fewer genes. The suggestion is ordinarily derived from quantitative analyses where it is impossible to isolate the individual units. If the usual estimates prove to be correct, however, then it is possible to apply classic genetic analyses in order to disjoin genetic aggregates. Once this is accomplished the strategy of each gene alone or their combinations can be studied. Behaviors in many cases appear complex and multivariate because of the interlocking of several gene effects and our inability to observe temporal variations of response in relatively simple environments. The investigator can gain insights into genetic control of behavior by constantly attempting to define individual elements of behavior, by simplifying the response requirements of the organism, and by seeking abrupt changes in ongoing response sequences. Each change in behavior from one element to another may represent a switch in genetic control.

Of course, as the number of influential genes increases, the problems of Mendelian decoding multiply. Whereas only nine genotypes appear in an F_2 population when two genes are acting, three times that many appear when three genes are involved. The comparable number of genotypes for backcrosses are four and eight. In other words, F_2 genotypes appear on the order of 3^n, where n is equal to the number of segregating gene pairs, and backcross genotypes appear on the order of 2^n. Clearly, the advantages of using a backcross are enhanced over that of an F_2 population as gene frequency increases. Beyond three genes, F_2 analyses are absolutely unwieldy. Table 4 sets out genetic and phenotypic expectations through four gene pairs and gives the general case for n gene pairs.

The recognition of more than two genes poses formidable problems. It bears repeating that our reference is to analyses that demonstrate the genotype of *individuals* and not to estimates of probability based on *population* parameters. Three gene pairs generate 27 genotypes and a minimum of eight phenotypes in an F_2 population (assuming the unlikely state of dominance at all three loci.) Of course behavior is generally too variable to pigeonhole accurately from small samples; hence, many individuals must be tested and classified. A rough estimate is that the least frequent phenotype must be represented at least 10 times in order to reliably reflect the genotype. In the best possible circumstances approximately 300 offspring meet this criterion for three gene pairs! A backcross of an F_1 to the recessive parent offers a better means of tackling the problem. A three gene situation would generate eight genotypes and phenotypes. Thus about 80 offspring would satisfy the needs of classification. Curiously, the power of the approach has not been exploited. In any case, these classic Mendelian techniques are still the most powerful and direct methods for resolving behaviors into the most refined elements possible. A rigorous analysis of each separate unit can follow.

COAT COLOR AS A DEFINING CHARACTERISTIC*

Coat color is perhaps the most thoroughly studied phenotype in *Mus musculus*. There are approximately 60 genes affecting pigmentation of hair (Deol, 1963; Wolfe and Coleman, 1966). Of these, at least 55 are nonlethal and easily identifiable. Obviously, color variations based on single genes present clear markers for the investigation of pleiotropic actions on behavior. Mice can easily be classified according to type and intensity of pigmentation, which in turn can be related to behavioral dimensions. Whenever an invariant association is found between coat color and behavior, the likelihood is high that the responsible gene is not specific to color but also penetrates into other developmental pathways.

It is the writers' opinion that much of the future work in behavior genetics will take advantage of the diverse stock of color genes available. Our studies have used coat-color labels almost exclusively. Hence, we are providing in this section some general information about the genetic determinants of coat color in the mouse. This will assist the reader in understanding our

* Our appreciation is extended to Dr. M. S. Deol, M. R. C. Experimental Genetics Research Unit, University College, London, who kindly commented on various aspects of this section. Any errors remain our own.

TABLE 4

Genetic and Phenotypic Expectations for Different Numbers of Gene Pairs

No. of Segregating Gene Pairs	Intercross ($F_1 \times F_1$)							Backcross ($F_1 \times$ recessive P)
	No. of Gametes	No. of Recombinants	No. of Genotypes	No. of Phenotypes[a]	Occurence of Most Frequent Genotype	Occurrence of Most Frequent Phenotype[a]	Occurrence of Least Frequent Genotype[b] and Phenotype	No. of Genotypes and Phenotypes
1 (Aa)	2	4	3	2	1/2	3/4	1/4	2
2 (Aa,Bb)	4	16	9	4	1/4 (Aa,Bb)	9/16 (A-,B-)	1/16	4
3 (Aa,Bb,Cc)	8	64	27	8	1/8 (Aa,Bb,Cc)	27/64 (A-,B-,C-)	1/64	8
4 (Aa,Bb,Cc,Dd)	16	256	81	16	1/16 (Aa,Bb,Cc,Dd)	81/256 (A-,B-,C-,D-)	1/256	16
n (general case)	$(2)^n$	$(4)^n$	$(3)^n$	$(2)^n$	$(1/2)^n$	$(3/4)^n$	$(1/4)^n$	$(2)^n$

[a]Dominance assumed at each locus.
[b]Homozygotes.

particular use of the genes and, at the same time, provide the interested investigator with knowledge about the most common gene alternatives.

Common Coat-Color Alleles

The pigment found in rodent hairs is melanin. Interestingly, only two kinds of melanin are found: phaeomelanin and eumelanin. The former is always yellow, while the latter may be black or brown. Hence, the wide variety of coat colors evident in the mouse are the result of genetic influences on only two kinds of pigment. This may prove to be a superlative research advantage, as any eventual description of behavior can be related to the parametric diversities of melanin. Once the biochemical chains of pigment formation are completely known, we will be close to uncovering the explanation for the associated behaviors. A great deal about the genetics and biochemistry of pigments is already known and can be used to advantage in behavioral studies.

The most common inbred strains of mice encountered in behavior genetic research are listed in Table 5. The genetic determinants of their coat color are also specified. The primary coat-color loci are *Agouti* (*A*), *Black* (*B*), *Albino* (*C*), and *Dilute* (*D*). They specify the distribution of black, brown, and yellow pigments in rodent hairs. *Pink Eyed* (*P*) and *Piebald* (*S*) are secondary loci which govern not only the amount of different pigments but the shape, size, and distribution of individual pigment granules. Generally two and often three or more alleles are present at each locus and the dominance

TABLE 5

Some Genetic Determinants of Mouse Coat Color at Common Loci

Strain	Locus					
	Agouti	Black	Albino	Dilute	Pink Eyed	Piebald
C57BL	*aa*	*BB*	*CC*	*DD*	*PP*	*SS*
C3H/2	*AA*	*BB*	*CC*	*DD*	*PP*	*SS*
DBA/2	*aa*	*bb*	*CC*	*dd*	*PP*	*SS*
I	*aa*	*bb*	*CC*	*dd*	*pp*	*ss*
BALB/c	*AA*	*bb*	*cc*	*DD*	*PP*	*SS*
A	*aa*	*bb*	*cc*	*DD*	*PP*	*SS*
R III	*AA*	*BB*	*cc*	*DD*	*PP*	*SS*
Linkage Group	V	VIII	I	II	I	III

and epistatic relations are complex. We shall be concerned with the alleles most often encountered and those that have the most immediate potential for behavioral research.

ALLELES AT THE A LOCUS.—The six genes in this series are located in Linkage Group V and govern the distribution of black, brown, and yellow pigments. *Yellow* (A^y) is the highest order of dominance and is lethal in the homozygous condition. The heterozygous effect on behavioral activity can be seen in Table 3. Mice of the designation have a rich yellow or orange fur and black eyes. They suffer from endocrine difficulties and generally become obese at a later age. *White-bellied agouti* (A^w) is recessive to A^y but dominant over the rest. The agouti appearance on the back takes its name after a South American mammal whose individual hairs are dark with yellow bands in the apical or subapical position. In A^w the back of the animal is the typical agouti appearance but the belly is lighter because the bases of the hairs are not so dark and the yellow bands are much wider and paler. *Grey-bellied agouti* (A) is recessive to A^y and A^w but dominant over other alleles. Mice with dominant expression of A are the typical wild type found in most mice. They differ from the white-bellied agouti in having a belly only slightly less dark than the back. *Black-and-tan* (a^t) is recessive to the foregoing but dominant over others. Mice with expression of a^t have glossy black backs and yellow or orange bellies. The border between the back and belly colors is sharp. No behavioral effect for this gene has been found in several observations (Les, 1959; and in this paper). *Nonagouti* (a) is recessive to those above but dominant over a^e (*Extreme nonagouti*). Mice of this type are entirely black except for a few light hairs around the ears, nipples, and genitals. The descending dominance relations of A^y, A^w, A, a^t, a, and a^e make the A locus ideal for behavioral study.

ALLELES AT THE B LOCUS.—Four alleles appear at this locus in Linkage Group VIII. They determine the type of eumelanin (black or brown) that is produced. The allele (*Black*) (B) which produces black eumelanin is dominant over b and b^c and semidominant over B^{lt}. The allele *brown* (b) is recessive to all others, and causes brown eumelanin to be produced. Behavioral studies show that b increases grooming (van Abeelen, 1963; 1965) and decreases water escape performance (this paper). The allele *cordovan* (b^c) is dominant over b but recessive to B and B^{lt}. The hairs are very light grey with darker tips, and in $B^{lt}B^{lt}$ the hairs are almost white with brownish tips. The alleles at this locus interact complexly with those at the A locus (Deol, 1963).

ALLELES AT THE C LOCUS.—The five alleles in Linkage Group I related to coat color determine the amount of phaeomelanin and eumelanin pigments. Full color (C) is completely dominant over other alleles. The allele *chin-*

chilla (c^{ch}) reduces the intensity of the yellow color by reducing the number of phaeomelanin granules. The effect is to alter an agouti mouse from brownish gray to a purer gray resembling that of a chinchilla. We have found that c^{ch} alters the mouse's sensitivity to a change in an environmental incline. *Extreme dilution* (c^e) suppresses the formation of phaeomelanin completely and reduces the amount of eumelanin significantly. Other pigment genes are therefore diluted in effect. *Himalayan* (c^h) produces a phenotype that is much like a Himalayan rabbit. The body is light and the nose, ears, tail, and scrotum are darker. *Albino* (c) suppresses the formation of any kind of pigment. The fur of such an animal is pure white, while the eyes are pink. The presence of this allele in the homozygous state prevents any other coat-color gene, regardless of locus, from being expressed. The allele has numerous effects on behavior (Table 3 and this paper). There is a decreasing intensity of color in the series C, c^{ch}, c^e, and c which corresponds to the level of dopa oxidase and tyrosinase activity in hair follicles (Wolfe and Coleman, 1966). The hierarchical relation between color and enzyme activity may prove to be a great advantage in explaining behavioral variations among alleles.

ALLELES AT THE D LOCUS.—The three alleles, D, d, and d^l are in Linkage Group II and specify the distribution of pigment granules in hair cells rather than the type or quantity of pigment produced. *Intense* (D) is necessary for the normal distribution of pigment and is dominant over the other two. *Maltese dilution* (d) results in the dilution of coat color because the pigment granules are clumped in irregular masses. The allele d decreases behavioral activity (Table 3). *Dilute-lethal* (d^l) affects fur in the same way as d but the homozygote suffers severe neurologic lesions and dies in early life (Sidman et al., 1965).

ALLELES AT THE P LOCUS.—Four alleles, with the dominance relation of $P > p^r > p > p^s$, exist in Linkage Group I. They regulate the amounts of phaeomelanin and eumelanin and also the shape, size, and distribution of individual granules. *Intense* (P) is necessary for normal production and distribution of pigment. *Japanese ruby* (p^r) dilutes the pigment effect on eyes and fur. The eyes may vary from pink to black. *Pink-eye* (p), like p^r, reduces the amount of eumelanin considerably and phaeomelanin slightly, but the effects are greater. The gene affects eye color especially, reducing eye pigments greatly. The allele *sterile* (p^s) resembles p in effect on pigments and, in addition, the mouse is severely retarded in development.

ALLELES AT THE S LOCUS.—*Piebald* (s) is a recessive allele found in Linkage Group III. It produces a spotting that is neither completely regular and symmetrical nor erratic in distribution. The degree of spotting is highly variable, ranging from almost regular coloration with a very few spots to entirely white forms. The extent of white depends on modifying genes. *Intense* (*S*) is necessary for the complete expression of other loci. No behavioral effects of this locus have been discerned.

Some idea of the interaction among loci can be gained from Table 6. The principal loci *A, B, D,* are represented, and the effects of the *P* locus imposed. All allelic changes are viewed in comparison to the agouti animal with dominance at all four loci. The alleles selected for illustration are the most common alternatives and those represented in the popular strains discussed in this paper (e.g., Table 5). In general, the substitution of *aa* for *AA* changes an agouti to a nonagouti (generally black), *dd* dilutes the primary phenotype, and *pp* adds pink eyes. Other variations are of course evident depending on the interacting qualities of the various loci. Clearly, the allelic substitutions are complex in scope and alter the phenotype qualitatively as well as quantitatively. Nevertheless these are common variations that an investigator will

TABLE 6

Effect of Recessive Genes on Coat Color in the Mouse

Segregating Alleles				Coat Color
A	B	D	P	
				Agouti
aa				Nonagouti
	bb			Dark-bellied Cinnamon
		dd		Dilute Agouti
			pp	Orange or Yellow–Pink Eyes
aa	bb			Chocolate
aa		dd		Blue Gray
aa			pp	Blue Lilac–Pink Eyes
	bb	dd		Dilute Cinnamon
	bb		pp	Orange or Yellow–Pink Eyes
		dd	pp	Dilute Agouti–Pink Eyes (pale)
aa	bb	dd		Dilute Chocolate
aa	bb		pp	"Champagne" or "Café-au-lait"–Pink Eyes
	bb	dd	pp	Dilute Cinnamon–Pink Eyes (pale)
aa	bb	dd	pp	Dilute Chocolate–Pink Eyes (pale)

TABLE 7
Segregating Alleles (SA) and Coat Color (C) in Selected Mouse Strains

Strain		Strain						
		C57BL	C3H/2	DBA/2	I	BALB/c	A	R III
C57BL	SA	—	A	BD	BDPS	ABC	BC	AC
	C	Nonagouti	Agouti	Nonagouti	Nonagouti	Agouti	Nonagouti	Agouti
C3H/2	SA		—	ABD	ABDPS	BC	ABC	C
	C		Agouti	Agouti	Agouti	Agouti	Agouti	Agouti
DBA/2	SA			—	PS	ACD	CD	ABCD
	C				Dilute Chocolate	Dark-bellied Cinnamon	Chocolate	Agouti
I	SA				—	ACDPS	CDPS	ABCDPS
	C					Dark-bellied Cinnamon	Chocolate	Agouti
BALB/c	SA					—	A	B
	C						Albino	Albino
A	SA						—	AB
	C							Albino

182

encounter in crossing strains to obtain an F_1 (Table 7) and in deriving F_2's from the strain crosses (Table 8).

The data contained in the four preceding tables provide the investigator with information about the genotype of popular research strains, the type of interaction expected at several prominent color loci, and the genetic and phenotypic expectations from crossing various strains and deriving segregating populations. The genotype can sometimes be deduced by knowing the phenotype alone; however, many times this cannot be relied on because several genotypes are responsible for the same phenotype. The advantages lie more in being able to construct any degree of pigment purity desired and any combination of gene forms. Thus each gene can be tested separately with regard to some standard type, and their interactions can be analyzed. The power of this technique for behavioral investigations cannot be overestimated.

Already several pigment genes have been implicated in behavioral proc-

TABLE 8

Expected Number of Genotypes and Phenotypes for Coat Color in Crosses Involving Seven Inbred Strains of Mice

Matings	Segregating Alleles	Number of F_2 Genotypes (3^n)	Number of F_2 Phenotypes (2^n)[a]
C57 X *C3H*	*A-----*	3	2
C57 X *DBA*	*-B-D--*	9	4
C57 X *I*	*-B-DPS*	81	16
C57 X *BALB*	*ABC---*	27	8
C57 X *A*	*-BC---*	9	4
C57 X *R III*	*A-C---*	9	4
C3H X *DBA*	*AB-D--*	27	8
C3H X *I*	*AB-DPS*	729	32
C3H X *BALB*	*-BC---*	9	4
C3H X *A*	*ABC---*	27	8
C3H X *R III*	*--C---*	3	2
DBA X *I*	*----PS*	9	4
DBA X *BALB*	*A-CD--*	27	8
DBA X *A*	*--CD--*	9	4
DBA X *R III*	*ABCD--*	81	16
I X *BALB*	*A-CDPS*	729	32
I X *A*	*--CDPS*	81	16
I X *R III*	*ABCDPS*	2187	64
BALB X *A*	*A-----*	3	2
BALB X *R III*	*-B----*	3	2
A X *R III*	*AB----*	9	4

[a]This is the theoretical number of phenotypes. In actual practice quite a few of them would look almost alike.

esses and several more will no doubt become evident. The masterful genetic analyses of coat color is of immense value to the clarification of pleiotropic influences on behavior, and will certainly be supported by biochemical analyses of pigment formation. Moreover, many rodent forms (and perhaps other orders) possess homologous color genes that can be used for comparative studies. The similar biochemical pathways involved, wherever the genes are found, suggest that the behavioral responsibilities will be similar.

EXPERIMENTAL STUDIES OF SINGLE GENES

Screening Program

The long-range task of our laboratory is to screen on a variety of behavioral problems as many known mutations in *Mus musculus* as possible, and then begin the arduous job of classifying gene effects according to the allele involved, linkage relation of loci of interest, and specific and general characteristics of the behavior influenced. Thus far 15 coat-color alleles have been investigated to a greater or lesser degree. These and others of immediate interest to our program are listed in Table 9. The albino allele (*cc*) has been investigated on 11 basic behaviors while segregating on an F_2 background, and 14 others have been investigated on four problems while segregating on the *C57BL/6J* inbred background.

The battery of 11 tasks measures important aspects of sensory-motor performance. The tasks include visual, auditory, tactual, olfactory, and temperature discriminations, as well as tests of vestibular responsiveness, general locomotion, and learning ability. We cannot, of course, include every task of interest or importance. What we have tried to do is bracket fundamental aspects of most sensory-motor systems, and choose those performances that appear to be of evolutionary significance to an animal when faced with a choice of habitat.

The behavioral apparatuses are seen in Fig. 3 and described in subsequent sections. In order for a test to be accepted into the battery, it must: (1) elicit a differential response based on the selected stimulus characteristics; (2) show significant test-retest reliability; and (3) separate inbred strains known to differ widely in genotype. All tests meet the above criteria with the possible exception of the temperature discrimination for which no strain differences have yet been found.

In most cases the mice are not forced to make a response, but are allowed to choose stimuli at their own discretion and speed. Thus, we are not measuring gene effects on absolute *capacity,* but rather we are observing environmental *preference* under a wide range of sensory cues. Presumably the performances chosen for study are those that direct habitat selection of animals in their normal environment. The evolutionary consequences of gene substitutions can thus be studied in a systematic fashion.

We are aware, of course, that genes of interest act in concert with background genotypes and have little absolute significance in isolation. We can obtain some idea, however, of the types of behavior involved, and eventually the genes can be transferred to other genotypes for study. We have developed a random-bred line for this purpose. Genes that act similarly on different backgrounds can be considered genome-specific. In any case, the controlled investigation of single genes offers a powerful tool for the study of relations of possible significance in phylogenetic and ontogenetic strategies.

The first part of this section describes the albino allele on the battery of 11 tasks and the latter part describes our preliminary findings for 14 genes on four of these problems. The albino allele is on an F_2 genome, and hence may not be entirely free of effective linkage. Other studies with this allele segregating on an inbred background (Table 3) suggest that linkage is relatively unimportant. The other 14 alleles involved in our studies segregate on the *C57BL/6J* genotype. For practical purposes each *C57* animal can be considered identical to any other; hence, the only gene allowed to vary in a systematic fashion is the particular gene of interest. Any behavioral difference between mutated and nonmutated forms may be assumed to be due to the action of the only gene allowed to vary.

Albinism and Behavior

ANIMALS.—F_2 animals were derived from intercrossing F_1's from the inbred lines *AKR/J* (*aaBBccDD*) and *DBA/2J* (*aabbCCdd*). The resulting offspring segregate at three coat-color loci, producing 27 genotypes ($3n$). The number of phenotypes would ordinarily be $8(2^n)$, but because *cc* prevents expression at other loci, only five result. The colors and black (*B_D_*), chocolate (*bb,D_*), dilute chocolate (*bb, dd*), blue gray (*B_dd*) and albino (*cc*).

Mice were weaned at 21 days, ear punched for identification, and distributed six to a cage. Three pigmented and three nonpigmented mice shared a 11¼ by 7 by 5 inch opaque cage. Purina Laboratory Chow and water were provided ad libitum at all times. Wood shavings and San-i-cel covered the bottom of each cage.

General Procedure

Testing over our battery of 11 problems began at 60 days of age and continued over a 2-month period. A total of 34 albinos (19 females and 15 males) and an equal number of nonalbinos were tested. The order of test presentation could not be systematically varied because of the relatively few animals involved. The sequence was the same as the order of apparatus

TABLE 9

Single Genes Selected for Testing

Strain Name or Genotype	Gene Name	Linkage Group	Biologic Phenotype
$C57BL/6J\text{-}a^t$	Black and tan	V	Glossy black backs and yellow or orange bellies
$C57BL/6J\text{-}a^{td}$	Tanoid	V	Light belly and very dark agouti back
$C57BL/6J\text{-}A^{i a}$	Intermediate agouti	V	Dark back and light belly
$C57BL/6J\text{-}A^y$	Yellow	V	Rich yellow or orange fur and black eyes
$C57BL/6J\text{-}A^{vy}$	Viable yellow	V	Various degrees of yellow; obese
$C57BL/6J\text{-}b^i(m)$	Brown-J	VIII	Brown instead of black pigmentation
$C57BL/6J\text{-}bf+(M)$	Buff	XVII	Khaki colored
$C57BL/6J\text{-}bg^J(M)$	Beige-J	XIV	Diluted coal color
$C57BL/6J\text{-}c^J(M)$	Albino-J	I	Absence of pigment in hair and eyes
$C57BL/6J\text{-}c^h$	Himalayan	I	Dilution of nonagouti color to dull black
$C57BL/6J\text{-}Ca^J+(M)$	Caracul-J	VI	Wavy coat and vibrissae
$C57BL/6J\text{-}ep+$	Pale ears	XII	Ears and tail are pale in color
$C57BL/6J\text{-}le$	Light ears	XVII	Dilution of coat color
$C57BL/6J\text{-}m$	Misty	VIII	Dilute coat color; tail and belly spots
$C57BL/6J\text{-}ma+$	Matted	XVI	Hairs erect and matted in clumps

TABLE 9 continued

Strain Name or Genotype	Gene Name	Linkage Group	Biologic Phenotype
C57BL/6J-mi^wh	White	XI	Light color; microphthalmic, slight eye pigmentation
C57BL/6J-Pt	Pintail	VIII	Short tail
C57BL/6J-Ra +	Ragged	V	Thin coat
C57BL/6J-ru	Ruby eye	XII	Reduced pigmentation of eyes and hair
C57BL/6J-Sl^d	Steel-Dickie	IV	White with black eyes; anemic
C57BL/6J-To	Tortoise shell	XX	Patches of light hair, males die in utero
C57BL/6J-W	Dominant spotting	XVII	White spotting; dilution of coat color; macrocytic anemia; sterility
C57BL/6J-W^v	Viable dominant spotting	XVII	White spotting; dilution of coat color; macrocytic anemia; sterility

description given below and indicated in Fig. 3. The choice was made on considerations of apparent noxiousness to the animal, the most noxious tests being run last. Males were always tested prior to females and testing began at the same time of day for any problem. Testing usually spanned several days on each problem. No two problems were run simultaneously and there was no overlap between the termination of one problem and the onset of another. Unless otherwise indicated, tests were run during the latter hours of the light half of a 12/12 light-dark cycle under dim red light to which the mice are apparently insensitive.

Apparatus and Specific Procedures

Testing equipment is shown in Fig. 3.

OPEN FIELD AND INCLINE PLANE.—The apparatus pictured in Fig. 3A was used to measure general exploration and vestibular sensitivity (Thiessen and Lindzey, 1967). The dimensions of the board are 36 by 28 inches. A ⅛ inch wire mesh screen is stretched tightly over the surface. The Plexiglas sheet is supported 3½ inches off the surface by four horseshoe brackets. Two drop leaves of this roof hinge upward and underneath so that the mouse can be trapped at the top edge after a response up the board. The board can then be tipped through the horizontal on a fulcrum to a new angle at which time the mouse is released and allowed to ascend the incline. The hinged sections are secured in the upward position by bull-dog clips and in the downward position by cork wedges. The angles of inclination are fixed by placing a dowel in holes beneath the board on the outer face of the fulcrum. With the angle settings of 20°, 40°, and 60° fixed on one side of the pivotal point and 30°, 50°, and 70° fixed on the other side, the animal need not be touched once placed on the board and can be tested on all six angles in a steadily increasing or decreasing sequence. Unpublished information from our laboratory shows that the sequence of testing has no effect on the overall function. The surface is lined off into 63 4-inch squares.

Two minutes of activity were recorded with the board at the horizontal. Each time an animal moved over a line its response was recorded. Following this the animal was trapped behind the retaining flap and the plane was tilted to 20°. The flap was then raised and the animal's progress up the plane was traced on the Plexiglas roof with a marking pen. At the upper edge the animal was again trapped, the board was tilted through the horizontal to 30° for the next trial. The procedure was repeated for each of the six inclines of 20°, 30°, 40°, 50°, 60° and 70°. After the run, the three longest line traversals for each incline were measured with a protractor and averaged. Thus, each animal was given an angle of orientation away from the horizontal for each of the six inclines. The validity and reliability of this procedure are discussed by Thiessen and Lindzey (1967).

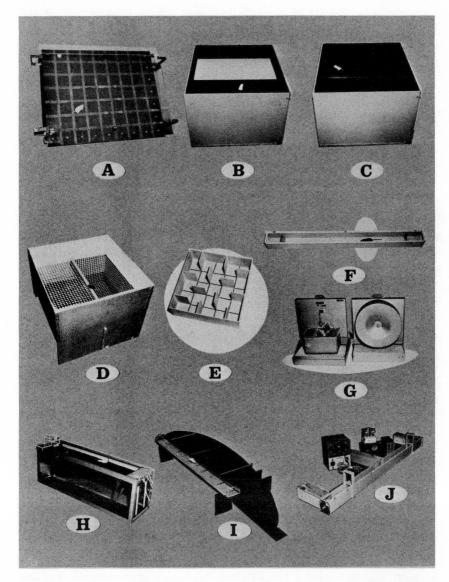

FIG. 3. Apparatuses used to screen single genes. A, open field (in horizontal) and incline plane; B, brightness plane; C, tactual plane; D, visual cliff; E, arena; F, olfactory alley; G, activity wheels; H, water escape; I, temperature gradient; J, auditory alley.

BRIGHTNESS PLANE.—This test measures the degree to which an animal prefers a light or dark environment. The apparatus is pictured in Fig. 3B. It consists of a plywood box 33 × 33 inches square and 22 inches deep. A vertical partition extending upward from the floor supports two panes of glass; one is painted flat black on the underside and the other is milky-white. Walls 6 inches high surround this floor. Beneath each pane is a 25 watt incandescent bulb which provides an even distribution of light through the milk glass and approximately equal temperatures on the two sides. Each animal was given a 5-minute preference test after being placed on the midline dividing the floor into smooth and rough halves. The amount of time spent on the rough side was recorded. The apparatus was cleaned and rotated 180° following each animal.

TACTUAL PLANE.—Tactual preference is observed in this task. The apparatus is pictured in Fig. 3C. The box is the same as that for the brightness plane. The only change is that the milk glass is replaced by a $\frac{1}{8}$ inch wire mesh screen. Each animal was given a 5-minute preference test after being placed on the midline dividing the floor into smooth and rough halves. The amount of time spent on the rough side was recorded. Cleaning and rotating procedures were identical to those of the last test.

VISUAL CLIFF.—Depth perception is assumed to be involved in this task (Walk, 1965). The apparatus is pictured in Fig. 3D. It is a box 30 inches square and 24 inches high. Inside the box and 8 inches from the top are two clear panes of glass. The walls and floor are covered with a vinyl cloth which has a $\frac{1}{2}$ inch black and white checkerboard pattern. The checkerboard floor is immediately beneath the glass on the "shallow" side of the apparatus and 16 inches beneath it on the deep side. A centerboard $1\frac{1}{2}$ inches high, 1 inch wide, and 30 inches long sits on top of the glass dividing the "shallow" from the "deep" side. An animal was placed on one end of the walkway and its choice and latency to step down to either side were recorded. General procedures of apparatus cleaning and rotation prevailed.

ARENA.—This is a measure of general activity under bright light conditions. The apparatus is pictured in Fig. 3E. The arena consists of a 20 by 20 inch floor lined off into 25 4-inch squares. Twelve barriers, 4 inches wide and 5 inches high, are erected at the base of every other square. The arena is enclosed by walls 5 inches high. The floor and barriers are painted gray and the squares are marked off in black. The apparatus was illuminated by two 75 watt bulbs suspended 2 feet above the arena. The test animal was placed in the center of the arena and the number of lines crossed in a 2-minute period was recorded. The apparatus was cleaned between tests.

OLFACTORY ALLEY.—Olfactory sensitivity to a noxious stimulus is involved in this problem. The apparatus is pictured in Fig. 3F. The straight alley wooden runway, measuring 48 by 4 by 4 inches is painted a flat grey and has a clear Plexiglas lid. Five inches from each end is a wire mesh barrier, behind which is an olfactory stimulus. A capful of water was placed at one end of the runway and a commercial solution of ammonia at the other. The position of the stimuli was alternated in a systematic manner. An animal was placed in the middle of the alley and the amount of time spent on the side of the alley containing water was recorded during a 5-minute test. The ammonia solution was replenished every 10 minutes and the apparatus was thoroughly cleaned with a 70 percent alcohol solution and ventilated between trials.

ACTIVITY WHEEL.—Long-term activity was recorded over days in activity wheels like those pictured in Fig. 3G. The wheels were obtained from Lafayette Instrument Co. Six days of activity were recorded for females in order that an entire estrous cycle would be included for each female, and 3 days of activity were recorded for males. Readings were made every 24 hours and an average computed for each animal and each genotype and sex.

WATER ESCAPE.—Learning and escape performance were assessed in this problem. The apparatus is pictured in Fig. 3H. It consists of an aquarium 36 by 12 by 10 inches filled with water at a temperature of approximately 25°C. A 3 by 3 inch restraining box with a sliding trap door in the bottom is positioned 2 inches above the water at one end. A wire-mesh ramp extending into the water provides an exit at the opposite end 33 inches from the restraining box. Each animal was given five trials with a 10-minute intertrial interval. On each trial the mouse was placed in the restraining box. When it oriented toward the exit ramp the trap door was quickly pulled free and the mouse dropped 2 inches into the water. Latency to swim to the ramp was recorded. Animals were placed in a drying cage between trials. Learning was estimated primarily by the change in latency from trial one to trial two, and performance was estimated from the latter four trials.

TEMPERATURE GRADIENT.—Temperature preference was measured on the thermal gradient pictured in Fig. 3-I. The gradient was imposed on a 58 inch metal runway 5 inches in width and painted gray, which was enclosed in a circular Plexiglas tube 5 inches in diameter and hinged along its length. Metal slides attached to the bottom of the floor at either end could be pushed in and out to alter the temperature of the runway. One slide was fitted with a heating tape attached to a rheostat obtained from Standard Electric Products and the other slide was cooled by chilled water circulating in a $\frac{3}{4}$ inch teflon tube attached to the underside of the slide. The water was chilled by a Blue M Electric Co. portable refrigerating unit and circulated through the tube by a Dayton Split Phase pump. By adjusting the length of the slides under the runway any desired gradient could be obtained. For this experiment the temperature at one end was $10 \pm 0.7°C.$ and $51.0 \pm 1.5°C.$ at the other. The gradient remained quite stable regardless of variations in ambient room temperature. The floor is divided into 11 equal intervals, each with a thermometer protruding through the cylinder and supported by brackets attached to the floor. Hence it was possible to record the temperature along the gradient every 5 inches. With the gradient turned off, mice spend most of their time at the two ends, but with the gradient turned on a distribution of responses around intermediate values is obtained.

Each animal was placed in the center of the gradient and allowed 15 minutes to adapt. Thereafter the animal's position on the gradient was recorded every 15 seconds for 5 minutes. The average number of responses along the 11 intervals was recorded for each individual. The apparatus was brushed clean after each trial. All testing was done between 7 A.M. and 2 P.M.

AUDITORY ALLEY.—Preference or aversion for an auditory stimulus was tested in the apparatus shown in Fig. 3J. Excluding the center unit which was not used in this experiment, the alley is 6 feet long with a 1-foot arm extending at a 90° angle at either end. The alley is 3 inches wide, 4 inches deep, and is covered by a hinged Plexiglas top. The alley is painted gray. At the end of each arm is a $3\frac{1}{2}$ inch Omaha speaker. A 14,000 Hz sine wave was fed into the speakers by an audio generator (Model WA-44C) and an audio stimulator (Model 1421) obtained from Lafayette Instrument Co. The sound level used was not accurately measured although it was clearly audible to the experimenter. Three photocells were positioned at each end of the alley. They were 1 inch from the floor and spaced 3 inches apart and fed into a Hunter Phototimer Mechanism. Preliminary observations indicated no differences between the actual time an animal spent at the two ends of the alley and photorelay recordings.

An animal was placed in the center of the alley and allowed to explore for 5 minutes. Sound was delivered to one or the other ends of the alley and the animal's time at either end was recorded. The position of the sound was systematically alternated. The time spent on the stimulus side was arbitrarily designated as positive and the time spent on the nonstimulus side as negative. A difference score was then derived by subtracting the two. Thus scores could range from primarily positive (indicating preference for the sound) to primarily negative (indicating aversion for the sound). The apparatus was thoroughly cleaned between tests.

Results of Albino Screening

The results are detailed in Table 10. The major findings for each problem can be summarized in the following way:

Open-field locomotion. Pigmented and nonpigmented mice did not differ in locomotion under red-light conditions. Sex differences were not apparent.

TABLE 10

Phenotypic Variations Between Albino and Nonalbino Genotypes

Problem	Measure		Mean and Standard Error				Probability Differences		
			Albino (A)		Nonalbino (NA)				
			Male	Female	Male	Female	A vs. NA	Sex	Interaction
Open Field	Line Crossings in 2 min.	$\bar{X}$	76.3333	80.4737	72.1333	72.9474	0.6443a	0.6964	0.7877
		S.E.	7.1159	5.0602	7.1767	5.9164			
Incline Plane	Overall Mean Orientation	$\bar{X}$	73.9403	73.3679	79.0477	80.0636	0.0000a	0.7725	0.3097
		S.E.	0.7372	0.8216	0.6971	0.7604			
Brightness Plane	Time Spent on dark area	$\bar{X}$	294.1333	253.0526	285.3333	264.5789	0.8484a	0.0002	0.1703
		S.E.	2.5521	8.9952	4.8074	8.4914	0.0708b		
Tactual Plane	Time spent on rough surface	$\bar{X}$	269.4667	214.6842	215.5333	210.9474	0.0047a	0.0038	0.0123
		S.E.	9.9222	9.4818	7.2151	10.7103			
Visual Cliff	Latency to descend	$\bar{X}$	25.0667	26.0474	22.3400	8.3263	0.2078a	0.5715	0.6404
		S.E.	6.9974	9.4818	11.3433	2.4135	0.0059b		
Arena	Line Crossings in 2 minutes	$\bar{X}$	48.0000	26.0000	83.7333	73.5789	0.0000a	0.0145	0.6358
		S.E.	9.1933	4.6372	6.0289	5.8732			
Olfactory Alley	Time spent on neutral side	$\bar{X}$	249.2667	238.2105	226.1333	247.5263	0.5578a	0.5668	0.0669
		S.E.	6.4893	9.4214	8.2504	9.2401	0.0032b		
Activity Wheel	Mean Daily Revolutions	$\bar{X}$	3183.2667	3352.3684	2938.0667	3779.8421	0.7760a	0.1195	0.3033
		S.E.	46.9037	63.4381	45.0272	54.2705			
Water Escape	Mean Latency to escape	$\bar{X}$	8.2516	5.3653	4.9750	4.9891	0.0696a	0.1418	0.1380
		S.E.	1.3543	0.7734	0.5003	1.0602			
Temperature Gradient	Mean Temperature Preference	$\bar{X}$	6.1800	6.2368	6.3786	6.5737	0.2427a	0.5889	0.7603
		S.E.	0.2107	0.2140	0.2018	0.2466			
Auditory Alley	Mean Time spent near sound source	$\bar{X}$	13.4667	16.1053	-16.6429	-15.6842	0.0012a	0.8339	0.9217
		S.E.	6.7146	12.5223	8.2037	3.7659			

aF test
bMann-Whitney U Test

193

Negative geotaxis. Pigmented animals appeared more sensitive to a change in environmental incline and responded at higher angles at every incline. This test proved to be highly sensitive for differentiating genotypes. No sex difference emerged.

Brightness preference. Albino mice spent more time on the bright surface than nonalbinos, although both genotypes preferred the dark. Males showed the stronger preference for the dark.

Tactual preference. Albino mice spent more time on the rough surface than nonalbinos, with both genotypes spending the majority of time on the rough surface. Males showed the stronger preference for the rough side.

Depth perception. Both genotypes preferred the shallow side of the visual cliff in approximately 63 percent of the responses, but the latency to descend from the center platform was longer for the albino animals. No sex differences occurred.

Arena activity. Under white-light conditions the albino form was less active than the pigmented form. Females of the albino strain were the least active of the four groups.

Olfactory choice. Albino mice avoided the noxious olfactory stimulus to a greater extent than nonalbino mice. It appeared that the difference is related to the albinos freezing on the nonstimulus side of the olfactory gradient. No sex differences were apparent.

Wheel activity. No gene or sex differences appeared in general wheel activity.

Water escape. Overall escape latencies between genotypes and sexes were similar, although when the major learning trial was eliminated from the total performance, the albino mice showed an increased latency to escape (see below).

Temperature preference. No gene differences emerged from the test, although it is our impression that a more refined temperature gradient might reveal differences.

Auditory preference. A clear qualitative difference appeared in this task. Albinos preferred the alley side containing the sound stimulus, whereas, non-albinos preferred the no-sound side. There were no apparent sex differences.

Of the 41 measures taken on these genotypes including stimulus selection, latencies, sex variations, and interactions, 20 revealed significant variations due to albinism. Quite an astonishing array of behaviors are affected. The albino gene lowers the normal sensitivity to a change in environmental incline, (incline plane) reduces activity under white-light conditions (arena, water escape, and visual cliff) but not under red-light conditions (open field) or when activity is measured primarily at night (activity wheel). The albino genotype remains in a lighted environment, or on a rough surface, longer than pigmented genotypes, although both genetic classes tend to avoid the light and remain on the rough surface. The albino responds *toward* a sound source while the nonalbino moves *away* from the source; the albino tends to avoid the olfactory stimulus more. Speed of *learning* to escape water is similar in the two types, although the albino is somewhat retarded in swimming performance (see Fig. 4). No gene variation occurs in temperature preference.

Explaining these variations is not easy. The increased hesitancy in a lighted environment, and the reduced water escape latency, but adequate active avoidance learning, corroborates earlier reports (Table 3). Photophobia would seem to be the basis for the light reaction (see chapter 2), although it cannot account for the other effects so easily. Incline behavior, tactual, olfactory and auditory preferences are gene-specific even under dim red light. These responses may, of course, be related to ontogenetic reactions to light. Russell Herbert, working in our laboratory, has found that the normal depression of activity in adult *A/J* albino animals is eliminated if the animals are raised in the dark! Further ontogenetic studies of this nature will be of critical interest.

One might take the position that the albino gene tends to disrupt sensitivity to environmental cues especially of a negative nature. Thus negative geotaxis is diminished at all levels of incline, as can be seen in Fig. 5. The decrease in sensitivity is approximately 10° slant, since the genetic differences

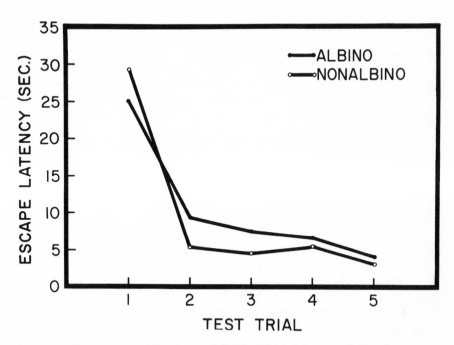

FIG. 4. Water escape latencies for albino and nonalbino mice. Change in latency between trials one and two denote "learning," while subsequent trials denote "performance."

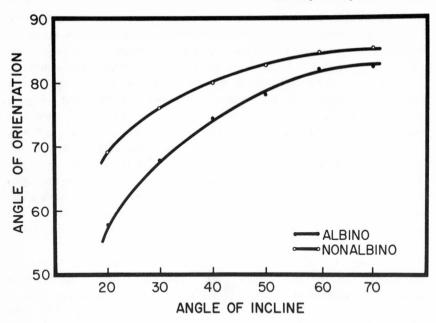

FIG. 5. Negative geotaxis of albino and nonalbino mice expressed as angle of orientation from horizontal over 6 levels of incline.

disappear when the albino animals at 30°, 40°, 50°, 60°, and 70° are compared to the nonalbino animals at 20°, 30°, 40°, 50°, and 60°, respectively. Likewise, the albino shows hesitancy to descend from heights, avoid light, escape from water, and avoid sound—all responses the pigmented animals perform more efficiently. In other words, when there is a change in the environment or an element of noxious stimulation the albinos are likely to hesitate in their reactions. Only the response to odorous stimuli seems to be an exception in this experiment; although even here the superior avoidance of the source by the albino may be interpreted as a diminished exploration of the gradient. While on the one hand the hypothesis about insensitivity to environmental changes may be too broad and need refining, on the other hand, it can account for the diverse effects of the gene on a variety of sensory-motor problems.

Screening of Fourteen Genes

This section presents the results of our preliminary investigation of 14 genes on open-field activity, negative geotaxis, water escape latency, and wheel revolutions. Procedures were the same as those already described for these tasks. The genes studied are indicated in Table 9. Both dominant and recessive genes are included, and there are genes from several different chromosome units. Only nine genes were available for testing in activity wheels. At the time these tests were made we had no nonmutated forms for a standard of comparison. We therefore tested each gene effect against every other for statistical significance and chose those with the greatest number of variations as important. We realize the possibility of statistical artifacts arising from so many statistical tests; therefore we choose to emphasize only the most prominent of the significant variations. In addition, we consider these results as tentative until a replication is performed.

The data on the various genes for the relevant behaviors are seen in the following figures. Even when the most conservative estimates are made there are many striking variations. Of the 14 genes studied, 10 show an influence

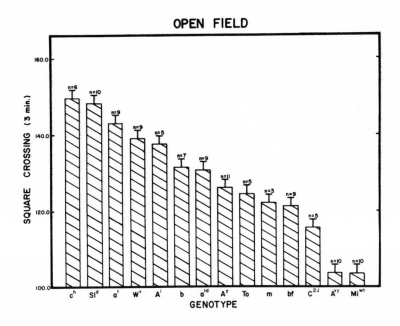

FIG. 6. Three-minute activity scores on open field for 14 single genes. Mean is equal to height of vertical bars. Lines over bars equal one standard error, and numbers indicate subjects.

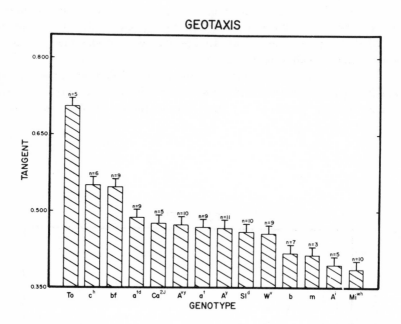

FIG. 7. Angle or orientation up an incline plane expressed as a tangent over 6 levels of incline for 14 single genes. Interpretation identical to Fig. 6.

on behavior. Relative to the other mutants two depress open-field activity (A^{vy}, Mi^{wh}), three increase the steepness of the geotaxic response (To, bf, ch), five depress water escape latencies (To, A^{vy}, A^i, Ca^{2j}, b), and two decrease wheel activity while one increases activity (Mi^{wh}, m, and W^v, respectively).

The salient points of these findings are:

1. Fully 71 percent of the genes tested modified some aspect of behavior. Clearly, it is not difficult to find single gene variations; the difficulty is knowing how to classify them.

2. As new tasks were added, the probability increased that genes affected some behavior. In other words, the larger the number of behaviors sampled the more likely that gene substitutions will be visible. Thus, 14 percent of

the genes affected open-field behavior, 36 percent of the genes affected open-field behavior and geotaxis, 57 percent of the genes affected open-field behavior, geotaxis, and water escape, and 71 percent of the genes affected open-field behavior, geotaxis, water escape, and wheel revolution. At this rate six or seven behavioral measurements should be sufficient to involve all 14 genes.

3. Genes that altered a behavior tended to act in the same direction. Of the effective genes, all either depressed open-field activity, decreased wheel running, reduced water escape latency, or increased the steepness of an animal's inclination up a slope. The one exception was for a gene (W^v) which increased wheel activity. The trends suggest that if a behavior is sensitive to gene substitutions, the direction of expression is predictable.

4. Gene effects probably have some behavioral specificity. Most genes influenced only one behavior, while only three genes influenced more than

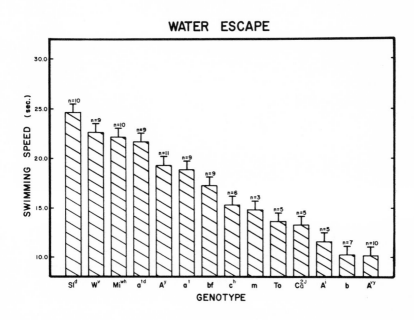

FIG. 8. Seconds to escape from water for six trials for 14 single genes. Interpretation identical to Fig. 6.

one behavior. Moreover, nearly 70 percent of the operative genes were dominant for their primary coat-color expression. Recessive effects were less common.

A FINAL COMMENT

Single gene screening appears to be a profitable approach to the study of behavior. Our investigations reveal the ease with which gene effects are found. Presumably a behavioral effect can be found for any allele if the search is extensive. Coat-color alleles in the mouse offer especially powerful means of tying gene variation to behavior. The genes are easily classified according to morphologic characters and can be considered to be normal in function. In

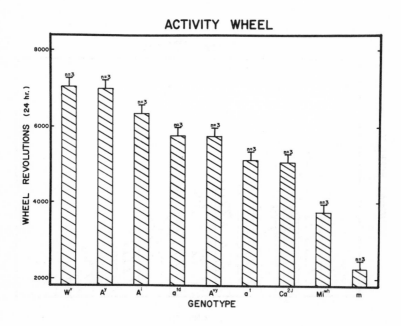

FIG. 9. Wheel activity for 18 days (3 mice per wheel) for 9 single genes. Interpretation identical to Fig. 6.

combination with an inbred background, like the *C57* mouse, and a rational battery of tests, coat-color genes assigned to known linkage groups give a means of clearly defining genetic units of behavior. The *gene unit of behavior* may be the most natural classification of behavior available to us. Moreover, it is capable of being readily manipulated.

Our results give some hope that any single gene can be fully analyzed for its effects on a wide range of behaviors, and assigned some general characteristic. On the other hand, the effect of several genes on a specific behavior can be studied with the aim of devising patterns of relations among alleles or linkage groups. Either approach should provide a means for studying the pathway from gene to behavior.

Mutation is the ultimate source of genetic variation (Mayr, 1963) and, hence, sets the possibilities for adaptation and evolution. Necessarily, therefore, our experimental questions must be addressed to these considerations. Genes that affect habitat selection, in particular, must be of experimental concern. Our battery of tests presents an array of stimulus fields which we consider analogs of those faced by animals in their natural environments. Sensitivity to stimulus change and stimulus preference are especially significant responses that can be studied in the laboratory and still have application in the field. In addition, several alleles, such as those for albinism or for blood groups, are chemically homologous across many species. The study of these alleles may lead to generalities that transcend species barriers.

REFERENCES

Altman, P. L., and D. S. Dittmer. 1964. Biology Data Book, Washington, D.C., Federation American Society for Experimental Biology.

Ashman, R. F. 1959. An attempt to discover behavioral effects of single gene differences in mice. Student report, The Jackson Laboratory, Bar Harbor, Maine.

Bartke, A., and G. L. Wolff. 1966. Influence of the lethal yellow (A^y) gene on estrus synchrony in mice. Science, 153:79–80.

Bastock, M. 1956. A gene mutation which changes a behavior pattern. Evolution, 10:421–439.

Brown, F. A., and V. A. Hall. 1936. The directive influence of light upon Drosophila melanogaster and some of its eye mutants. J. Exp. Zool., 74:205–220.

Bundy, R. E. 1950. A search for some effects of the "short-ear" gene on behavior in mice. Amer. Nat., 84:393–399.

Cattell, R. B., H. B. Young, and J. D. Hundleby. 1964. Blood groups and personality traits. Amer. J. Hum. Genet., 16:397–402.

Crossley, S. 1963. An experimental study of sexual isolation within a species of Drosophila. Unpublished doctoral dissertation, University of Oxford.

Crow, J. F. 1967. Genetics and medicine. *In* Brink, R. A., ed., Heritage from Mendel, Madison, University of Wisconsin Press, pp. 351–374.

Crozier, W. J., and G. Pincus. 1936. Analysis of the geotropic orientation of young rats. X. J. Gen. Physiol., 20:111–144.

DeFries, J. C., J. P. Hegmann, and M. W. Weir. 1966. Open-field behavior in mice: Evidence for a major gene effect mediated by the visual system. Science, 154:1577–1579.

Denenberg, V. H., S. Ross, and M. Blumenfield. 1963. Behavioral differences between mutant and nonmutant mice. J. Comp. Physiol. Psychol., 56:290–293.

Deol, M. S. 1963. Inheritance of coat colour in laboratory rodents. *In* Lane-Petter, W., ed., Animals for Research, New York, Academic Press, Inc., pp. 177–197.

Elens, A. A. 1965a. Studies of selective mating using the melanistic mutants of Drosophila melanogaster. Experientia, 21:145–146.

———— 1965b. Studies of selective mating using the sex-linked mutants white and bar of Drosophila melanogaster. Experientia, 21:594–595.

Elston, R. C., and E. Glassman. 1967. An approach to the problem of whether clustering of functionally related genes occurs in higher organisms. Genet. Res., 9:141–147.

Erlenmeyer-Kimling, L., and W. Paradowski. 1966. Selection and schizophrenia. Amer. Nat., 100:651–665.

Ewing, A. W., and A. Manning. 1967. The evolution and genetics of insect behavior. Ann. Rev. Entom., 12:471–494.

Fingerman, M. 1952. The role of the eye pigments of Drosophila melanogaster in photic orientation. J. Exp. Zool., 120:131–164.

Fuller, J. L. 1967. Effects of the albino gene upon behavior of mice. Anim. Behav., 15:467–470.

Green, M. C. 1966. Mutant genes and linkages. *In* Green, E. L., ed., Biology of the Laboratory Mouse, New York, McGraw-Hill Book Co., pp. 87–150.

Geer, B. W., and M. M. Green. 1962. Genotype, phenotypes, and mating behavior of Drosophila melanogaster. Amer. Nat., 96:175–181.

Grüneberg, H. 1952. The Genetics of the Mouse, 2nd ed., The Hague, Nijhoff.

Gutherz, K., and D. D. Thiessen. 1966. Albinism and audiogenic seizures in the mouse. Psychon. Sci., 6:97–98.

Hadorn, E. 1961. Developmental Genetics and Lethal Factors, London, Methuen.

Hamburg, D. A. 1967. Genetics of adrenocortical hormone metabolism in relation to psychological stress. *In* Hirsch, J., ed., Behavior-Genetic Analysis, New York, McGraw-Hill Book Co., pp. 154–175.

Hawkins, J. D. 1965. Effects of the A^y substitution on mouse activity level. Psychon. Sci., 3:279–270.

———— 1966. Developmental effects of the A^y substitution on mouse activity level. Psychon. Sci., 4:105–106.

Henry, K. R., and K. Schlesinger. 1967. Effects of the albino and dilute loci on mouse behavior. J. Comp. Physiol. Psychol., 63:320–323.

Hirsch, J. 1967. Behavior-genetic analysis at the chromosome level of organization. *In* Hirsch, J., ed., Behavior-Genetic Analysis, New York, McGraw-Hill Book Co., pp. 258–269.

Kalmus, H. 1965. Diagnosis and Genetics of Defective Colour Vision. Oxford, Pergamon Press, Inc.

Keeler, C. E., and H. D. King. 1942. Multiple effects of coat color genes in the Norway rat, with special reference to temperament and domestication. J. Comp. Psychol., 34:241–250.

———— S. Ridgway, L. Lipscomb, and E. Fromm. 1968. The genetics of adrenal size and tameness in colorphase foxes. J. Hered., 59:82–84.

Les, E. P. 1959. A study of the effects of single locus heterozygosity on traits which may have survival value in eight stocks of laboratory mice. Unpublished doctoral dissertation, Ohio State University.

Levine, L., and P. L. Krupa. 1966. Studies on sexual selection in mice. III. Effects of the gene for albinism. Amer. Nat., 100:227–234.

Lewis, E. B. 1967. Genes and gene complexes. *In* Brink, R. A., ed., Heritage from Mendel, Madison, University of Wisconsin Press, pp. 17–47.

McKusick, V. A. 1966. Mendelian Inheritance in Man. Baltimore, The Johns Hopkins Press.

Manning, A. 1965. Drosophila and the evolution of behaviour. Viewpoints Biol., 4:125–169.

———— 1967. Genes and the evolution of insect behaviour. *In* Hirsch, J., ed., Behavior-Genetic Analysis, New York, McGraw-Hill Book Co., pp. 44–60.

Martin, P. G., and H. G. Andrewartha. 1962. Success in fighting of two varieties of mice. Amer. Nat., 96:375–376.

Mayr, E. 1963. Animal Species and Evolution. Cambridge, Belknap Press.

Medioni, J. 1959. Sur le role des yeux composés de Drosophila melanogaster meigen dans la perception du proche ultraviolet: Experiences sur le phototrophisme de races mutantes. Comp. Rend. Soc. Biol. 153:164–167.

Merrell, D. J. 1965. Methodology in behavior genetics. J. Hered., 56:262–266.

———— 1949. Selective mating in Drosophila melanogaster. Genetics, 34:370–389.

Muller, H. J. 1932. Further studies on the nature and causes of gene mutations. Proc. 6th Internatl. Cong. Genet., 1:213–255.

Murphy, R. M. 1967. Instrumental conditioning of the fruit fly. Anim. Behav., 15:153–161.

Petit, C. 1959. Les facteurs genetiques de la compétition sexuelle entre une forme et son allelomorphe sauvage chez Drosophila melanogaster. Amer. Genet., 1:83–87.

Reed, E. W., and S. C. Reed. 1965. Mental Retardation: A Family Study, Philadelphia, W. B. Saunders Co.

Reed, S. C., and Reed, E. W. 1950. Natural selection in laboratory populations of Drosophila. II. Competition between white-eye gene and its wild type allele. Evolution, 4:34–42.

Rendel, J. M. 1951. Mating of ebony, vestigial, and wild type Drosophila melanogaster in light and dark. Evolution, 5:226–230.

Robertson, A. 1967. The nature of quantitative genetic variation. *In* Brink, R. A., ed., Heritage from Mendel. Madison, University of Wisconsin Press, pp. 265–280.

Rothenbuhler, W. C. 1967. Genetic and evolutionary considerations of social behavior of honeybees and some related insects. *In* Hirsch, J., ed., Behavior-Genetic Analysis. New York, McGraw-Hill Book Co., pp. 61–106.

Scott, J. P. 1943. Effects of single genes on the behaviour of Drosophila. Amer. Nat., 77:184–190.

Sidman, R. L., M. C. Green, and S. H. Appel. 1965. Catalog of the Neurological Mutants of the Mouse. Cambridge, Harvard University Press.

Spuhler, J. N. 1951. Genetics of three normal morphological variations: Patterns of superficial veins of the anterior thorax, peroneus tertius muscle, and number of vallate papillae. Cold Springs Harbor Symp. Quant. Biol., 15:175–189.

Sturtevant, A. H. 1915. Experiments in sex recognition and the problem of sexual selection in Drosophila. J. Anim. Behav., 5:351–366.

Thiessen, D. D. 1967. Chromosome mapping of mutations in the mouse. Paper presented at the American Psychological Association, Washington, D.C., September.

———— 1966. Pleiotropism in behavior genetics: The mouse as a research instrument. Percept. Motor Skills, 23:901–902.

———— and G. Lindzey. 1967. Negative geotaxis in mice: Effect of balancing practice on incline behaviour in C57BL/6J male mice. Anim. Behav., 15:113–116.

United Nations. 1958. Report of the United Nations Scientific Committee on the effects of atomic radiation. General Assembly, Official Records, 13th Session, Suppl. 17 (A/3838).

van Abeelen, J. H. F. 1963. Mouse mutants studied by means of ethological methods. Genetics, 34:270–286.

———— 1964–1965. Behavioral and pharmacological studies of mutants and inbred strains of mice. *In* The Jackson Laboratory 36th Annual Report, 90.

———— 1965. An ethological investigation of single-gene differences in mice. Unpublished doctoral dissertation, The Netherlands, University of Nyinegen.

Walk, R. D. 1965. The study of visual depth and distance perception in animals. *In* Lehrman, D. S., R. A. Hinde, and E. Shaw, eds., Advances in the Study of Behavior. New York, Academic Press, Inc., pp. 99–154.

Williams, C. N., and S. C. Reed. 1944. Physiological effects of genes: The flight of Drosophila considered in relation to gene mutations. Amer. Nat., 78:214–223.

Winston, H. D., and G. Lindzey. 1964. Albinism and water escape performance in the mouse. Science, 144:189–191.

———— G. Lindzey, and J. Connor. 1967. Albinism and avoidance learning in mice. J. Comp. Physiol. Psychol., 63:77–81.

Wolfe, H. G., and D. L. Coleman. 1966. *In* Green, E. L., ed., Biology of the Laboratory Mouse. New York, McGraw-Hill Book Co., pp. 405–425.

Gene-Physiologic Determination of Behavior

8

Mechanism-Specific Behavior: An Experimental Alternative

DAVID A. RODGERS
Cleveland Clinic,
Cleveland, Ohio

INTRODUCTION

Although the present volume is about research on the house mouse, it is not likely to be read by rodentologists interested in the natural behavior of this species. Most of the studies, based on highly "artificial" inbred strains of laboratory animals (see McClearn, this volume), concern behaviors that are often unique to particular strains and of little general ecologic relevance. They demonstrate, convincingly, that genetic factors can critically influence a wide variety of behaviors; and they often elegantly detail the specific characteristics of these genetically stabilized patterns of behavior. The elegance and genetic precision of these studies, however, pose a problem if the avowed goal of behavior genetics research is, as Meissner stated it, to determine "... the genotypes corresponding to the psychological phenotypes" (1965, p. 206). In the present state of the art, genotype is essentially defined by the laboratory strain in which it is demonstrated or by a subsequent precise cross of that strain; and generalizations are therefore usually strain specific. For example, knowledge of the precise sequence and timing of mountings, thrusts, ejaculations, and refractory period that characterize the copulatory behavior of the *DBA/2J* mouse (see McGill, this volume and 1962) cannot be generalized to the wild mouse or even to another laboratory strain such as the *C57BL* or *BALB/c*. This behavior, therefore, is essentially the copulatory phenotype of the *"DBA/2J"* genotype; and the very precision of this phenome-genome relationship tends to prevent its generalizability in any useful way beyond this one strain.

Until techniques are developed for identifying specific genomes in animals of diffuse genetic backgrounds, extrapolation of this present trend in behavior genetics research would, as Hirsch stated, ". . . require experimental analysis for each behavior we wish to study for all populations in which we wish to study it" (1962, p. 14), with each population being defined in highly precise genetic terms. To the extent that the research in the present volume is proto-typical, as the title implies, determination of comprehensive phenome-genome relationships in natural populations would in fact be impossible. The fantastic genetic variability that characterizes many natural populations tends to limit practical rigorous genetic analyses either to a relatively few unusual pheno-types or to laboratory populations specially bred for genetic uniformity. In contrast to the essential genetic identity of one inbred mouse to another of the same strain, it is unlikely that any two existing human beings other than monozygous twins are genetically identical. Other than monozygous multiple births, the probability that any two human siblings from typical parents will be genetically identical is no more than approximately one in 70 trillion, and the probability is vanishingly small that offspring of two unrelated couples will be genetically identical (Hirsch, 1962; Rodgers, 1967). Thus, naturally occurring populations do not provide the genetic uniformity on which the present studies are largely dependent. Nor do the present studies furnish prototypic examples of research that could practically be carried out on the human species, even under laboratory conditions. Systematic inbreeding to establish homozygous strains, the basis for stabilized genotype in most of the mouse studies (see McClearn, this volume), is obviously unfeasible with humans and would require too many years' lead time to be immediately practical with many other species. Indeed, the lack of prototypicality that makes the mouse an ideal species for genetic manipulation constitutes much of its utility for behavior genetics studies.

It might be concluded, not entirely without justification, that the present volume is written by the authors for themselves and a few colleagues who find satisfaction in exquisitely studying species that have no reason for existing in psychology laboratories except to be the object of these studies, the results of which studies have no generalizability beyond the confines of these laboratories. I think there is an alternative rationale for this volume and these studies, but it is one that may escape recognition even (or especially) by behavior geneti-cists unless it is explicitly detailed. It seems to me that the utility of the behavior genetics analyses reported here is to demonstrate the way in which specific mechanisms underlying behaviors that can, in some sense, be abstracted from (rather than limited to) their relationship to a unique genotype can be studied most effectively by explicitly utilizing the tools of genetic analysis and behavior genetics. I thus would suggest that the goal of behavior genetics be not to relate phenotype to genotype, but to use genotype to understand the essential mechanisms underlying any given phenotype of interest. The resulting mechanism-specific analyses can then be generalized broadly, independent of genome.

HISTORICAL SETTING

The argument has been made in a previous paper (Rodgers, 1967) that behavior genetics historically is a reaction against overgeneralizations associated with the "black box" fiction of behaviorism that ignores the organism which intervenes between stimulus and response in S-R analyses. The black box generalization was, to a large degree, a reaction against the overparticularization that attributed (primarily for theologic reasons) uniqueness to one genetic species, homo sapiens, without the empirical evidence to support such attribution of uniqueness. More specifically, many of man's complex problem-solving and adaptive behaviors were attributed to his unique "higher" mental processes, which were confused theologically with man's soul and which were formally excluded for mechanistic study by a mind-body dualism philosophy. The mind-body dualism argument ran as follows: "Man is unique because he is man. His uniqueness consists in having a soul that is transcendent over the corporeal body. He is also unique in being a reasoning, thinking, exquisitely problem-solving, adaptive animal; and therefore these thinking capacities must be manifestations of his unique soul." These conceptualizations were well formulated by Aristotle and were revitalized by Thomas Aquinas, in the context of reformation thinking. They entered psychology directly as a mind-body dualism that still muddies conceptual waters. Psychology indeed originated as a science for the study of the mind, which was conceived of as immaterial and nonmechanistic (e.g., Boring, 1929). A brief perusal of the writings of animal experimentalists prior to 1920 provides ample evidence of the attempt to "explain" behavior in terms of immaterial mind dimensions rather than in terms of physiology. The mind was, of course, known to be correlated with and in some sense equatable to the brain, but perhaps both because of theologic considerations and because of the complexity of analyzing a 10^{10} unit computer constructed of less than $1\frac{1}{2}$ kilograms of rather gelatinous material, few early psychologists attempted to relate mind function to specific brain structure. As a consequence, speculation about the nature of the mind was unanchored to observable empirical structure. Indeed, the anchoring points, or observable data, of early psychologists were essentially limited to stimulus on the input side and gross behavior on the output side. John Watson (1913) and the early behaviorists placed their empiric bets on the study of observable variables rather than an immaterial or unknowable mind, which emphasis on the observable necessarily largely meant examination of regularities between stimulus and response without much regard for organismic variables as such.

Even though the brain as a structure has remained complex, although by no means too complex to yield to empirical analysis, behavioral data have not allowed the nature of organism to remain indefinitely unexamined.

Geneticists and experimenters utilizing genetic variables have been especially successful in emphasizing the essential role of organismic factors. The ethologists have demonstrated numerous complex behavior patterns that are problem-solving in nature and that are unique to specific species. Herring gull chicks, for example, respond with begging responses to the mandible spot on adult herring gulls (Tinbergen, 1961). Game chicks scurry for cover at the first view of a moving hawk silhouette, but do not scurry for cover when the identical silhouette is moved in the opposite direction and has the functional characteristics of a flying goose (Tinbergen, 1951). These and other behaviors are demonstrably not related to learned stimulus-response associations, and therefore explicitly disprove the universal applicability of the popular stimulus-response-reinforcement theory of problem-solving.

Tolman (1924) and Tyron (1940) provided the first deliberate behavior genetics study of stimulus-organism-response relationships in their pioneering selective breeding of "bright" and "dull" strains of rats that showed systematic differences in ability to learn a maze, the differences being determined by genetic-organismic variables rather than by stimulus-experiential variables. The subsequent numerous studies that have been done in behavior genetics, including those detailed in the present volume, amply demonstrate the relevance of the genetic-organismic factor as a unique modifier of the stimulus-response relationships. Historically, therefore, part of the vitality and importance of the behavior genetics movement can be understood as an implosion into the vacuum that the mind-body dualism conception and its subsequent replacement by behaviorism has left in an understanding of the role of organismic variables in behavior.

At least one important function of the present volume, therefore, can be to demonstrate unequivocally the relevance of organismic factors to an understanding of behavior. Hopefully, young psychologists will find this demonstration anachronistic and unnecessary. An older generation of psychologists, however, and the psychologic layman may still be impressed by these demonstrations of this crucially important principle that should be accepted as a truism, but often is not (e.g., compare Breland and Breland, 1961; and Bitterman, 1965).

Important as the demonstration of relevance of organismic structure is, it would hardly justify the present volume, since this principle is well documented elsewhere. For example, McClearn observed in 1960 that "the fact of genetic influence on behavior has been amply demonstrated ..." (p. 157); and Casperi observed as far back as 1963 that simply correlating behavioral differences with genetic differences has limited promise as an academic or experimental procedure (p. 97).

GENETIC-BEHAVIORAL CORRELATES

Fig. 1 schematically presents the somewhat remote relationship between genetic material and behavior. The starting point for any ultimate behavior is a zygote that, subject to numerous modifying influences, eventually becomes a behaving organism. The zygote, which is itself a consequence of behavior, does not directly give rise to what we normally regard as behavior. It, instead, is a template for enzyme production that is dependent for activation on intracellular nutritional inputs. The enzymes that can be produced by this intracellular environment are limited to those which are templated in the information code of the particular zygote, just as realization of the information potential of the zygote is dependent on appropriate stimulation and raw materials from the intracellular environment. The enzymes themselves act on other nutrients and their biochemical surround to produce organisms that are capable of acquiring behavioral experience. Again, the nutritional environment is limited to the production of organismic potentials contained in the enzyme patternings, just as realization of the potential in the enzymes is dependent on the nutritional nature of the surround. For example, we assume that no manipulation of nutritional surround can convert the enzymatic output of a mouse zygote into a rat or rabbit embryo. On the other hand, we know that the nutritional surround can have profound influence on the nature of the developing embryo, as when the presence of thalidomide produces grotesque structural "abnormalities" that are obviously expressions of the information potential contained in the human zygote, but that find expression only under ecologically unusual circumstances. Organisms, in turn, do not inevitably "produce" the complex behavior patterns in which

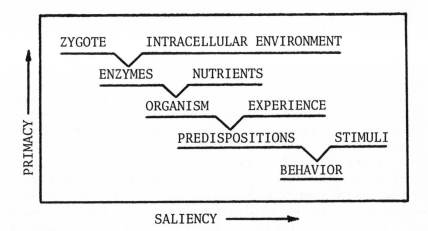

FIG. 1. Determinants of behavior.

we are normally interested until they have first been subjected to shaping experiences. Again, no amount of experience can train a horse to fly or a fish to write scientific papers. Thus, behavioral potential is limited by the nature of the organism. Conversely, however, the allegiance that a mallard chick shows to its mother is dependent on an imprinting experience of a specific sort (Hess, 1959), just as is the ability of a human to communicate in English instead of Spanish or Hindi. Furthermore, as both the behaviorists and the ethologists have been at great pains to demonstrate, even a well-trained organism is not the equivalent of behavior as such. The organism is simply a behavioral potential that awaits release by appropriate triggering stimuli. The experienced male beta spendens may attack viciously a color display resembling another male beta spendens, but becomes quiet as soon as this visual stimulus is removed.

This somewhat tedious review of the obvious, but often ignored, is simply to emphasize the many stages at which external influence can modify the relationship between genetic code and its final expression in the behavior of an experienced organism. Each of these points of influence represents a potential source of uncertainty in the relationship between genome and behavior. The genetically determined regularities that exist in the studies reported in the present volume are therefore as dependent on the careful control of all intervening factors as they are on the nature of the initial genome and the final stimulus that releases the "genetically determined" behavior. Numerous examples can be cited in which "expected" genetic behavioral expressions are aborted by intervening events. For example, the phenylpyruvic oligophrenia associated with phenylketonuria is now routinely aborted by carefully controlling the diet of babies with genetically determined "errors" of phenylalanine metabolism. Similarly, cretinism of the genetically athyroid individual can be avoided if thyroid hormone is exogenously supplied soon enough. Man is not genetically predisposed to fly, but artificial "prostheses" in the form of airplanes have overcome this genetic deficiency. Man was not born with visual capacity in the infrared spectrum, but he nevertheless can routinely rely on infrared visual cues, by use of "snooper-scopes" and other infrared sensing devices.

Not only can a given genome give rise to diverse phenotypes, depending on modifying influences of prostheses and other environmental manipulations, but also diverse genotypes can result in essentially identical phenotypic expression. Turning again to the phenotype of cretinism, this is a condition of structural and intellectual impairment that can result from any one of the following: genetically determined athyroidism with normal nutrition, genetically determined inability to incorporate iodine into tyrosine, genetically determined production of an iodinated amino acid that is less physiologically active than thyroxin, dietary iodine deficiency in "normal" genomes (although there are probably genetic differences in vulnerability to the effects of such deficiency), exposure of normal genomes to thyroxin-suppressing drugs such as propylthiouracil, and surgical removal of the thyroid gland at birth. The condition, recognizable as deficit in ability to perform those intellectual tasks that typify the normal members of a species, can be found in humans, rats, mice, and many other if not all species with a thyroid gland. The condition can be avoided, regardless of genome, by exogenous administration of thyroxin during a critical developmental period.

MECHANISM-SPECIFIC BEHAVIOR

What the previously discussed condition of cretinism clearly demonstrates is that genome is not the appropriate antecedent structure from which to predict cretinism. The presence or absence of physiologic levels of thyroid hormone during a critical period of brain development is a much more consistent and parsimonious predictor of this condition. Any factor, genetic or otherwise, that influences thyroxin level, will influence the development of cretinism. Eayrs (1960) and others have carried the structural analysis still further and have identified differences in complexity of dendritic structure in the brain cortex (along with other morphologic variables) as a still more direct antecedent to cretinism. If thyroxin is available during dendritic proliferation, then its absence after these structures have developed does not result in cretinism (instead, the largely reversible condition of myxedema will develop). Conversely, if sparse dendritic structure is established during the early postpartum period, then no amount of thyroxin at a later age can eliminate the cretinism that results. It thus seems reasonable to conclude that cretinism *is* a characteristic sparsity of dendritic structure, with the functional manifestation of cretinism being simply the inevitable corollary of the structural condition.

It is such "inevitable," or what I shall call *essential* structure-function relationships that constitute the ultimately most parsimonious descriptive units in science. When reduced to the condition of failure of cerebral dendritic proliferation during the neonatal period because of lack of thyroid hormone, the condition of cretinism is abstractable from the particular species in which it occurs. Thus, factors affecting this condition in man can be and have been effectively studied in subhuman species such as mice and rats. The identification of structure-function regularities that can be studied independent of the particular host organism is as applicable to behavioral phenotypes as it is to anatomic ones. For example, the effect of early stimulation on subsequent complexity of behavior and learning ability has been most rigorously examined in the rat, by Krech et al., (e.g., 1960), but appears to be equally applicable in the human species. In the practical sense, therefore, laboratory work with genetically inbred strains of rats is being directly translated into

Headstart programs in impoverished human environments, with acceptably satisfactory results.

To emphasize the structure-function behavioral units that are thus abstractable from the host genome, I have suggested labeling such units mechanism-specific behavior (Rodgers, 1966, 1967).

MECHANISM-SPECIFIC BEHAVIOR AND BEHAVIOR GENETICS

Genetic techniques are uniquely suitable for mechanism-specific analyses, and indeed many such analyses have been made within the behavior genetics tradition. The already mentioned studies of Krech et al., (1960) on effects of early handling on behavior, and the identification of brain cholinesterase levels and dendritic mass as the structures through which such effects are mediated, is one example. The identification of these biochemical differences associated with differences in experience was made possible only because these investigators used genetic controls to reduce intracondition variability of brain enzyme levels, so that intercondition variability could be economically identified.

Another example of generalizable behavior-genetic analysis is Herter's early studies of thermotactic optima of grey and white mice (see Fuller and Thompson, 1960, p. 118). These studies demonstrated that genetic factors control ventral hair density and that ventral hair density correlated highly with a preference for a warm or cool environment. It seems plausible that the important finding was the relationship between ventral insulation and temperature preference, not between a given genotype and thermotaxia. The genome-phenotype relationship would be of relevance only to the specific strains of grey and white mice with which Herter worked. The mechanism-specific relationship between insulation and temperature preference can be generalized to diverse situations and diverse species, often even without regard for genetically influenced variables. For example, it would suggest the conclusion that men wearing business suits in public buildings will prefer cooler temperatures than will women wearing light dresses, because of the difference in exogenously derived insulation. This generalization and corollary conclusion is a correct one that is of significant relevance in the industrial setting, where a common complaint is that if the boss controls the thermostat, the secretary freezes, whereas if the secretary controls it, the boss swelters. Similarly, it would suggest the probably correct conclusion that building temperatures will be kept warmer in countries where most people drive to work in their own heater-equipped cars than in countries where most people walk or wait on street corners for public transportation and therefore dress warmly in cold weather. These are perhaps trivial examples, but they do illustrate the greater generalizability of mechanism-specific findings over findings that simply establish genetic-behavior correlates or that are limited to the reference group on which the studies were done. Especially when behavioral phenotypes are specifically so chosen that results can be generalized, the consequences need not be trivial. My own work on alcohol preference of mice, as a way of studying certain dimensions of the human alcoholism problem, has been previously discussed from this point of view (Rodgers, 1966).

The utility of behavior genetics studies for mechanism-specific analyses lies in the potential offered for stabilizing structure-function relationships in a way that they normally cannot be stabilized in genetically heterogeneous material. In one study, for example, we demonstrated that 97 percent of total variance in quantified alcohol intake of a group of mice was attributable to controlled genetic differences, with error of measurement, sex difference, and litter difference contributing only 3 percent to total variance (Rodgers and McClearn, 1962). Since the genetic contribution to variance is presumably mediated entirely by structural differences, the use of genetic control in this case would make available 97 percent of measured behavioral variability for correlation with structural factors. Conversely, by negative implication, this same finding implies that only 3 percent of variability would have been associated with environmental manipulations carried out on genetically heterogeneous stock. With genetic control, thus, the noise-to-signal ratio could be 3 to 97 whereas without genetic control, it could be no better than 97 to 3, a ratio in which signal would almost inevitably be obscured by noise.

I submit that the primary implication of the studies reported in this present volume is the demonstration that genetic control allows much more penetrating and precise identification of structure-function regularities than do alternative experimental approaches that ignore the genetic variable. Mechanism-specific behavior units, the identification of which can be made possible by behavior genetics techniques, can then be either generalized or tested in species remote from the one in which the behavior was initially studied. Within this context, therefore, the mouse is not simply a prototype animal, but can also be a primary animal for behavioral research. Laboratory strains are useful precisely because they are carefully developed to provide behavioral characteristics which naturally occurring species do not as effectively provide. It is not, or need not be, our intent to study such animals because we wish to generalize to those animals which they demographically represent. Rather, we study such animals to understand the mechanism-specific behavior segments which we wish to generalize much more broadly than to a single genome.

If mechanism-specific analyses were the primary goal of most behavior genetics studies, this chapter would not be necessary. Such a goal would be self-evident in the nature of the studies reported. In fact, however, it seems to me that relatively little behavior genetics work is specifically directed toward structure-function analyses. Most would agree with Meissner's already mentioned observation that the emphasis of behavior genetics work is "...on the determination of the genotypes corresponding to the psychological phenotypes. The overriding concern [is] to identify the genetic factors to which the observed variance in psychological abilities and performance could be ascribed" (1965, p. 206). From a mechanism-specific point of view, this emphasis ties behavior specifically to genotype and does not allow cross-strain or cross-species generalizations. It also predisposes the experimenter to stop short of extending genetic analyses to structural analyses. Such stopping short can result in grossly misleading inferences. For example, several studies have demonstrated that *C57BL* mice show better learning ability than do *C3H* mice under conditions of positive reinforcement on discrimination problems (McClearn, 1962; Lindzey and Winston, 1962; Winston, 1964), but worse ability in escaping from noxious stimulation (Royce and Covington, 1960; Winston, 1964). It would be tempting to conclude that two different processes are involved in escape versus positive reinforcement learning, since the *C3H* strain consistently shows inferiority on one kind of problem and superiority on the other. The genetic complexity of these differences is undoubtedly precisely analyzable by the usual breeding procedures, and, in fact, may well be of single-gene simplicity. Such a conclusion, that positive reinforcement and escape learning differ, would have major theoretical implications, of course, for one-factor versus multifactor learning theories. A more likely explanation for these strain differences in positive versus escape learning, however, is provided by the demonstration that the *C3H* mouse is functionally blind (Sidman and Green, 1965). The positive reinforcement discrimination problems in these studies were primarily visual and could not easily be solved by blind animals. The escape problems, in contrast, were less dependent on the visual system and did not differentially penalize the "learning" ability of the *C3H* strain. Thus, what initially appeared to be a genetic demonstration of differences between positive reinforcement and escape learning is more parsimoniously interpreted as a demonstration that a nonfunctional visual system impairs the learning of visual problems. This latter mechanism-specific behavioral relationship seems safely generalizable across numerous genotypes, whereas the gene-specific difference in learning ability of the mouse substrains was not generalized beyond the particular laboratory strains on which it was demonstrated; and the relationship between the essential structure and resulting phenotype—blindness and poor visual learning—was in danger of being obscured rather than highlighted by concentration on the genetic rather than essential structural antecedents. Parallels to this example are still common in the current research literature.

Until psychology assumes the burden of trying to establish detailed structure-function regularities in the behavioral sciences, we still will not have completely laid to rest the ghost of mind-body interactionism that thinks of behavior as being under the immaterial control of a nonmechanistic homunculus. When we do accept the burden of making structure-function analyses, then behavior genetics will be a primary tool available to us; and the inbred laboratory mouse will be an invaluable research animal for such analyses. Hopefully, this volume will provide some prototypic suggestion of the potential in behavior genetics work for mechanism-specific analyses.

REFERENCES

Bitterman, M. E. 1965. Phyletic differences in learning. Amer. Psychol., 20:396–410.
Boring, E. G. 1929. A History of Experimental Psychology, New York, Appleton-Century-Crofts.
Breland, K., and M. Breland. 1961. The misbehavior of organisms. Amer. Psychol., 16:681–684.
Casperi, E. W. 1963. Vice-presidential address: Genes and the study of behavior. Amer. Zool., 3:97–100.
Eayrs, J. T. 1960. Influence of the thyroid on the central nervous system. Brit. Med. Bull., 16:122–127.
Fuller, J. L., and W. R. Thompson. 1960. Behavior Genetics, New York, John Wiley and Sons, Inc.
Hess, E. H. 1959. Imprinting. Science, 130:133–141.
Hirsch, J. 1962. Individual differences in behavior and their genetic basis. *In* Bliss, E. L., ed., Roots of Behavior, New York, Harper and Row, Publishers, pp. 3–23.
Krech, D., M. R. Rosenzweig, and E. L. Bennett. 1960. Effects of environmental complexity and training on brain chemistry. J. Comp. Physiol. Psychol., 53:509–519.
Lindzey, G., and H. Winston. 1962. Maze learning and effects of pretraining in inbred strains of mice. J. Comp. Physiol. Psychol., 55:748–752.
McClearn, G. E. 1960. Behavior and heredity. *In* McGraw-Hill Encyclopedia of Science and Technology, New York, McGraw-Hill Book Co., pp. 148–157.
——— 1962. The inheritance of behavior. *In* Postman, L., ed., Psychology in the Making, New York, Alfred A. Knopf, Inc., pp. 144–252.
McGill, T. E. 1962. Sexual behavior in three inbred strains of mice. Behaviour, 19:341–350.
Meissner, W. W. 1965. Functional and adaptive aspects of cellular regulatory mechanisms. Psychol. Bull., 64:206–216.
Rodgers, D. A. 1966. Factors underlying differences in alcohol preference among inbred strains of mice. Psychosom. Med., 28:498–513.
——— 1967. Behavior genetics and overparticularization: An historical perspective. *In* Spuhler, J. N., ed., Genetic Diversity and Human Behavior, Chicago, Aldine Publishing Co., pp. 47–59.
——— and G. E. McClearn. 1962. Mouse strain differences in preference for various concentrations of alcohol. Quart. J. Stud. Alcohol, 23:26–33.
Royce, J. R., and M. Covington. 1960. Genetic differences in the avoidance conditioning of mice. J. Comp. Physiol. Psychol., 53:197–200.

Sidman, R. L., and M. C. Green. 1965. Retinal degeneration in the mouse: location of the rb locus in linkage group XVIII. J. Hered., 56:23–29.

Tinbergen, N. 1951. The Study of Instinct, London, Oxford University Press.

———— 1961. The Herring Gull's World, New York, Basic Books Inc., Publishers.

Tolman, E. C. 1924. The inheritance of maze-learning ability in rats. J. Comp. Physiol. Psychol., 4:1–18.

Tryon, R. C. 1940. Genetic differences in maze-learning ability in rats. Yearb. Nat. Soc. Stud. Educ., 39(1):111–119.

Watson, J. B. 1913. Psychology as a behaviorist sees it. Psychol. Rev., 20: 158–177.

Winston, H. D. 1964. Heterosis and learning in the mouse. J. Comp. Physiol. Psychol., 57:279–283.

9

The Genetics and Biochemistry of Audiogenic Seizures*

KURT SCHLESINGER AND BARBARA J. GRIEK
*Institute for Behavioral Genetics,
University of Colorado, Boulder*

INTRODUCTION

Interest in audiogenic seizures has resulted in the publication of an enormous research literature over the last 40 years. This literature has been reviewed by Finger (1947) who covered the period up to 1947, by Bevan (1955) who covered the period 1947 to 1954, and by Henry (1966) who covered the period 1958 through 1965. Over these years several interesting trends have occurred which characterize the work on audiogenic seizures. For example, Finger found 84 percent, Bevan 47 percent, and Henry 13 percent of the work on seizures reported in psychology journals. Finger found the literature exclusively North American, Bevan found it predominantly so (83 percent), while Henry found a more even distribution (51 percent North American). Finally, research on audiogenic seizures has shifted away from an almost exclusive reliance on the laboratory rat as the experimental animal of choice. The results of these developments have been the following: (1) Research on audiogenic seizures has become truly interdisciplinary, often involving the collaborative efforts of biochemists, geneticists, pharmacologists, psychiatrists, and psychologists. (2) Research on audiogenic seizures has taken on a decidedly cosmopolitan flavor which is perhaps best exemplified by a recent international symposium on this topic (Busnel, 1963). (3) Finally, the use of different species and especially of different strains of mice in this research has led to a greater emphasis on the genetic determinants of susceptibility to audiogenic seizures.

Two questions are repeatedly asked with respect to audiogenic seizure

* This research was supported by grant number MH-13026 from The National Institute of Mental Health.

research: (1) What is the interest of researchers in this phenomenon? (2) What is known of the mechanism(s) underlying seizure susceptibility? The first question can be rephrased as follows: Is susceptibility to audiogenic seizures merely a biologic curiosity or is the phenomenon of a more general interest? It can be stated fairly that the phenomenon is of more general interest for two reasons: First, audiogenic seizures have been observed in a variety of species including the mouse, rat, guinea pig, *Peromyscus,* and rabbit. Second, strains of mice susceptible to audiogenic seizures have, in general, also been found to be more susceptible to other seizure-inducing agents than mice resistant to audiogenic seizures (see a later section of this paper). If it is true that seizure susceptibility is not a biologic curiosity and that certain genotypes within a given species are more seizure susceptible regardless of the methods used to induce the seizures, then the following argument seems plausible: Seizure susceptible genotypes might be used as experimental prototypes to determine the mechanisms underlying the variation known to exist in central nervous system (CNS) excitability. Viewed in this light susceptibility to audiogenic (and other) seizures becomes merely a special case of CNS excitability and the seizures per se are simply tools used to investigate genetically determined differences in CNS function. It is this possibility which has excited great interest in seizure susceptible animals, as well as the fact that certain seizure prone strains of mice resemble human phenylketonuria (PKU) in some respects (Coleman, 1960), and it is hoped that these animals will come to serve as models useful in studying the CNS correlates of this hepatic pathology.

With respect to the second question, i.e., what do we know of the mechanisms underlying seizure susceptibility, we must agree with Chance (1963) who points out that only a very limited insight has been reached into the mechanisms determining seizure susceptibility and that a great discrepancy exists between the amount of effort expended and the limited gain in understanding. Although the reasons for this state of affairs are probably quite complex, it is true that work on audiogenic seizures has suffered from the lack of a single model to which all researchers in the field could relate. This situation is aggravated by the fact that these individuals come from such diverse backgrounds as biochemistry, genetics, pharmacology, psychiatry, and psychology. Several potentially useful models have been suggested in the past and these will be discussed in greater detail in a later section of this paper. More recently, a more comprehensive model has been proposed by Roberts (1966a, 1966b, 1967). As Roberts points out in his discussion, it is undoubtedly true that the genetic constitution of the organisms used in seizure research is extremely important; but, as Deol's (1964) work on the *kreisler* mutant in the mouse has illustrated, even single gene differences in otherwise isogenic organisms can be reflected in developmental and structural differences of the whole nervous apparatus. Since audiogenic seizures are in all probability not under single gene control, the situation with respect to this phenotype becomes very complex; nevertheless, the ideas presented by Roberts appear comprehensive enough to accommodate both such an order of complexity and such divergent information as is provided by this inter-disciplinary research. It is for this reason that this model holds great promise. Roberts' model is too complex to be restated here in its entirety. Basically, however, the model suggests that the excitability of the central nervous system depends on the interaction of excitatory and inhibitory factors acting on CNS tissue. In other words, seizure susceptibility is a function of the sum total of all excitatory and inhibitory factors, neurohumors and neuromodulators, acting on a particular nervous system at a particular moment in time.

A Model of Seizure Susceptibility

Our attempts at understanding seizure susceptibility are based on an observation reported by Coleman (1960). Coleman found that certain dilute mice (d/d and d^1/d^1) were deficient in phenylalanine hydroxylase activity in liver. Similar results have also been reported by other investigators (Rauch and Yost, 1963). These observations indicate that mice of certain dilute genotypes, specifically *DBA/1J, DBA/2J* and dilute lethals, have only 14 to 50 percent as much phenylalanine hydroxylase activity as nondilute (wild type) mice. Phenylalanine hydroxylase, of course, is the enzyme which converts phenylalanine to tyrosine and is the enzyme missing in human phenylketonuria. Several important differences between the enzyme defect in dilute mice and in man should be pointed out. First, in the mouse the enzyme deficiency is only partial and the remaining activity is sufficient to metabolize dietary phenylalanine adequately. No excess phenylalanine can be demonstrated in dilute mice maintained on standard laboratory diets. Second, the defect in dilute mice does not appear to be a failure to make the enzyme; if crude liver homogenates from dilute mice (deficient in enzyme activity) are centrifuged at $17,000 \times g$ enzyme activity in the supernatant fraction becomes normal. It has, therefore, been suggested that in dilute mice an endogenous inhibitor of phenylalanine hydroxylase is associated with the mitochondrial fraction.

The enzyme deficiency in dilute mice, nevertheless, has certain demonstrable in vivo consequences. Given excess phenylalanine, dilute mice excrete this excess amino acid more slowly than mice of nondilute genotypes. Under normal dietary conditions dilute mice excrete certain abnormal phenylalanine metabolites such as phenylacetic acid. As Coleman (1960) pointed out, phenylacetic acid has been shown to inhibit decarboxylating reactions in a number of tissues (Hanson, 1958; Sandler and Close, 1959), and it is possible that dilute strains of mice are, therefore, deficient in certain products of decarboxylating reactions. Specifically, it is possible that these animals are deficient in GABA, NE, and 5-HT in brain tissue and that these deficiencies, in turn, account for the high proportion of seizures typically observed in dilute strains of mice (Ginsburg and Miller, 1963; Coleman and Schlesinger, 1965).

Several assumptions must be made if these speculations are to be useful as a working model. It is first of all necessary to assume that GABA, NE, and 5-HT are inhibitory in their action in the CNS, or at least in certain critical regions. The evidence with respect to this point is not complete, and is based primarily on observations reported by Roberts et al. (1958) with respect to GABA and by Marrazi (1961) with respect to 5-HT. It is also necessary to assume that phenylacetic acid, in the amounts present in dilute mice, inhibits decarboxylation. In our laboratory we have been able to show that 5-hydroxytryptophan decarboxylase is inhibited by phenylacetic acid, but that the degree of inhibition is not great (approximately 40 percent) and requires rather large concentrations of the inhibitor. Finally, a serious complication arises from the fact that hydroxylation rather than decarboxylation is the rate-limiting reaction in the synthesis of, for example, NE (Spector et al., 1965). In human PKU, however, abnormal phenylalanine metabolites have been shown to be excreted conjugated with pyridoxine (vitamin B_6) (Yen and Ritman, 1964). Pyridoxine is the cofactor in decarboxylating reactions and if similar excretions occur in dilute mice, and this is entirely speculative, then these animals might also be functionally deficient in this vitamin. In such an instance it is possible that decarboxylation becomes more rate limiting.

In summary, then, we speculate that dilute genotypes have lower seizure thresholds because these animals might be deficient in inhibitory factors controlling CNS functioning in the manner detailed above. It should be emphasized that this model is not complete and that many remaining problems must be worked out. Nevertheless, we have used this model in all of our work although other attractive explanatory hypotheses have been advanced (Abood and Gerard, 1955; Ginsburg, 1963). These investigators have demonstrated differences in oxidative phosphorylating mechanisms in dilute versus nondilute strains of mice. They were able to show that seizure susceptible mice are deficient in ATPase activity and have lower P/O ratios, but only at ages which correspond to the period of maximal seizure sensitivity in these mice. Ginsburg (1963) has shown that the administration of substances related to the tricarboxylic acid cycle and/or to energy turnover (e.g., lactic acid, oxoloacetic acid, succinic acid, etc.) protects mice against audiogenic seizures. It could also be hypothesized that the uptake of metabolic precursors into the brain tissue of dilute mice is slower than that of nondilute mice. Indeed, McKean et al. (1962) have shown that high levels of phenylalanine inhibit the transport of 5-hydroxytryptophan into brain and that this, in turn, leads to reduced levels of 5-HT in this organ.

Description of the Phenotype

Audiogenic seizures as observed in our laboratory are a series of psychomotor reactions to an intense acoustic stimulus. The complete syndrome, typically, consists of (1) a latency period of variable duration during which the mouse may huddle while appearing to be attending to the stimulus, or appear to ignore the stimulus while washing and grooming excessively; (2) a wild running phase characterized by frenzied running around the boundary of the container; (3) a clonic convulsion during which the animals fall on their sides while drawing up their rear legs towards their chins; (4) a tonic seizure during which all four legs are extended caudally; and (5) death due to respiratory failure.

Extensive variations in this pattern have been observed. The duration of the latency period ranges considerably. The wild running phase, which clearly differentiates audiogenic seizures from other convulsive patterns, may be accompanied by a change of gait which can best be described as a series of stiff-legged bounds. It may terminate without becoming a clonic seizure; tonic seizures may or may not be fatal, and death may or may not be prevented by artificial resuscitation. Nevertheless, the more or less characteristic and discrete phases of this pattern have made scoring of the response relatively easy. Each phase of the response may be graded one point (Plotnikoff, 1958), the latent period may be timed (Cooke et al., 1964), the severity of the seizure may be rated (Lehman and Busnel, 1963), the incidence of fatal seizures may be recorded (Cooke, 1964), or other criteria may be used.

All the experiments on audiogenic seizures reported below were performed using the procedure described by Schlesinger et al. (1965). Animals are taken from their home cages one at a time, brought into another room, placed into a large chromatography jar, and given 30 seconds to adapt. A 5 inch electric bell generating approximately 102 db's of noise at the level of the mouse is then sounded for 90 seconds. During this interval the animals are observed and records made of the wild running phase, clonic, and tonic seizures. In some experiments, in which we were interested in whether or not the seizures were fatal, a slight modification in this procedure was introduced. The experiments were performed as before except that the sound was discontinued during the clonic phase of the seizure response precisely when the animal's hind legs were drawn up maximally towards its chin.

On the basis of some genetic analyses of the response to sound (Schlesinger et al., 1966), it was found that each individual phenotype making up the complex audiogenic seizures is under different genetic control. Further, different drugs affect different components of this syndrome differentially. For these reasons it is extremely important that complete records be kept and reported of each phase of the response and that detailed explanations be given as to the exact criteria used in judging whether or not a seizure had occurred. Unfortunately, this suggestion has not always been followed in audiogenic seizure research.

STRAIN DIFFERENCES IN SEIZURE SUSCEPTIBILITY

Audiogenic Seizures

The importance of genetic factors in determining susceptibility to audiogenic seizures has repeatedly been demonstrated; strain comparison and selective breeding experiments, mostly of mice and rats, have shown time and again that the genetic constitution of the organisms used in audiogenic seizure research is of great importance (see, e.g., Fuller and Thompson, 1960). Dilute strains of mice, i.e., mice of diluted pigmentation homozygous for the autosomal recessive allele d in linkage group II, seem to be particularly susceptible to audiogenic seizures (Ginsburg and Miller, 1963; Coleman and Schlesinger, 1965), although nondilute genotypes have also been reported to be susceptible to sound-induced convulsions (Fuller and Sjursen, 1967). Using *DBA/2J* and *C57BL/6J* as examples of dilute and a nondilute genotype respectively we have been able to replicate these results which are summarized in Table 1.

TABLE 1

Audiogenic Seizures in Mice

Genotype	Age	N	Percent Seizures		
			Wild and Running	Clonic Seizures	Tonic Seizures
DBA/2J	14 days	15	73%	27%	13%
DBA/2J	21 days	15	100%	87%	87%
DBA/2J	28 days	15	87%	40%	13%
DBA/2J	35 days	15	67%	40%	7%
DBA/2J	42 days	15	27%	7%	0%
C57BL/6J	14 days	15	27%	0%	0%
C57BL/6J	21 days	15	7%	0%	0%
C57BL/6J	28 days	15	0%	0%	0%
C57BL/6J	35 days	15	13%	0%	0%
C57BL/6J	42 days	15	0%	0%	0%
F_1	14 days	15	0%	0%	0%
F_1	21 days	15	80%	20%	20%
F_1	28 days	15	87%	0%	0%
F_1	35 days	15	60%	13%	7%
F_1	42 days	15	13%	0%	0%

These results show that *DBA/2J* mice are susceptible to audiogenic seizures, whereas *C57BL/6J* mice are resistant to such seizures. F_1 hybrid mice are intermediate in seizure susceptibility, although phenotypically closer to the nonsusceptible parent strain. Age was found to be a major variable affecting susceptibility. *DBA/2J* mice, for example, had a 90 percent seizure risk at 21 days of age (using full clonic-tonic seizures as the index of susceptibility), but only a 13 percent risk at 14 and 28 days of age. *C57BL/6J* mice were resistant to clonic-tonic seizures at all ages. F_1 hybrid mice were found to have a developmental pattern similar to that observed in *DBA/2J* mice. These results, both with respect to genotypic and ontogenetic determinants of seizure susceptibility, are similar to those reported by other investigators (see e.g., Fuller and Thompson, 1960). The exact age of maximal seizure risk, however, varies somewhat from laboratory to laboratory. This discrepancy points to the importance of some, as yet unspecified, environmental factors, such as diet, conditions of housing, temperature, and/or time of day at which the seizure tests are conducted, interacting with genetic factors to give slightly different developmental patterns.

Metrazol Seizures

Fig. 1 summarizes the results obtained in an experiment designed to determine metrazol seizure thresholds in *DBA/2J, C57BL/6J* and F_1 hybrid mice, either 21, 28, or 35 days of age. ED_{50}'s were calculated using the procedure described by Karman and Spearman (Finney, 1952). These results indicate that *DBA/2J* mice had significantly lower metrazol seizure thresholds than *C57BL/6J* animals, when the occurrence of muscle spasms, clonic, or tonic seizures is used as an index of susceptibility. These thresholds tend to increase with age in all three genotypes, although this effect was not statistically significant. It should be pointed out that these results are consistent with those reported by Busnel and Lehman (1961) and at odds with those reported by Swinyard et al. (1963).

Electroconvulsive Seizures

Electroconvulsive seizure (ECS) thresholds were also determined in 21- and 28-day-old *DBA/2J, C57BL/6J* and F_1 hybrid mice. Table 2 gives the results of this experiment. ECS thresholds were determined in 80 *DBA/2J*, 80 *C57BL/6J*, and 80 F_1 hybrid mice. Equal numbers of 21- and 28-day-old mice were used. Five or six mice per genotype per age were tested at each of seven current settings between 3.9 and 9.5 milliamperes (mA). It is important to point out that we used ear clips and that the mice were stimulated with a 833.3 cycles/second square wave of 0.2 second duration and a delay of 0.6 milliseconds between each cycle, giving a total period of 1.2 milliseconds.

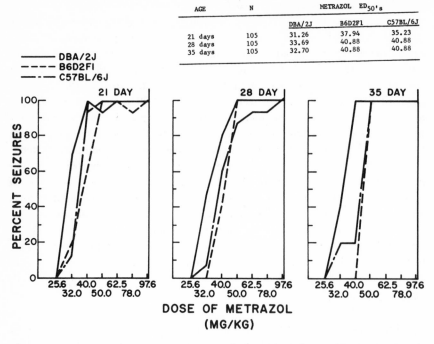

FIG. 1. Metrazol seizure thresholds in DBA/2J, C57BL/6J and F₁ hybrid mice.

Current was held constant at each particular setting. Such control of the stimulus parameters was necessary since seizures were found to be contingent not only upon current, but upon frequency, duration, and pulse width.

On the basis of these data ECS thresholds were obtained by interpolation. The results clearly indicate that *DBA/2J* mice have lower ECS seizure thresholds than either *C57BL/6J* or *F₁* hybrid mice. In *DBA/2J* and *C57BL/6J* mice, ECS thresholds increase with age; in *F₁* hybrid animals the opposite results were obtained.

We have interpreted these results as preliminary evidence suggesting that *DBA/2J* mice are more susceptible to seizures regardless of the agents used to induce the convulsion. Although contradictory evidence exists, other investigators have reported similar results; Hamburgh and Vicari (1960) have reported that animals susceptible to audiogenic seizures are also more susceptible to ECS. Busnel and Lehman (1961) report that animals susceptible to audiogenic seizures are also more susceptible to metrazol- and strychnine-induced seizures. It is clear, however, that the type of generality which we would like to establish can never be shown to exist unequivocally since it is impossible to test every known convulsant. Nevertheless, a strong case can now be made that *DBA/2J* mice are more susceptible to seizures

TABLE 2

Electroconvulsive Seizure Thresholds in Mice[a]

Genotype	Age	N	Electroconvulsive Seizure Threshold (mA)
DBA/2J	21 days	40	4.6
DBA/2J	28 days	40	4.9
C57BL/6J	21 days	40	7.0
C57BL/6J	28 days	40	7.6
F_1	21 days	40	7.2
F_1	28 days	40	5.9

[a]Electroconvulsive seizure thresholds are defined as that current setting at which 50 percent of the animals showed clonic-tonic seizures. These values were obtained by interpolation.

than animals of some other genotypes and as other convulsants are tested in our and other laboratories the evidence will, perhaps, become more conclusive.

STRAIN DIFFERENCES IN LEVELS OF 5-HT AND NE IN BRAIN

Whole Brain Assays

As has been stated previously, our hypothesis of seizure susceptibility depends, in part, on our ability to demonstrate differences in levels of GABA, 5-HT, and NE in brains of *DBA/2J* and *C57BL/6J* mice. Specifically, the hypothesis predicts that seizure susceptible genotypes would have lower levels of these biogenic amines and of GABA in the brain tissue than mice of seizure resistant genotypes. Having shown that both genotype and age affect seizure susceptibility it was now easier to test whether these differences were correlated with differences in levels of 5-HT and NE; but, if such correlations exist they must vary with both genotype and age, and such a result is less likely to be due to chance. With this in mind, levels of 5-HT and NE were determined in 14-, 21-, 28-, 35-, and 42-day-old *DBA/2J,* *C57BL/6J* and F_1 hybrid mice. All analyses were performed under code numbers using a modification of the spectrophotofluorometric technique described by Mead and Finger (1961). These results are summarized in Table 3.

Table 3 gives brain weights and brain 5-HT and NE levels in these three genotypes at five ages. The data are presented separately for total and specific (per gram wet weight) 5-HT and NE levels. No significant differences in brain weight were observed. The difference in total 5-HT in brain

TABLE 3

Brain Serotonin and Norepinephrine Levels as a Function of Strain and Age

Age	Strain	N	Brain Weights (g)		Total 5H-T (mμg)		5-HT Total (mμg/g)		Total NE (mμg)		NE Total (mμg/g)	
			x̄	SD	x̄	SD	x̄	SD	x̄	SD	x̄	SD
14 days	DBA/2J	6	0.340	±0.15	154	±25	456	±88	121	±25	365	±17
"	B6D2F1	6	0.374	±0.01	145	±18	388	±46	117	±11	312	±24
"	C57BL/6J	6	0.371	±0.17	185	±19	498	±54	125	±13	337	±36
21 days	DBA/2J	7	0.328	±0.02	155	±33	466	±104	105	±31	317	±73
"	B6DF1	6	0.413	±0.01	201	± 9	485	±29	167	±17	397	±47
"	C57BL/6J	7	0.359	±0.21	228	±10	638	±28	189	±16	526	±36
28 days	DBA/2J	6	0.418	±0.27	214	±38	593	±93	152	±25	427	±82
"	B6D2F1	7	0.402	±0.01	230	±15	574	±39	196	±23	489	±60
"	C57BL/6J	7	0.398	±0.03	235	±30	589	±82	185	±55	465	±144
35 days	DBA/2J	6	0.376	±0.02	229	±14	612	±43	163	±51	438	±146
"	B6D2F1	7	0.396	±0.01	246	±10	622	±34	188	±14	474	±14
"	C57BL/6J	7	0.398	±0.01	231	±53	588	±94	189	±45	481	±81
42 days	DBA/2J	6	0.390	±0.14	262	±33	683	±62	212	±15	555	±41
"	B62DF1	6	0.434	±0.03	282	±27	652	±74	208	±39	480	±33
"	C57BL/6J	5	0.421	±0.10	255	±20	605	±46	231	±14	548	±25

between $DBA/2J$ and $C57BL/6J$ mice is approximately 32 percent at 21 days of age; the value for the F_1 hybrid mice is intermediate. The difference in total NE in brain is approximately 42 percent; the value for the F_1 hybrid group is intermediate.

If differences in levels of 5-HT and NE in brain are important in determining seizure susceptibility then we would have expected to find differences in levels of these amines but *only* at ages at which significant differences in seizure susceptibility are found. These data point to the importance of levels of 5-HT and NE in brain in determining seizure susceptibility since $DBA/2J$ mice have significantly lower levels of these compounds in brain, but *only* at 21 days of age which corresponds to the period of maximal seizure risk in these animals.

Subcellular Distribution of 5-HT

These results led us to investigate subcellular distribution of 5-HT in $DBA/2J$ and $C57BL/6J$ mice. Since 5-HT found in the particulate fraction has (not with entire accuracy) been called "free" serotonin, the ratio of free to total 5-HT is of some interest. The data obtained in these experiments (Table 4) indicate that $DBA/2J$ mice have less 5-HT in both the particulate and supernatant fraction and that the ratio of free to total 5-HT is not significantly different in $DBA/2J$ and $C57BL/6J$ mice. It should be pointed out, however, that these results in no way preclude differences which might be obtained with finer techniques such as, for example, density gradient centrifugation or analyses of 5-HT in isolated free nerve endings.

PHARMACOLOGIC CORRELATES OF SEIZURES

Reserpine and Tetrabenazene Depletion of 5-HT and NE

If, as the hypothesis predicts and as the correlation studies reported above suggest, levels of 5-HT and NE determine susceptibility to seizures, then

TABLE 4

Subcellular Distribution of 5-HT in Brains of Mice

Genotype	Age	N	5-HT (mμg/whole brain ± 1 SD)			
			Particulate	Supernatant	Total	Percent Supernatant
C57BL/6J	21 days	12	212 ± 33	50 ± 11	264 ± 41	19% ± 2.5
DBA/2J	21 days	12	169 ± 19	40 ± 12	210 ± 23	19% ± 2.5
C57BL/6J	42 days	11	263 ± 46	84 ± 23	346 ± 68	24% ± 3.0
DBA/2J	42 days	11	214 ± 44	58 ± 16	272 ± 56	21% ± 4.0

manipulations of levels of these amines should have predictable results in terms of altered seizure susceptibility. However, before studying the effects of some drugs on seizures, the effects of these drugs on levels of 5-HT and NE in the nervous system of *DBA/2J* and *C57BL/6J* mice were examined. Specifically, reserpine depletion, tetrabenazene depletion, and repletion of levels of 5-HT and NE were investigated.

Twenty-one and 42-day-old *DBA/2J* and *C57BL/6J* mice were injected intraperitoneally with either vehicle, 1.0, 2.0, or 5.0 mg reserpine/kg body weight. Twenty-four hours later animals were sacrificed and levels of 5-HT and NE in brain were determined by fluorescence spectrometry. The results of these experiments are detailed in Fig. 2. These results indicate the following: (1) Reserpine depletes 5-HT and NE in both genotypes. (2) The extent of the depletion is a function of both genotype and age. Reserpine had a quantitatively *smaller* effect in older animals and a significantly larger effect in *DBA/2J* mice. The mechanisms underlying this quantitatively differential effect in mice of these genotypes remain to be explained.

In order to verify and extend these results, tetrabenazene, a synthetic, reserpine-like drug, was studied for its effects on 5-HT and NE depletion and repletion in brains of *DBA/2J* and *C57BL/6J* mice, 21 days of age. The results of this experiment are summarized in Fig. 3 and indicate that both amines are depleted to a greater extent and at a faster rate in *DBA/2J* than in *C57BL/6J* mice. Further, rates of repletion are significantly slower in *DBA/2J* than in *C57BL/6J* animals. Experiments are in progress to determine the mechanism(s) underlying this genotype by drug interaction.

Depletion of 5-HT and NE and Seizure Susceptibility

The effects of reserpine on susceptibility to audiogenic seizures were studied in 42-day-old *DBA/2J, C57BL/6J,* and F_1 hybrid mice. Animals well past the period of maximal seizure risk were chosen since the predicted effect of reserpine was to enhance seizure susceptibility. All animals were injected intraperitoneally with either vehicle, 1.0 or 2.0 mg reserpine/kg body weight. The results of these experiments are given in Table 5. These results indicate that reserpine does indeed enhance susceptibility to audiogenic seizures, *except* when the drug was allowed to deplete too much 5-HT and NE, in which case no enhancement of seizure susceptibility was observed. These data are in agreement with those reported by other investigators: Bevan and Chinn (1957), for example, reported that seizure susceptibility increased with increasing doses of reserpine up to 0.9 mg reserpine/kg, whereas Plotnikoff (1960) reported that 5 mg reserpine/kg protects animals against audiogenic seizures. These data also indicate that the degree of enhancement of seizure susceptibility by reserpine depends on the genotype of the animals used; in *DBA/2J* mice, for example, 1 mg/kg of reserpine increased sus-

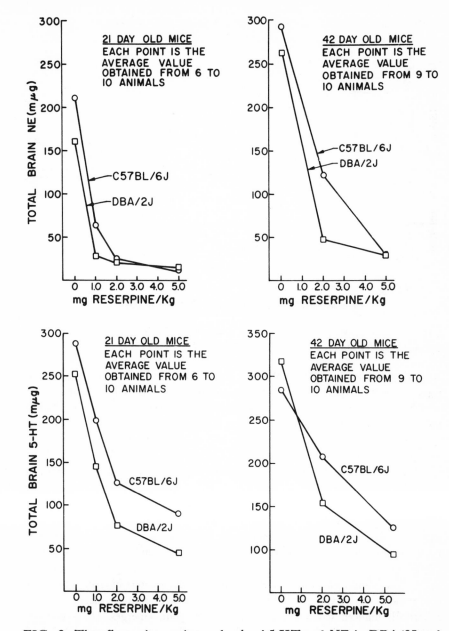

FIG. 2. The effects of reserpine on levels of 5-HT and NE in DBA/2J and C57BL/6J mice.

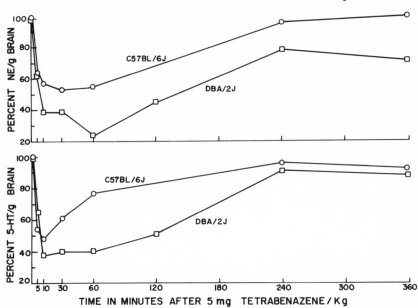

FIG. 3. The effects of tetrabenazene on levels of 5HT and NE in DBA/2J and C57BL/6J mice.

ceptibility to the clonic-tonic phase of the seizure syndrome, but 2 mg/kg of reserpine were required to increase susceptibility to the wild running phase of the seizure response in *C57BL/6J* mice. The response of the F_1 hybrid mice appeared to be intermediate since 1 mg/kg of reserpine increased susceptibility to the wild running phase and 2 mg/kg were required to increase susceptibility to the clonic-tonic phase.

The effects of reserpine on metrazol- and electrically-induced seizures were also studied. Animals were injected with either vehicle or 2.0 mg/kg of reserpine, placed back into their home cages, and then either injected again with metrazol or used in the electroconvulsive seizure tests. The results of these experiments, which are summarized in Table 6, are clear evidence that reserpine enhances susceptibility to both metrazol- and electrically-induced seizures.

These results indicate that the balance of the various biogenic amines is not without significance in terms of susceptibility to sound-, metrazol-, and electrically-induced seizures, but an additional point emerges. If, as our hypothesis and our preliminary results led us to believe, low levels of 5-HT and NE are correlated with seizure susceptibility, then alterations of levels of these amines should have predictable results in terms of increasing or decreasing seizure susceptibility. In other words, using the appropriate drugs one should be able to make a phenocopy of "susceptible" by lowering levels of 5-HT and NE in "nonsusceptible" animals and vice versa. However, in order

TABLE 5

Effects of Reserpine on Susceptibility to Audiogenic Seizures[a]

Treatment	Time of[b] Test	N	DBA/2J			F₁			C57BL/6J		
			WR	C	T	WR	C	T	WR	C	T
Control	1 hour	5	40%	0%	0%	0%	0%	0%	0%	0%	0%
Reserpine (1 mg)	1 hour	5	60%	20%	20%	60%	0%	0%	0%	0%	0%
Reserpine (2 mg)	1 hour	5	60%	20%	20%	60%	0%	0%	60%	20%	0%
Control	2 hour	5	40%	20%	0%	20%	0%	0%	0%	0%	0%
Reserpine (1 mg)	2 hour	5	100%	80%	60%	40%	0%	0%	0%	0%	0%
Reserpine (2 mg)	2 hour	5	20%	0%	0%	60%	40%	40%	20%	0%	0%
Control	3 hour	5	20%	0%	0%	20%	0%	0%	0%	0%	0%
Reserpine (1 mg)	3 hour	5	80%	60%	60%	40%	0%	0%	0%	0%	0%
Reserpine (2 mg)	3 hour	5	20%	0%	0%	20%	20%	20%	0%	0%	0%
Control	4 hour	5	20%	0%	0%	20%	0%	0%	0%	0%	0%
Reserpine (1 mg)	4 hour	5	60%	20%	20%	20%	0%	0%	0%	0%	0%
Reserpine (2 mg)	4 hour	5	40%	40%	0%	0%	0%	0%	0%	0%	0%

[a]All mice were 42 days of age.
[b]Time after drug administration at which seizure tests were conducted.
[c]WR = wild running phase; C = clonic seizures; T = tonic seizures.

to make accurate phenocopies of "susceptible," 5-HT and NE must be lowered to approximately the same endogenous levels as occur in susceptible lines, i.e., to approximately 60 percent of normal. For example, 5 mg reserpine/kg, deplete 5-HT and NE almost completely and such treatment does not lead to a true phenocopy of "susceptible" nor to increased seizure susceptibility. The point is, of course, while not explaining the mechanisms involved in seizure susceptibility, accurate dose and time experiments are of the greatest importance.

Tetrabenazene was also utilized to determine if susceptibility to audiogenic seizures is increased following the administration of this compound. The results of these experiments are summarized in Table 7 and indicate that this drug, too, increases seizure susceptibility. It should be pointed out that while the effects of this drug were quite small, only one dose and one time were tried, and the results might have been more convincing had more parameters been investigated.

Experimental Phenylketonuria and Audiogenic Seizures

Phenylketonuria (PKU), an inborn error of metabolism, is caused by the lack of the liver enzyme phenylalanine hydroxylase. Several behavioral and neurologic symptoms are commonly associated with the disease. These in-

TABLE 6

Effect of Reserpine on Seizure Susceptibility[a]

	Genotype	Treatment	N	Seizure Susceptibility		χ^2	p
				Number of Animals Not Seizing	Number of Animals Seizing		
Electroconvulsive seizures	DBA/2J	Control	10	8	2	7.27	<0.01
	DBA/2J	Reserpine	10	1	9		
	C57BL/6J	Control	15	11	4	11.25	<0.001
	C57BL/6J	Reserpine	15	1	14		
Metrazol Seizures	DBA/2J	Control	10	10	0	30.47	<0.001
	DBA/2J	Reserpine	30	1	29		
	C57BL/6J	Control	10	10	0	14.40	<0.001
	C57BL/6J	Reserpine	30	0	30		
	F_1	Control	10	10	0	27.83	<0.001
	F_1	Reserpine	30	2	28		

[a]For metrazol seizures the occurrence of either muscle spasms, clonic or tonic seizures was used as an index of susceptibility. For electroconvulsive seizures the occurrence of the clonic-tonic seizures was used as an index of susceptibility. χ^2 were computed comparing control and reserpine groups for each genotype. Yates' correction for continuity was used in all cases.

TABLE 7

Effects of Tetrabenazene on Susceptibility to Audiogenic Seizures

Genotype	Age	Treatment[a]	N	Percent Seizures		
				Wild Running	Clonic Seizures	Tonic Seizures
DBA2J	35 days	Control	30	96%	36%	6%
DBA/2J	35 days	Tetrabenazene	30	87%	50%	27%
DBA/2J	42 days	Control	6	17%	0%	0%
DBA/2J	42 days	Tetrabenazene	24	27%	4%	0%
C57BL/6J	35 days	Control	30	0%	0%	0%
C57BL/6J	35 days	Tetrabenazene	30	27%	0%	0%
C57BL/6J	42 days	Control	8	0%	0%	0%
C57BL/6J	42 days	Tetrabenazene	20	30%	0%	0%
F_1	35 days	Control	30	27%	3%	0%
F_1	35 days	Tetrabenazene	30	50%	0%	0%
F_1	42 days	Control	10	10%	0%	0%
F_1	42 days	Tetrabenazene	26	12%	0%	0%

[a]Animals were injected with either vehicle or 5 mg tetrabenazene/kg. Seizure tests were conducted 6 or 7 minutes after drug administration.

clude mental retardation, perhaps the most interesting and certainly the best known symptom, accentuation of superficial and deep reflexes (Harris, 1962), microcephaly, purposeless movements and severe temper tantrums (Hsia, 1967), and a high incidence of epileptic seizures (Partington, 1961). Research on the behavioral symptoms associated with this disease in human phenylketonurics and in animals made experimentally phenylketonuric has focused on the "mental retardation" aspects of this syndrome, while the other symptoms have not been studied as intensively. The mechanisms underlying these neurologic and behavioral symptoms are not fully understood, although several interesting explanatory hypotheses have been advanced (Hsia, (1967). It was with a view toward explaining how a deficiency in a liver enzyme could have such profound effects on the central nervous system that attempts were first made to mimic this disease in animal populations. As Karrer and Cahilly (1963) have pointed out, attempts to produce PKU experimentally have been of the following types: (1) administration of excess phenylalanine in diets or by other means, either alone or in combination with other amino acids; (2) pharmacologic manipulation of phenylalanine hydroxylase activity, as with para-chlorophenylalanine (p-ChPhe); and, (3) use of genetic stocks deficient in phenylalanine hydroxylase activity, such as dilute strains of mice. In numerous experiments, investigators have now been able to show that animals on phenylketogenic diets either learn more slowly or make more errors in learning tasks than animals maintained on control diets (see, e.g., Mc-Kean et al., 1967). There remains the question, however, of whether or not the learning deficit outlasts the dietary treatment; Schalock and Klopfer (1967) report that it does, whereas Palidora (1967) reports that it does not.

Inasmuch as (1) PKU individuals are deficient in levels of 5-HT (Harris, 1962), (2) unaffected heterozygous carriers of PKU have lower metrazol seizure thresholds than normal individuals (Nakai et al., 1966) and, (3) we have found that lowering levels of 5-HT increases seizure susceptibility, it is possible that the high incidence of seizures associated with PKU is related to the low levels of this amine. Several experiments, comparing commonly employed techniques for producing experimental phenylketonuria in animals, were performed to determine if these treatments enhance susceptibility to audiogenic seizures. Specifically, the effects of p-ChPhe, injections of phenylalanine, and diets containing either an excess of, or no, phenylalanine and tyrosine, were studied. In some of these experiments a-methyl tyrosine (a-MT) was given in combination with the phenylketogenic treatment. p-ChPhe is a compound which inhibits the tryptophan- and phenylalanine-hydroxylating capabilities of liver tissue; this leads to an increase in levels of phenylalanine and a decrease in levels of 5-HT (Koe and Weissman, 1966). No effects on levels of NE were observed. a-MT inhibits the enzyme tyrosine hydroxylase and decreases levels of NE in brain and other tissue without affecting levels of 5-HT (Spector et al., 1965).

The results of single p-ChPhe injections and chronic injections of phenylalanine on levels of 5-HT in brain are summarized in Table 8. The data indicate that these treatments lower levels of 5-HT in brain, but do not affect NE. Phenylalanine administration had only a small, marginally sig-

TABLE 8

Effects of p-ChPhe and Phenylalanine on Levels
of 5-HT and NE in Brain

Genotype	N	Treatment	Percent Change	
			5-HT	NE
DBA/2J	11	Control	100%	100%
DBA/2J	16	p-ChPhe	49%	98%
DBA/2J	10	Control	100%	100%
DBA/2J	10	Phenylalanine	80%	107%
C57BL/6J	13	Control	100%	100%
C57BL/6J	15	p-ChPhe	49%	114%

nificant, effect; however, the magnitude of the effect was similar to that reported by other investigators (Yuwiler and Louttit, 1961).

Table 9 summarizes the results obtained when p-ChPhe, a-MT, and p-ChPhe plus a-MT were tested for their effects on audiogenic seizure susceptibility. In *DBA/2J* mice, treatment with either p-ChPhe or a-MT increased susceptibility to all phases of the seizure response. Treatment with both drugs increased susceptibility further. In *C57BL/6J* mice, only treatment with p-ChPhe increased susceptibility to these seizures, and the effect was slight; a-MT had no effect and the combination of the two drugs did not increase seizure susceptibility. In F_1 hybrid mice, treatment with p-ChPhe and with a-MT increased seizure susceptibility; combination of the two compounds increased susceptibility more than the administration of either drug alone. These data were tested statistically and the results of these analyses are included in Table 9: Each mouse was given a seizure severity score based on its responses. Wild running was scored 1 point, clonic seizures were scored 2 points, tonic seizures 3 points, and lethal seizures 4 points. The average seizure severity score was calculated for each group and is shown in Table 9.

Table 10 summarizes the results of experiments in which *DBA/2J* mice were injected chronically with phenylalanine. Treatment was started when these animals were 14 days of age, daily injections of 1.2 mg phenylalanine/kg were given and seizure tests were conducted approximately 2 hours after the last injection at 29 days of age. This treatment was also combined with a-MT administration. It can be seen from these results that phenylalanine increased susceptibility only slightly; seizure severity was, however, further increased when this treatment was combined with a-MT injections.

Finally, the effects of manipulating dietary levels of phenylalanine and tyrosine on susceptibility to audiogenic seizures are summarized in Table 11. In *DBA/2J* mice, diets rich in phenylalanine and tyrosine as well as diets containing no phenylalanine or tyrosine were observed to increase susceptibility to audiogenic seizures significantly. In *C57BL/6J* animals, a single instance of seizures was observed; otherwise, these animals failed to exhibit seizures on any dietary regimen.

Several interesting conclusions can be reached on the basis of these data: First, decreasing levels of 5-HT with p-ChPhe and with injections of phenylalanine increases seizure susceptibility. These results are similar to those obtained with reserpine and tetrabenazene depletion of 5-HT and NE. Diets rich in phenylalanine increased seizure susceptibility. In most instances a-MT depletion of NE, when combined with phenylketogenic treatments, increased seizure susceptibility more than either treatment alone. These findings support our hypothesis that low levels of 5-HT and NE increase seizure susceptibility. Since PKU individuals (1) have low levels of 5-HT and (2) are susceptible to seizures, these results suggest that this particular symptom may be caused by the amine deficiency associated with this disease. The drug by strain inter-

TABLE 9

Effects of p-ChPhe and α-MT on Audiogenic Seizures[a]

Genotype	Treatment	N	Percent Seizures				Seizure Severity		
			Wild Running	Clonic Seizure	Tonic Seizure	Lethal Seizure	x̄ ± 1 SD	t	p
DBA/2J	Control	46	91%	44%	22%	15%	1.72 ± 1.21		
DBA/2J	p-ChPhe	44	96%	55%	39%	27%	2.16 ± 1.35	1.64	NS
DBA/2J	α-MT	19	100%	53%	37%	32%	2.21 ± 1.36	1.38	NS
DBA/2J	p-ChPhe+α-MT	20	100%	64%	46%	27%	2.36 ± 1.25	2.01	<0.05
C57BL/6J	Control	10	20%	10%	0%	0%	0.30 ± 0.68		
C57BL/6J	p-ChPhe	10	50%	30%	10%	10%	1.00 ± 1.33	1.48	NS
C57BL/6J	α-MT	11	18%	0%	0%	0%	0.18 ± 0.41	0.05	NS
C57BL/6J	p-ChPhe+α-MT	11	46%	27%	9%	9%	1.00 ± 1.33	1.48	NS
F_1	Control	11	91%	9%	0%	0%	1.00 ± 0.45		
F_1	p-ChPhe	10	100%	60%	20%	20%	2.00 ± 1.15	2.57	<0.01
F_1	α-MT	16	100%	44%	13%	0%	1.56 ± 0.73	2.49	<0.01
F_1	p-ChPhe+α-MT	10	100%	90%	90%	60%	3.40 ± 0.31	7.16	<0.001

[a]All mice were 21 days of age. p-ChPhe = 320 mg/kg; α-MT = 80 mg/kg. Animals were injected with p-ChPhe or vehicle 3 days before seizure test. α-MT was given 2 hours before the test. All drug groups were compared statistically to their respective control groups.

238

TABLE 10

Effects of Phenylalanine and α-methyl tyrosine on Audiogenic Seizures

Genotype	Treatment	N	Percent Seizures				Seizure Severity	t	p
			Wild Running	Clonic Seizures	Tonic Seizures	Lethal Seizures	$\bar{x} \pm 1$ S D		
DBA/2J	Control	13	69%	38%	15%	8%	1.30 ± 1.60	–	–
DBA/2J	Phenylalanine	13	77%	62%	46%	0%	1.85 ± 1.28	0.96	NS
DBA/2J	Phenylalanine+α-MT	10	100%	90%	50%	0%	2.40 ± 0.70	2.22	<0.05

239

TABLE 11

Effects of Diets Varying in Phenylalanine and Tyrosine on Audiogenic Seizures

Genotype	Diet[a]	Age[b]	N	Percent Seizures				Seizure Severity		
				Wild Running	Clonic Seizure	Tonic Seizure	Lethal Seizure	$\bar{x} \pm 1\,SD$	t	p
DBA/2J	1	7 days	17	88%	41%	12%	12%	1.53 ± 1.13		
DBA/2J	2	7 days	11	91%	73%	73%	36%	2.73 ± 1.42	2.36	<0.05
DBA/2J	3	7 days	—	—	—	—	—	—	—	—
DBA/2J	1	14 days	8	88%	50%	25%	13%	1.75 ± 1.28		
DBA/2J	2	14 days	10	100%	80%	80%	10%	2.80 ± 0.63	2.12	<0.05
DBA/2J	3	14 days	8	88%	88%	88%	25%	2.88 ± 1.24	1.78	<0.05
DBA/2J	1	20 days	7	88%	52%	12%	12%	1.65 ± 1.11		
DBA/2J	2	20 days	13	100%	78%	54%	25%	2.54 ± 1.13	2.16	<0.05
DBA/2J	3	20 days	23	100%	87%	87%	74%	2.48 ± 1.04	5.65	<0.05
C57BL/6J	1	7 days	10	0%	0%	0%	0%	—	—	—
C57BL/6J	2	7 days	5	20%	20%	20%	20%	—	—	—
C57BL/6J	3	7 days	—	—	—	—	—	—	—	—
C57BL/6J	1	14 days	6	0%	0%	0%	0%	—	—	—
C57BL/6J	2	14 days	8	0%	0%	0%	0%	—	—	—
C57BL/6J	3	14 days	9	0%	0%	0%	0%	—	—	—
C57BL/6J	1	20 days	15	0%	0%	0%	0%	—	—	—
C57BL/6J	2	20 days	17	0%	0%	0%	0%	—	—	—
C57BL/6J	3	20 days	22	0%	0%	0%	0%	—	—	—

[a]Diet 1 = control diet; diet 2 = supplemented with 7% phenylalanine and 7% tyrosine; Diet 3 = no phenylalanine and no tyrosine.
[b]All seizure tests were conducted when these animals were 28 days of age. The mice were weaned on to these diets at 20 days of age. Dietary treatments were, however, began when the animals were either 7, 14, or 20 days of age, i.e., the mothers of the pups were placed onto these diets. No animal survived one condition.

actions observed in some of these data remain to be explained as does the finding that diets deficient in phenylalanine and tyrosine increase susceptibility to audiogenic seizures.

Effects of Increasing Levels of 5-HT and NE on Seizures

If, as we have shown, a lowering of levels of 5-HT and NE increases seizure risk, then raising the levels of these compounds in brain should protect animals against seizures. In our first experiments we injected 5-hydroxytryptophan (5-HTP), the metabolic precursor of 5-HT; such treatment raises levels of 5-HT in brain. The effects of this treatment on audiogenic, metrazol, and electroconvulsive seizures were determined. Twenty one-day-old *DBA/2J* mice were used; these animals were injected intraperitoneally with either 0.0 or 150 mg 5-HTP/kg and tested for seizures after 3 hours. The results of these experiments are given in Table 12; it is quite clear that increasing levels of 5-HT in brain protects these animals against seizures.

In order to extend and verify these observations, iproniazid, a monoamine oxidase inhibitor which raises levels of 5-HT and NE in brain, was tested for its effects on audiogenic and electroconvulsive seizures. Again, 21-day-old *DBA/2J* as well as *C57BL/6J* mice were used. Animals were injected intraperitoneally with either 0.0 or 150 mg iproniazid/kg and seizure tests were conducted 6 hours after drug administration. The data obtained in these ex-

TABLE 12

Effects of 5-HTP on Seizure Susceptibility

	Genotype	Treatment[a]	N	Seizure Susceptibility	
				Number of Animals Not Seizing	Number of Animals Seizing
Audiogenic	*DBA/2J*	Control	10	1	9
Seizures	*DBA/2J*	5-HTP	9	7	2
Metrazol	*DBA/2J*	Control	21	2	19
Seizures	*DBA/2J*	5-HTP	21	10	11
	C57BL/6J	Control	15	0	15
	C57BL/6J	5-HTP	15	10	5
	F_1	Control	15	1	14
	F_1	5-HTP	15	7	8
Electroconvulsive	*DBA/2J*	Control	10	3	7
Seizures	*DBA/2J*	5-HTP	10	10	0
	C57BL/6J	Control	11	0	11
	C57BL/6J	5-HTP	10	9	1

[a]All animals were 21 days of age. Animals were injected with either vehicle or 150 mg 5-HTP/kg. Seizure tests were conducted 3 hours after drug administration.

periments are summarized in Table 13. Iproniazid was found to protect these animals against seizures and these results were interpreted to mean that this effect is due to increased levels of 5-HT and NE in brain of iproniazid-treated animals.

This experiment was repeated using another monoamine oxidase inhibitor, catron. Since this compound also increases levels of 5-HT and NE we predicted that it should protect animals against audiogenic seizures. Twenty-one-day-old *DBA/2J* mice were injected with 0.0 or 10 mg catron/kg and tested for audiogenic seizures 6 hours later. The results of these experiments are also summarized in Table 13 and indicate that this drug does protect these mice against seizures.

Effects of Intracranial 5-HT and NE on Seizures

Next, certain experiments were conducted to provide more direct evidence for the effects of 5-HT and NE on seizure susceptibility, since it could be argued that the drugs which we had used previously have some side-effects; in other words we wished to study directly the action of 5-HT and NE on seizure susceptibility. These amines do not readily cross the blood-brain barrier; therefore, to accomplish our purpose they were injected directly into the brain and the effects of this treatment on seizure susceptibility were noted. For these experiments mice were removed from their home cages and weighed to the nearest 0.1 g. Each was then given an intracranial injection of (1) vehicle, (2) 800 mμg 5-HT and 800 mμg NE, (3) 800 mμg NE, or (4)

TABLE 13

**Effects of Iproniazid and Catron on
Susceptibility to Seizures[a]**

	Genotype	Treatment	N	Seizure Susceptibility	
				Number of Mice Not Seizing	Number of Mice Seizing
Audiogenic	*DBA/2J*	Control	10	0	10
Seizures	*DBA/2J*	Iproniazid	10	6	4
	DBA/2J	Control	10	0	10
	DBA/2J	Catron	10	10	0
Metrazol	*DBA/2J*	Control	10	3	7
Seizures	*DBA/2J*	Iproniazid	10	9	1
	C57BL/6J	Control	11	0	11
	C57BL/6J	Iproniazid	10	6	4

[a]Clonic-Tonic seizures were used as an index of seizure susceptibility.

800 mµg 5-HT. Injections were given into the frontal lobe to a depth of 3 mm; at a volume of 4.0 µl/mouse; similar injections with trypan blue showed that the dye was distributed throughout the subarachnoid space almost immediately after injection. Animals were tested for seizure susceptibility 10 minutes later: either for audiogenic seizures by the method described above, or for metrazol seizures by being injected with 78 mg metrazol/kg.

Table 14 shows that, in *C57BL/6J* and *DBA/2J* mice, intracranial injections of 5-HT and NE protected against metrazol-induced seizures; in F_1 hybrid mice this effect was not statistically significant. Administration of 5-HT and NE was also observed to protect *DBA/2J* mice against audiogenic seizures.

The separate intracranial administration of 5-HT was without effect, except that the incidence of lethal seizures was reduced by approximately 43 percent from control values ($N=7$/group). Intracranial administration of NE reduced the incidences of clonic seizures by approximately 70 percent, of tonic seizures by 30 percent, and of lethal seizures by 72 percent from control values ($N=7$/group).

These results were interpreted as further evidence that increasing levels of 5-HT and/or NE has a general protective effect against seizures.

Effects of Manipulating Levels of GABA on Seizures

Our speculations also involve differences between levels of GABA in brains of susceptible and in brains of nonsusceptible strains of mice. We have not as yet measured endogenous levels of this compound and, therefore, do not know whether they are correlated with seizure susceptibility. We have, however, manipulated levels of GABA pharmacologically and investigated the effects on seizure susceptibility in a series of four experiments; the agents used were: (1) amino oxyacetic acid; (2) glutamic acid; (3) glutamine; and (4) the direct intracranial administration of GABA.

Administration of amino oxyacetic acid results in increased levels of GABA in brain as shown in Table 15; GABA levels continue to rise for approximately 7 hours after drug treatment and then slowly return to control values over the next 30 hours. As can also be seen from Table 15, amino oxyacetic acid decreases susceptibility to audiogenic seizures approximately as it increases levels of GABA. As we had predicted, increasing levels of GABA protect *DBA/2J* mice against audiogenic seizures.

The effects of amino oxyacetic acid on seizures induced by other agents were also studied, and are summarized in Table 16. Amino oxyacetic acid was found to protect *DBA/2J* and *C57BL/6J* mice against electroconvulsive seizures as well as against metrazol seizures; in F_1 hybrid mice, however, this compound was not observed to have a significant effect.

Glutamic acid and glutamine, injections of which should increase levels

TABLE 14

Effects of Intracranial 5-HT *and* NE on Seizure Susceptibility

	Genotype	N	Percent Seizures[a]				Seizure Severity Score ± 1 SD	t	p
			WR	C	T	L			
Audiogenic Seizures	DBA/2J	21	100%	100%	90%	52%	3.43 ± 0.64	–	–
	DBA/2J	10	90%	40%	30%	20%	1.80 ± 1.33	4.60	<0.001
Metrazol Seizures	DBA/2J	36	100%	92%	64%	58%	3.14 ± 1.08	–	–
	DBA/2J	15	87%	67%	40%	33%	2.27 ± 1.44	2.12	<0.05
	C57BL/6J	40	100%	95%	62%	58%	3.15 ± 1.04	–	–
	C57BL/6J	20	100%	55%	25%	25%	2.05 ± 1.20	3.65	<0.001
	F_1	27	100%	92%	37%	30%	2.56 ± 1.03	–	–
	F_1	18	100%	78%	44%	17%	2.39 ± 1.01	0.54	NS

[a]WR= wild running for audiogenic seizures; muscle spasms for metrazol seizures. C= clonic seizures. T= tonic seizures. L= lethal seizures.

TABLE 15

Effects of Amino Oxyacetic Acid on Audiogenic Seizures in Mice

Dose[a] (μg/g)	N	Time of test (hours)	Percent Increase[b] in GABA	Percent Seizures		
				Wild Running	Clonic Seizures	Tonic Seizures
0.0	15	–	–	100%	87%	87%
20.0	10	1	100%	50%	10%	10%
20.0	10	2.5	220%	20%	0%	0%
20.0	10	5	300%	0%	0%	0%
20.0	10	7.5	270%	0%	0%	0%
20.0	10	15	70%	80%	10%	0%
20.0	10	30	20%	100%	60%	60%

[a]The control data (0.0 μg/g) are those reported by Schlesinger et al.(1965). All mice used in these experiments were 21-day-old *DBA/2J* animals.
[b]The data on levels of GABA were kindly supplied by Dr. E. Roberts, City of Hope Medical Center, Duarte, California.

of GABA, were also studied for their effects on susceptibility to audiogenic seizures. Ginsburg and Roberts (1951) have reported that glutamic acid protects mice against audiogenic seizures; our experiments summarized in Table 17, replicate these findings. Ginsburg (1963) has also reported that administration of glutamine not only failed to protect mice against audiogenic seizures but actually increased seizure susceptibility. In this respect our results do not agree, since glutamine was found to decrease seizure susceptibility slightly. The discrepancy between these results is probably attributable to (1) differences in dose and, (2) differences in time of seizure test following the administration of glutamine.

Finally, Table 18 shows the effects of intracranial injections of GABA on susceptibility to audiogenic and metrazol-induced seizures. The three different doses of GABA that we used were all found to protect 21-day-old *DBA/2J* mice against audiogenic seizures; GABA was also observed to protect *C57BL/6J* and *DBA/2J* mice against metrazol-induced seizures. It should be pointed out, however, that in *DBA/2J* animals only high doses of GABA significantly reduced seizure susceptibility. GABA was not found to protect F_1 hybrid mice against metrazol-induced seizures.

Summary of Pharmacologic Experiments

The results of all of these pharmacologic experiments are summarized in Table 19. This abstract attempts to show, in a qualitative sense, the effects of the various drugs used in the experiments discussed above on levels of

TABLE 16

The Effects of Amino Oxyacetic Acid on Metrazol- and Electrically-Induced Seizures[a]

	Genotype	Treatment	N	Number of Animals Not Seizing	Number of Animals Seizing	χ^2	p
Metrazol Seizures	DBA/2J	Control	15	1	14	5.71	<0.02
	DBA/2J	AOAA	15	8	7		
	C57BL/6J	Control	16	3	13	9.30	<0.01
	C57BL/6J	AOAA	15	12	3		
	F_1	Control	11	0	11		Not Significant
	F_1	AOAA	10	0	10		
Electrical Seizures	DBA/2J	Control	10	3	7	5.20	<0.05
	DBA/2J	AOAA	10	9	1		
	C57BL/6J	Control	11	0	11	11.02	<0.001
	C57BL/6J	AOAA	10	8	2		

[a]χ^2 were calculated separately for each genotype and drug group and Yates' correction for continuity was used in all cases.

246

TABLE 17

**Effects of Glutamic Acid and Glutamine
on Audiogenic Seizures[a]**

Response	Percent Seizures	
	Glutamic Acid (N = 8)	Control (N = 10)
Wild Running	50%	100%
Clonic Seizures	25%	90%
Tonic Seizures	25%	90%
Lethal Seizures	12%	40%
	Glutamine (N = 15)	Control (N = 5)
Wild Running	100%	100%
Clonic Seizures	80%	100%
Tonic Seizures	80%	100%
Lethal Seizures	40%	80%

[a]All animals were 21-day-old *DBA/2J* mice; 2 mg/kg of glutamic acid or 20 mg/kg of glutamine were injected.

5-HT, NE, and GABA in brain and on susceptibility to audiogenic, metrazol- and electrically-induced seizures. It does not take into account the size or statistical significance of the various effects, nor the drug by genotype interactions observed in some of these experiments. Nevertheless, the overall picture which emerges indicates that, in general, drugs which lower levels of 5-HT, NE, and GABA increase susceptibility to seizures, whereas drugs which raise levels of these amines and of GABA protect mice against seizures. These data are reinforced by our findings that the intracranial administration of 5-HT, NE, and GABA also protects against audiogenic- and metrazol-induced seizures.

THE EFFECTS OF PYRIDOXINE ON SEIZURE SUSCEPTIBILITY

Yen and Ritman (1964) have reported that human phenylketonurics excrete abnormal phenylalanine metabolites conjugated with vitamin B_6. Since *DBA/2J* mice excrete abnormal phenylalanine products and are similar to human phenylketonurics in other aspects as well, it is possible that these mice are functionally deficient in pyridoxine. This deficiency could be related to our hypothesis of seizure susceptibility since vitamin B_6 is the cofactor in decarboxylating reactions; this might be the cause of the low levels of 5-HT and NE in 21-day-old *DBA/2J* mice (see below). If pyridoxine is related to seizure susceptibility then diets deficient in vitamin B_6 should increase sus-

TABLE 18

Effects of Intracranial Injections of GABA on Audiogenic and Metrazol-Induced Seizures

	Genotype	N	Treatment		Percent Seizures		
				Wild Running	Clonic Seizures	Tonic Seizures	Lethal Seizures
Audiogenic Seizures	DBA/2J	21	Control	100%	100%	90%	52%
	DBA/2J	10	GABA (2 mg/kg)	100%	30%	30%	10%
	DBA/2J	10	GABA (4 mg/kg)	70%	30%	20%	10%
	DBA/2J	10	GABA (8 mg/kg)	70%	50%	20%	0%
Metrazol Seizures	C57BL/6J	40	Control	100%	95%	62%	58%
	C57BL/6J	20	GABA (2 mg/kg)	100%	50%	35%	35%
	DBA/2J	36	Control	100%	92%	64%	58%
	DBA/2J	21	GABA (2mg/kg)	100%	95%	67%	57%
	DBA/2J	11	GABA (8 mg/kg)	100%	73%	9%	9%
	F_1	27	Control	100%	92%	37%	30%
	F_1	10	GABA (2 mg/kg)	100%	70%	20%	10%
	F_1	10	GABA (8 mg/kg)	100%	70%	50%	30%

TABLE 19

Summary of the Effects of Various Drugs on Seizures[a]

Drug	5-HT	NE	GABA	Audiogenic Seizures	Metrazol Seizures	Electroconvulsive Seizures
Reserpine	↓	↓	−	↑	↑	↑
Tetrabenazene	↓	↓	−	↑	−	−
α-MT	−	↓	−	↑	−	−
p-ChPhe	↓	−	−	↑	−	−
5-HTP	↑	−	−	↓	↓	↓
Iproniazid	↑	↑	−	↓	−	↓
Catron	↑	↑	−	↓	−	↓
AOAA	−	−	↑	↓	↓	↓

[a]Under the 5-HT, NE, and GABA headings the direction of the arrows have the following meaning: ↑ = increases levels; ↓ = decreases levels; − = no known effect. Under the seizure headings the direction of the arrows mean the following: ↑ = increases seizure susceptibility; ↓ = decreases seizure susceptibility; − = experiment not carried out.

ceptibility to audiogenic seizures and administration of supplemental B_6 should protect mice against these seizures.

Diets Deficient in Pyridoxine

The results of these experiments have been reported previously (Coleman and Schlesinger, 1965). *DBA/2J, BDP/J,* and *P/J* mice (examples of dilute genotypes) as well as *C57BL/6J* and *B6D2F1* mice (examples of nondilute genotypes) were maintained on diets containing either 30, 10, or 0 mg pyridoxine/kg of diet for either 3, 5, or 7 weeks starting at 21 to 28 days of age. At the end of this treatment, the animals were tested for susceptibility to audiogenic seizures; the results indicate that mice of dilute strains were much more susceptible to audiogenic seizures on the pyridoxine deficient diets than on control diets, while in mice of non-dilute genotypes the effects were extremely small. It should also be pointed out that body weights of the animals on these diets, which were recorded every 3 to 4 days, were affected as follows: Dilute strains (*DBA/2J, BDP/J,* and *P/J*) and the hybrid *B6D2F1* showed smaller increments in body weight on pyridoxine deficient diets than did nondilute *C57BL/6J* mice.

Supplemental Pyridoxine

If diets deficient in pyridoxine increase susceptibility to audiogenic seizures, and if one of the reasons for susceptibility in dilute mice is a functional deficiency of this vitamin, then it seems reasonable to predict that administration of supplemental B_6 should result in a lower incidence of seizures. Accordingly, *DBA/2J* mice were injected daily with 2 mg pyridoxine/mouse. Injections were started at 14 days of age and continued until the animals were 21 days of age; tests for seizure susceptibility were conducted approximately 3 hours after the last injection. The incidences of wild running, clonic, and tonic seizures were not affected by this treatment, but the number of lethal seizures was reduced by approximately 40 percent. This effect, although quantitatively quite small, was statistically significant ($p<0.05$).

Interaction of Pyridoxine Deficiencies and Drug Manipulation of Levels of 5-HT and NE

It has been demonstrated that pyridoxine deficient diets affect body weight gains, food intake, organ levels of pyridoxine, and susceptibility to audiogenic seizures (Lyon et al., 1958; Coleman and Schlesinger, 1965). Pyridoxine is the essential cofactor in decarboxylation and diets deficient in this vitamin might, therefore, be expected to lower levels of 5-HT, NE, and GABA. Experiments in our laboratory have shown that mice maintained on a B_6 deficient diet for 4 weeks do not have significantly less 5-HT or NE in brain tissue; Tews and Lovell (1967), however, have demonstrated that such treatment significantly lowers levels of GABA in brain. The increased susceptibility to audiogenic seizures observed in B_6 deficient mice could, therefore, be attributed to the effects of this treatment on levels of GABA, rather than on levels of the amines. The purpose of the experiments discussed below was to determine if the effects of pyridoxine deficiency and the pharmacologic depletion of levels of 5-HT and NE are additive or synergistic as far as susceptibility to audiogenic seizures are concerned. 5-HT was depleted by injections of p-ChPhe and NE was depleted by injection of a-MT.

One hundred and ninety-two F_2 mice (derived from *C57BL/6J* and *DBA/2J* animals) were used as experimental subjects. Animals were maintained on either control diets (Purina Mouse Breeder Chow) or diets deficient in pyridoxine (Nutritional Biochemical Corporation). Animals were weaned onto these respective diets at 21 days of age and maintained on these diets till they were 49 days of age. Some animals within each dietary condition were used as controls, or were injected with p-ChPhe, a-MT or a combination of both drugs. The results of these experiments are summarized in Table 20.

These results indicate the following: (1) Pyridoxine deficient diets significantly increase seizure susceptibility ($t = 2.29$, df $= 46$, $p < 0.025$). (2) On the control diet no drug treatment significantly increases seizure severity. (3) The pyridoxine deficient diet combined with p-ChPhe and a-MT increased susceptibility significantly over that obtained with the dietary regimen alone. In summary, then, these results suggest that pyridoxine deficiencies increase seizure susceptibility and that if, in addition to this dietary treatment, levels of 5-HT and NE are reduced, further increases in seizure susceptibility are observed.

TABLE 20

Effects of B$_6$ Deficiency, α-MT and p-ChPhe on Audiogenic Seizures[a]

Treatment	N	Percent Seizures				Seizure Severity Score		
		Wild Running	Clonic Seizures	Tonic Seizures	Lethal Seizures	$\bar{x} \pm 1$ SD	t	p
Control Diet								
Control	24	33%	13%	4%	0%	0.50 ± 0.83	–	–
α-MT	24	21%	0%	0%	0%	0.21 ± 0.41	1.55	NS
p-ChPhe	24	42%	8%	4%	4%	0.54 ± 0.78	0.173	NS
α-MT + p-ChPhe	24	37%	13%	8%	4%	0.58 ± 0.91	0.316	NS
B$_6$ Deficient								
Control	24	67%	25%	25%	17%	1.17 ± 1.17	–	–
α-MT	24	63%	37%	21%	8%	1.21 ± 1.18	0.373	NS
p-ChPhe	24	67%	33%	29%	21%	1.29 ± 1.23	1.091	NS
α-MT + p-ChPhe	24	87%	63%	54%	33%	2.04 ± 1.16	8.208	<0.001

[a]All mice used in these experiments were F_2 animals derived from mating F_1 mice from *DBA/2J* and *C57BL/6J* matings.

SUMMARY AND DISCUSSION

The research which we have reported above can be summarized as follows:

1. *DBA/2J* mice are significantly more susceptible to audiogenic seizures than *C57BL/6J* mice, especially at 21 days of age.
2. *DBA/2J* mice have lower metrazol- and electrically-induced seizure thresholds than *C57BL/6J* mice.
3. Endogenous levels of 5-HT and NE are significantly lower in *DBA/2J* than in *C57BL/6J* mice, especially at 21 days of age.
4. Reserpine and tetrabenazene deplete 5-HT and NE more in *DBA/2J* than in *C57BL/6J* mice. Rates of repletion of 5-HT and NE following tetrabenazene treatment are significantly slower in *DBA/2J* mice.
5. Depletion of 5-HT and NE by drugs or diets increases susceptibility to audiogenic, metrazol- and electrically-induced seizures.
6. Increasing levels of 5-HT, NE, and GABA, by drugs, diets, or direct intracranial injection of these compounds, protects animals against audiogenic, metrazol- and electrically-induced seizures.
7. Phenylketogenic treatments, which lower levels of 5-HT in brain, increase susceptibility to audiogenic seizures.
8. Diets deficient in pyridoxine increase susceptibility to audiogenic seizures; this effect is further enhanced when levels of 5-HT and NE are also lowered. Supplemental pyridoxine tends to decrease seizure severity.

The purpose of these experiments was to investigate the importance of 5-HT, NE, and GABA in determining the excitability of nervous tissue, measured in terms of susceptibility to audiogenic, metrazol- and electrically-induced seizures. More specifically, we have attempted to answer two questions: (1) Are mice, known to be susceptible to audiogenic seizures, also more susceptible to other seizure-inducing agents? Using *DBA/2J* and *C57BL/6J* mice as our experimental prototypes, we have found this to be true with respect to metrazol- and electrically-induced seizures. Our tentative conclusion, on the basis of our own data and that of other researchers, is that *DBA/2J* mice, in general more susceptible to seizures, have an extremely excitable nervous system. (2) Are levels of 5-HT, NE, and GABA, which we postulate to have inhibitory functions with respect to seizure susceptibility, lower in *DBA/2J* than in *C57BL/6J* mice and is this a possible explanation for the differing seizure thresholds of these mice? *DBA/2J* mice were found not only to have less 5-HT and NE in their brain tissue than mice more resistant to seizures but also to have lower levels of these amines only at 21 days of age, the period of maximal seizure risk for these animals. It is possible that seizure-susceptible mice also have higher levels of excitatory "transmittors", e.g., acetylcholine. Such a phenomenon has been observed in *ep* mice, which are susceptible to convulsions after loss of postural equilibrium; these mice have approximately 40 to 60 percent more acetylcholine in brain than other mice (Kurokawa et al., 1961, 1963). Preliminary experiments performed in collaboration with Dr. M. Bowers, Department of Psychiatry, Yale University School of Medicine, and summarized in Table 21, indicate that this is not likely to be the case in our animals, since our dilute mice synthesize less acetylcholine than the nondilute.

Having obtained a correlation between levels of 5-HT and NE, and seizure susceptibility, we reasoned that pharmacologic and dietary manipulation of levels of these amines should have predictable effects in terms of either increasing or decreasing this susceptibility, regardless of the agent used to induce the seizure; our results support our predictions. Similarly, we have observed that increasing levels of GABA protects mice against seizures, whereas diets deficient in pyridoxine increase risk of seizure, presumably because of the effects of this vitamin deficiency on levels of GABA. On the basis of these experiments, we would like to argue that 5-HT, NE, and GABA are inhibitory in function in the CNS, at least with respect to seizure susceptibility.

The role of the *d* locus in seizure susceptibility has been tested directly. In F_2 and $F_1 \times DBA/2J$ cross animals four coat color phenotypes are recovered. These are black, chocolate brown, dilute black, and dilute brown; the first two phenotypes are genetically either D/d or D/D, and the second two phenotypes are genetically d/d. Single gene mutants (reversions) to full coat-color in $DBA/2J$ stocks have also been obtained. These mice, on $DBA/2J$ background, i.e., either homozygous dilute (d/d) or homozygous or heterozygous nondilute $(D/D$ or $D/d)$, can be studied, and in a number of experiments we have observed that the dilute locus does not contribute importantly to seizure susceptibility (Schlesinger et al., 1966). Since it is more reasonable to assume that these mutant mice have normal phenylalanine hydroxylase

TABLE 21

Acetylcholine Synthesis in *DBA/2J* and *C57BL/6J* Mice

Acetylcholine Synthesis (µg/g wet weight/hour)

	Incubation Mixture		Tissue Slices	
	DBA/2J	*C57BL/6J*	*DBA/2J*	*C57BL/6J*
Experiment 1	4.28	6.16	–	1.80
	4.00	5.88	1.78	2.10
	5.27	6.84	1.56	2.45
Experiment 2	3.02	4.93	2.21	2.28
	2.89	6.66	1.18	3.21
	2.35	4.37	1.18	0.91
Average	3.64	5.81	1.58	2.13

activity there arises the question of whether or not the data which we have presented above can account for the high incidence of seizures typically observed in dilute mice in the manner which we have hypothesized. Several alternatives seem plausible: (1) That dilute mice are indeed more seizure susceptible because of deficiencies in levels of 5-HT, NE, and GABA, but that the reason for the deficiency of these amines and of GABA is other than the one we have described, i.e., other than abnormal phenylalanine hydroxylase activity. (2) That pharmacologic manipulation of levels of 5-HT, NE, and GABA does change seizure thresholds but that endogenous levels of these compounds do not explain the high degree of seizure susceptibility observed in dilute strains. (3) That the dilute locus is not involved in seizure susceptibility, but that a closely linked locus is, one which may or may not determine levels of 5-HT, NE, and GABA.

All of these experiments are aimed at elucidating the mechanisms underlying seizure susceptibility. Our research is, of course, not conclusive, but it strongly suggests that gene-controlled synthesis and/or destruction of 5-HT, NE, and GABA are important factors in seizure susceptibility.

REFERENCES

Abood, L. G., and R. W. Gerard. 1955. A phosphorylation defect in the brain of mice susceptible to audiogenic seizures. *In* Waelsch, H., ed., Biochemistry of the Developing Nervous System, New York, Academic Press, Inc., pp. 467–472.

Bevan, W. 1955. Sound-precipitated convulsions: 1947–1954. Psychol. Bull., 52:473–504.

―――― and R. Chinn. 1957. Sound-induced convulsions in rats treated with reserpine. J. Comp. Physiol. Psychol., 50:311–314.

Busnel, R. G. 1963. Psychophysiologie Neuropharmacologie et Biochemie de la Crise Audiogène, Paris, Centre National de la Recherche Scientifique.

―――― and A. Lehman. 1961. Action de convulsivants chimigues sur les souris de lignées sensible et résistance à la crise audiogène, Part III. Cafeine. J. Physiol., 53:285–286.

Chance, M. R. A. 1963. A biological perspective on convulsions. *In* Busnel, R. G., ed., Psychophysiologie Neuropharmacologie et Biochemie de la Crise Audiogène, Paris, Centre National de la Recherche Scientifique, pp. 15–43.

Cooke, J. P. 1964. An investigation of audiogenic seizure prone mice and rats as indicators of low intensity chronic gamma irradiation effects. Dissertation Abstr., 24:4263–4264.

―――― S. O. Brown, and G. M. Krise. 1964. Prenatal chronic gamma irradiation and audiogenic seizures in rats. Exp. Neurol., 9:243–248.

Coleman, D. L. 1960. Phenylalanine hydroxylase activity in dilute and non-dilute strains of mice. Arch. Biochem., 91:300–306.

―――― and K. Schlesinger. 1965. Effects of pyridoxine deficiency on audiogenic seizure susceptibility in inbred mice. Proc. Soc. Exp. Biol. Med., 119:264–266.

Deol, M. S. 1964. The abnormality of the inner ear in kreisler mice. J. Embryol. Exp. Morph., 123:111–119.

Finger, F. W. 1947. Convulsive behavior in the rat. Psychol. Bull., 44:201–248.

Finney, D. J. 1952. Statistical Methods in Biological Assay, New York, Hafner Pub. Co., Inc.

Fuller, J. L., and F. H. Sjursen, Jr. 1967. Audiogenic seizures in eleven mouse strains. J. Hered., 58:135–140.

———— and W. R. Thompson. 1960. Behavior Genetics, New York, John Wiley and Sons, Inc.

Ginsburg, B. E. 1963. Causal mechanisms in audiogenic seizures. *In* Busnel, R. G., ed., Psychophysiologie Neuropharmacologie et Biochemie de la Crise Audiogène, Paris, Centre National de la Recherche Scientifique, pp. 228–240.

———— and D. S. Miller. 1963. Genetic factors in audiogenic seizures. *In* Busnel, R. G., ed., Psychophysiologie Neuropharmacologie et Biochemie de la Crise Audiogène, Paris, Centre National de la Recherche Scientifique, pp. 218–225.

———— and E. Roberts. 1951. Glutamic acid and central nervous system activity. Anat. Rec., 111:492–493.

Hamburgh, M., and E. Vicari. 1960. A study of some physiological mechanisms underlying susceptibility to audiogenic seizures in mice. J. Neuropath. Exp. Neurol., 19:461–472.

Hanson, A. 1958. Inhibition of brain glutamic acid decarboxylase by phenylalanine metabolites. Naturwissenschaften, 45:423.

Harris, H. 1962. Human Biochemical Genetics, Cambridge University Press.

Henry, K. R. 1966. Audiogenic Seizures. Unpublished thesis, University of North Carolina.

Hsia, D. Y. 1967. The hereditary metabolic diseases. *In* Hirsch, J., ed., Behavior-Genetic Analysis, New York, McGraw-Hill Book Co., pp. 176–193.

Kurokawa, M., M. Kato, and Y. Machiyama. 1961. Choline acetylase activity in a convulsive strain of mouse. Bioch. Biophys. Acta, 50:385–386.

———— Y. Machiyama, and M. Kato. 1963. Distribution of acetylcholine in the brain during various states of activity. J. Neurochem., 10:341–348.

Karrer, R., and G. Cahilly. 1963. Experimental attempts to produce phenylketonuria in animals: A critical review. Psychol. Bull., 64:52–64.

Koe, B. K., and A. Weissman. 1966. p-Chlorophenylalanine: A specific depletor of brain serotonin. J. Pharm. Exp. Ther., 154:499–516.

Lehman, A., and R. G. Busnel. 1963. Le métabolisme de la sérotonine cérébrale dans ses rapports avec la crise audiogène de la Souris et ses variations sous l'influence de divers composés psychotropes. *In* Busnel, R. G., ed., Psychophysiologie Neuropharmacologie et Biochemie de la Crise Audiogène, Paris, Centre National de la Recherche Scientifique, pp. 454–469.

Lyon, J. B., Jr., H. L. Williams, and E. A. Arnold. 1958. The pyridoxine-deficient state in two strains of inbred mice. J. Nutr., 66:261–275.

Marrazi, A. S., 1961. Inhibition as a determinant of synoptic and behavioral patterns. Ann. N. Y. Acad. Sci., 92:990–1003.

McKean, C. M., S. M. Acharnberg, and N. J. Giarman. 1962. A mechanism of the indole defect in experimental phenylketonuria. Science, 137:604–605.

———— S. M. Schanberg, and N. J. Giarman. 1967. Aminoacidemias: Effects on maze performance and cerebral serotonin. Science, 157:213–215.

Mead, J. A. R., and K. F. Finger. 1961. A single extraction method for the determination of both norepinephrine and serotonin in brain. Biochem. Pharm., 6:52–53.

Nakai, K., H. Ida, Y. Kakinoto, Y. Hishikawa, I. Sano, and Z. Kaneko. 1966. A low metrazol threshold in the heterozygote of phenylketonuria. J. Nerv. Ment. Dis., 144:436–442.

Palidora, V. J. 1967. Behavioral effects of "phenylketonuria" in rats. Proc. Nat. Acad. Sci., 57:102–106.

Partington, M. W. 1961. The early symptoms of phenylketonuria. Pediatrics, 27:465–473.

Plotnikoff, N. P. 1958. Bioassay of potential tranquilizers and sedatives against audiogenic seizures in mice. Archives Internationales de Pharmacodynamie et de Thérapie, 116:130–135.

———— 1960. Ataractics and strain differences in audiogenic seizures in mice. Psychopharmacologia, 1:429–432.

Rauch, H., and M. T. Yost. 1963. Phenylalanine metabolism in dilute-lethal mice. Genetics, 48:1487–1495.

Roberts, E. 1966a. The synapse as a biochemical self-organizing cybernetic unit. *In* Rodahl, K., ed., Nerve as a Tissue, New York, Harper and Row, Publishers, pp. 212–265.

———— 1966b. The synapse as a biochemical self-organizing microcybernetic unit. Brain Res., 2:117–166.

———— 1967. Synaptic neurochemistry: A projection. *In* Hirsch, J., ed., Behavior-Genetic Analysis, New York, McGraw-Hill Book Co., pp. 194–210.

———— M. Rothstein, and C. F. Baxter. 1958. Some metabolic studies of gamma-aminobutyric acid. Proc. Soc. Exp. Biol. Med., 97:796–802.

Sandler, M. and H. Close. 1959. Biochemical effects of phenylacetic acid in a patient with 5-hydroxytryptophan secreting carcinoid tumor. Lancet, 277:316–318.

Schalock, R. L., and F. D. Klopfer. 1967. Phenylketonuria: Enduring behavioral deficits in phenylketonuric rats. Science, 156:1033–1035.

Schlesinger, K., W. Boggan, and D. X. Freedman. 1965. Genetics of audiogenic seizures: I. Relation to brain serotonin and norepinephrine in mice. Life Sci., 4:2345–2351.

———— R. C. Elston, and W. Boggan. 1966. The genetics of sound induced seizures in inbred mice. Genetics, 54:95–103.

Spector, S., A. Sjoerdsma, and S. Udenfriend. 1965. Blockage of endogenous norepinephrine synthesis by a-methyl-tyrosine, an inhibitor of tyrosine hydroxylase. J. Pharm. Exp. Ther., 147:86–95.

Swinyard, E. A., A. W. Castellion, G. B. Fink, and L. G. Goodman. 1963. Some neurophysiological and neuropharmacological characteristics of audiogenic seizure-susceptible mice. J. Pharm. Exp. Ther., 140:375–384.

Tews, J. K., and R. A. Lovell. 1967. The effect of a nutritional pyridoxine deficiency on free amino acids and related substances in mouse brain. J. Neurochem., 14:1–7.

Yen, H. L., and P. Ritman. 1964. New metabolites of phenylalanine. Nature, 203:1237–1239.

Yuwiler, A., and R. T. Louttit. 1961. Effects of phenylalanine diet on brain serotonin in the rat. Science, 134:831–852.

SECTION V

The Evolution
of Behavior

10

Behavioral Population Genetics and Wild *Mus musculus**

JAN H. BRUELL
Behavior Genetics Laboratory,
Department of Psychology,
The University of Texas at Austin

INTRODUCTION

This chapter deals with two rather unfamiliar subjects: behavioral population genetics, an emerging field of behavioral research, and wild *Mus musculus,* an animal that could become important for psychology. The two are treated in the same chapter because a suitable research animal is needed for population-genetic studies of behavior, and the wild mouse may turn out to be just the animal for that purpose. This at least will be the thesis of this chapter.

Behavioral population genetics is not entirely foreign to psychology. Studies dealing with racial differences in human behavior (cf. Spuhler and Lindzey, 1967), selective breeding experiments (cf. Fuller and Thompson, 1960), and studies of behavioral isolating mechanisms (see Spieth, 1958; Ehrman, 1964b) fall into its province. But the field of behavioral population genetics awaits systematic development. More psychologists must become familiar with its concepts through reading and research. The purpose of this chapter iş to serve as a guide to some of the pertinent literature, and to suggest population-genetic research that could be carried out now with natural populations of mice. Since wild mice have seldom been used by psychologists, a section of the chapter deals with the races and world distribution of the species *Mus musculus.*

* Preparation of this paper was aided by Grant No. HD-052589 from the National Institute of Child Health and Human Development. U.S. Public Health Service, granted to the Department of Psychology, Case Western Reserve University, Cleveland, Ohio 44106.

POPULATION GENETICS

Population genetics has not yet had an appreciable impact on psychology. Introductory textbooks of psychology do not mention the field. Psychologists engaged in research rarely test its theories. Even some of its basic concepts are not well known to psychologists. In this section I will, therefore, discuss some concepts of population genetics. My discussion will be short and in no way exhaustive. I am including it here as background for the subsequent discussion of "behavioral population genetics." More explicit statements of the theory of population genetics can be found in many other places, for example, in Dobzhansky's highly informative "Review of some Fundamental Concepts and Problems of Population Genetics" (Dobzhansky, 1955; see also Dobzhansky, 1958, 1959, 1962; Mayr, 1959, 1963; Simpson, 1958; Spiess, 1968).

Mendelian Populations

The term "population" means one thing to a psychologist, and something entirely different to a population geneticist. For a psychologist the term denotes the "universe" of all college students, females over forty, children in orphanages, or whatever. Such populations are artificial groups, sharing some arbitrary characteristic. By contrast, the populations of geneticists are biologic entities, groups of related individuals sharing hereditary material. Dobzhansky (1955) calls such groups Mendelian populations. In Dobzhansky's words, a Mendelian population is "a reproductive community of individuals who share in a common gene pool."

In theory, all members of a species can interbreed; they form one large reproductive community, one large Mendelian population. Actually, however, mostly because of geographic barriers to free interbreeding, species tend to break up into infraspecific Mendelian populations, *subspecies* or *races*. These, in turn, are often composed of numerous, relatively small, local breeding communities, or *demes*. A completely panmictic Mendelian population, that is, one within which the chances of mating of any female with any male are equal, is sometimes referred to as an *isolate*.

Within a species the flow of genes is hampered primarily by geographic barriers, often simply the distance separating demes or racial groups. Yet, within a species there is always some gene diffusion. "Although flies native to California do not usually mate with flies from Texas, the genes of the California population may percolate, though slowly, through a chain of geographically intermediate populations, to Texas, or anywhere else in the species area" (Dobzhansky, 1958, p. 27). Thus a valuable new gene or an advantageous gene constellation can eventually spread throughout the range of the species. In time, such genetic novelties can become the property of the whole species.

However, genetic innovations arising in one species cannot spread to other species. Demes and races are genetically open systems; species are not. Species are genetically closed Mendelian populations. They do not share in a common gene pool with other species.

The Gene Pool

The term "gene pool" appears in Dobzhansky's definition of a Mendelian population. I will use an illustration to explain it. Much has been written lately about the plight of whooping cranes: the species seems doomed, only a few members of the species still survive. Each egg a whooping crane lays is greeted enthusiastically by conservationists and reporters. Suppose the last whooping crane died and a few eggs were preserved. These eggs would carry all the genetic information about whooping cranes and their ways. They would contain the entire gene pool of the species. If these eggs were incubated, and a new generation of cranes emerged, these birds in turn would carry all the genes of whooping cranes. And so it could go on. We could speak of a next "generation" of eggs and another generation of cranes, and then a generation of eggs again. Generations of eggs and birds would not follow each other neatly; eggs and adults would often coexist. Not every bird and not every egg would carry all the genes of cranes, but together they would hold the entire set of genes, the gene pool, the species possesses. In the case of whooping cranes, a nearly extinct species, the entire gene pool of the species is possessed by a single population of birds. But most species consist of many Mendelian populations, and thus the gene pools of species tend to be subdivided into numerous smaller pools.

Factors Determining the Contents of a Gene Pool

Each of the many gene pools of a species may contain a somewhat different set of genes. Two factors, chance and natural selection, are responsible for such differences. Chance affects the gene pool in various ways: (1) It operates whenever new territories are occupied. Local populations are often founded by a few colonists, sometimes one pregnant female. The colonists carry with them only a small sample of the genes present in the gene pool of the population from which they originated. Such samples are subject to "errors of sampling," and thus gene frequencies in the pool of the new colony may differ "by chance" quite considerably from the gene frequencies in the ancestral gene pool. (2) Chance operates when genes are added to a local pool by individual stragglers from neighboring demes. Gene flow from one deme to another is a random event. (3) Chance also operates when mutations occur, adding new alleles to the gene pool of a population. But in the long run, random events are of minor significance. Natural selection, so it is believed today (see Mayr, 1963), is the major factor shaping the gene pool of a population.

Natural selection affects the contents of a gene pool through a simple feedback mechanism. Individuals differ in viability and fertility. Therefore, they contribute different numbers of offspring to the next generation. Some individuals are reproductively fit, others less fit. But the relative fitness of individuals depends on the genes they carry. Carriers of fitness-enhancing genes have more offspring than carriers of fitness-reducing genes. Thus, in each generation, the proportion of fitness-enhancing genes added to the gene pool of a population is larger than the proportion of fitness-reducing genes. This progressively changes the contents of the gene pool: the proportion of fitness-promoting genes rises, while the proportion of fitness-reducing genes decreases. "Reproductive success rather than survival is stressed in the modern definition of natural selection. A superior genotype has a greater probability of leaving offspring than has an inferior one. Natural selection, simply, is the differential perpetuation of genotypes" (Mayr, 1963, p. 183).

Factors Determining the Selective Value of Genes

Terms like "fitness-enhancing" or "fitness-decreasing" genes can be misleading. Such terms may suggest that there exist intrinsically "good" and "bad" genes. But the value of a gene depends on the environment in which it occurs. This point is vividly illustrated by the widely publicized phenomenon of industrial melanism. Many species of moths fly at night and spend the day resting on the trunk of a tree. One such moth is the peppered *Biston betularia* and its melanic form *B. carbonaria*. The melanic form of the moth was rare one hundred years ago; now it is the light form that is rare. In the past, the peppered form, resting on the lichen-encrusted trunks of trees, was well protected against predation by birds. Today, the dark *B. carbonaria* is protected when it rests on trees blackened by the polluted air of industrial cities. In earlier times it was the gene conferring lightness on *Biston* that contributed to the reproductive success of the moth, and so it was the prevalent gene in the gene pool of the species. Now, in an environment changed by industrial soot, the melanic gene does for the moth what the light gene did for it previously. Today, dark moths are the more successful reproducers, and the melanic gene has almost completely replaced the light gene in the gene pool of *Biston* populations that inhabit industrial areas (Kettlewell, 1959, 1961; Ford, 1964). The case of industrial melanism illustrates the principle of "genic relativity": genes do not have absolute values. Their selective value depends on the environment in which they occur. As the environment changes, the selective value of genes changes also.

The selective value of a gene depends not only on the external environment in which it occurs, but also on the internal, physiologic environment provided by other genes with which it interacts. This fact is well illustrated by the case of the sickle-cell gene and its normal allele (see Dobzhansky, 1962). Individuals carrying the sickle-cell gene in double dose, that is, homozygous carriers of this gene, are likely to become severely anemic and to die at an early age. The homozygous carriers of the normal allele of the sickle-cell gene are similarly handicapped. If they happen to live in malaria infested regions of Africa they are likely to succumb to malaria. But heterozygotes, individuals carrying both the sickle-cell gene and its normal allele, are rather resistant to malaria, and they do not become anemic. Thus the selective value of the sickle-cell gene depends on the gene environment in which it occurs, as does the value of its "normal" allele.

MAINTENANCE OF GENETIC HETEROGENEITY IN POPULATIONS

The obvious genetic heterogeneity of populations has puzzled evolutionists since Darwin. One could assume that adaptation of a population to its environment is accomplished by the gradual elimination of all less favorable alleles of a gene, and the creation of a monopoly for one "best" gene. If this happened, most individuals in a population would become homozygous for most genes. Since each gene would be represented only by its best possible allele, heterozygosity would, in time, be eliminated. Such is obviously not the case in nature, and one must explain how populations are protected "against the restless ravages of natural selection" (Mayr, 1963, p. 215). Many, not mutually exclusive, explanations of the phenomenon have been advanced.

1. Even if genetic homogeneity had been achieved by a population, mutations would continue to occur, thus adding allelic variants to the gene pool. Though the mutants would be eliminated again, the lag between the appearance of a mutant and its elimination by natural selection would result in unavoidable temporary heterogeneity.
2. Secular changes in the environment occupied by a population would create similar conditions for inescapable variation, as illustrated by industrial melanism. In time, *Biston carbonaria* will replace *B. betularia,* but a transitional period during which both morphs of the moth coexist cannot be avoided.
3. The transitory polymorphism of *Biston* is caused by a changing environment. Persistent polymorphism can be caused by an environment sufficiently heterogeneous to provide niches for several morphs, each adjusted to the particular microhabitat in the macroenvironment of the population.
4. Genetic variants could persist in a population if they were selectively neutral, that is, inconsequential for the reproductive fitness of their bearers. Such variants would survive in a population because they would neither

be selected for nor selected against. Some evolutionists doubt that entirely "neutral" genes actually exist (see Mayr, 1963).

5. Both alleles of a gene would persist in a population if the heterozygote *Aa* were more fit than either of the homozygotes, *AA* and *aa*. That this does occur is illustrated by the case of the sickle-cell gene and its normal allele. The phenomenon is referred to as "balanced polymorphism" (Dobzhansky, 1955) because, in the population, the low fitness of the homozygous geno- types or morphs is balanced by the superior fitness of the heterozygous morph; thus all three morphs survive. What applies to pairs of genes is known also to apply to sections of chromosomes, or even entire chromo- somes. Bearers of homozygous chromosome structures are often less fit than their heterozygous counterparts in the population. Today the phenomenon of heterozygote superiority or balanced polymorphism is considered to be one of the most important reasons why genetic heterogeneity persists in populations (Dobzhansky, 1955; Mayr, 1963).

Polymorphism, Polytypism, and Continuous Variation

Polymorphism and polytypism are two terms often used in the discussion of population genetics. Polymorphism is an intrapopulation phenomenon. It refers to the occurrence of several different, heritable, discontinuous types within a single interbreeding population, for example, the melanic and white forms of *Biston*. Polytypism is an interpopulation phenomenon. The term polytypic is used to point to the existence within a species of several distin- guishable populations, each differing from the others. "Mankind is a *polytypic* species composed of a cluster of races, Mendelian populations with more or less different gene pools. And mankind is *polymorphic,* each population being genetically variable" (Dobzhansky, 1962, p. 221).

Much of the phenotypic variation within populations is continuous rather than discrete. Members of a population are not necessarily either black or white; they are apt to come in all shades of grey. But there is widespread agreement among geneticists that continuous, "quantitative" variation differs from discontinuous, "qualitative" variation only in the number of factors af- fecting the particular aspect of the phenotype. If one or few genes have a large effect on the phenotype, variation tends to become manifestly discontinuous; if many genes with small effects cooperate in determining the phenotype, dif- ferences tend to become blurry, and we speak of continuous variation. In either case, however, so it is believed today, the hereditary material underlying pheno- typic variation is discontinuous, or particulate.

Coadapted Gene Pools and Speciation

A gene must prove its worth in several environments: the geographic environment in which it occurs, as the example of *Biston carbonaria* has shown; the environment provided by all the other genes with which it forms a genotype, as illustrated by the phenomenon of heterozygote superiority; and the genic environment of the gene pool itself, as the following considerations will suggest. Individuals, ephemeral holders of genes, die, but genes are passed on from individual to individual, from generation to generation. On their journey through time, genes enter into multiple unions with other genes of the pool, forming ever-new genotypic combinations. In order to survive, genes must produce viable genotypes in cooperation with most, not just some of the other genes of the pool: they must be "good mixers" (Dobzhansky, 1955). Selection creates cohesion among the genes of a pool by removing from it genes that are "rugged individualists" (Dobzhansky, 1955) and retaining those that mix well. Natural selection is not only responsible for the adaptedness of genotypes to their environment, but also for the coadaptedness of the genes of a gene pool.

Progressive increase in the cohesiveness of a local gene pool can culminate in speciation, the irreversible splitting off of a local pool from the other gene pools of the species. In geographically isolated Mendelian populations, relatively undisturbed by intrusion of genes from other populations of the species, the internal cohesion of a gene pool often nears perfection. The local genes "know" each other, they cooperate harmoniously, they form a perfect team creating a rich variety of genotypes that are well adapted to local conditions. The pool reaches an "adaptive peak" (S. Wright); things can hardly be improved. Any influx of genes, carried by immigrants, can only disturb the well attuned harmony attained by the gene pool. Genotypes put together from resident and immigrant genes must be inferior to those assembled from resident genes only. The hybrid offspring from such incompatible unions are selected against. The process is self-reinforcing. As fewer and fewer strange genes are admitted to the pool, chances for new arrivals become slimmer and slimmer. Finally isolation becomes complete. What began as a population sharing, though infrequently, in the gene pool of other populations of the species, has now become a population posing impenetrable barriers to the inflow of genes from the outside; the population has evolved into a separate species.

BEHAVIORAL POPULATION GENETICS

In their research, population geneticists deal almost exclusively with qualitative traits. Most often, they choose for study discontinuous morphologic traits that are detectable without time-consuming measurement, are well known genetically, and are not readily changed by the environment. Qualitative morphologic traits offer many advantages. Individuals displaying such traits are easily classified. The numbers of the various morphs occurring in a given population can be determined by proper sampling. The relative frequencies of the genes responsible for the polymorphism can then be estimated. This, in turn, permits one to calculate whether the genes occur in the population in the "right" proportions, conforming to expectations based on the Hardy-Weinberg law, the basic law of population genetics. Gene frequencies in one population can be compared with gene frequencies observed in other populations, using simple statistical tests. There can be no doubt that the substantial advances made by population geneticists are due to a great extent to their selection of qualitative morphologic traits for study. The choice of such traits for population-genetic research was felicitous for practical reasons, but it was not strictly demanded by theoretical considerations. The theory of population genetics applies to any heritable trait, whether the trait be qualitative or quantitative, morphologic or nonmorphologic.

The theory of population genetics certainly also applies to behavioral traits, a fact that was apparent to many previous writers (see Hirsch, 1963; King, 1967; Spuhler and Lindzey, 1967). A new area of behavioral research, behavioral population genetics, awaits development. The goals of the behavior-genetic study of populations would not have to be in any way ambitious. Psychologists would not undertake population-genetic studies of behavior to contribute to the "genetics" of populations (see Thompson, 1967, p. 344). The aim of such studies could simply be to focus attention on some untenable beliefs of contemporary psychology, such as the belief in species-typical behavior, the belief in the species-wide uniformity of behavioral adaptations, and the belief that the behavioral characteristics of species are static and rather immutable. In short, the goal of behavioral population genetics could be to make us more aware of the out-dated "typological thinking" (see Mayr, 1963; Dobzhansky, 1967) still prevalent among psychologists.

The Belief in Species-Typical Behavior

B. F. Skinner has recently stated that "no reputable student of animal behavior has ever taken the position ... that species differences are insignificant" (Skinner, 1966). Whatever the merits of this assertion, it can be proposed with equal assurance that even today many reputable psychologists tacitly assume that intraspecies differences do not really matter. This "counterfactual uniformity assumption" (Hirsch, 1963) seems to pervade much of contemporary psychology. Gudmund Smith, in a twin study, had difficulties with one of his twins: "What do you want to test me for? Haven't you tested my brother already?" (Smith, personal communication). Some psychologists remind one of this twin: "Since all rats behave more or less alike, test any rat. Behavioral variation between your subjects is mostly of your own doing. Be more rigorous, use better controls. Test one rat well, and you know them all." Actually, I may have overdrawn the picture. Today, under the influence of behavior-genetic research, the assumption of intraspecies uniformity appears to be giving way to a more sophisticated view. The fact that some behavioral variation among subjects, even human subjects, may be due to heredity seems more acceptable now than it was a decade ago (see Skinner, 1966).

While acknowledging hereditary variation within species, psychologists persist in assuming that such variation is "normally distributed" throughout a species. Consequently, psychologists believe that one can obtain species-typical values by testing an adequate sample of rats, or pigeons, or men. The sample mean is thought to indicate what is typical for the species as a whole. Many psychologists still adhere to this monotypic species concept. They mistakenly assume that the concept of species implies a single genetically homogeneous population. They are not sufficiently aware of the fact that most species consist of numerous genetically distinct Mendelian populations, and that such genetic distinctness often must entail behavioral distinctness.

Although today psychologists do not think in terms of behaviorally poly-typic species, this could change. There is nothing to suggest that, as a group, psychologists nurture strong beliefs about the behavioral unity of animal species. For example, they certainly are not upset by the fact that breeds of dogs, Mendelian populations kept separated by man, display different heritable "personalities." Thus, psychologists will not be upset by the idea that what man has done for dogs, geographic isolation often could have done for other species. Psychologists would accept a polytypic view of animal behavior if it were supported by the results of behavior-genetic studies of Mendelian populations.

Turesson, a botanist, has created the model after which population-genetic studies of behavior could be patterned. Turesson dealt, essentially, with one aspect of the nature-nurture problem (see Mayr, 1963, p. 351). Fellow bota-nists, believing in the genetic uniformity of plant species, attributed pheno-typic differences between local populations of plants to nurture, that is, soil conditions, climatic factors, and other environmental influences. For Turesson this was only one part of the story. He believed that local deviations from what botanists regarded to be the "species-type" were due, in part, to genetic differences between local populations of a plant species. A unitary "species-type" was an abstraction; species were polytypic. To test his views, Turesson transplanted samples of individual plants from various localities within the species range to a uniform garden or greenhouse environment. When the transplanted specimens remained different under the conditions of identical cultivation, Turesson had made his point. Behavior-genetic studies of popula-tions could be conducted in an analogous manner. Samples of individuals, drawn from different populations of an animal species, would have to be brought to the laboratory. From them a new generation would have to be bred and raised under uniform laboratory conditions. The behavior of this laboratory-raised generation could then be tested. Behavioral differences between animals derived from different populations would point to heritable behavioral differences between populations.

Populations for such studies would be selected using a simple geographic criterion: the relative spatial separation of populations. On the one hand, one would choose series of populations from adjacent territories. On the other hand, one would select populations that are separated by practically unsurmountable geographic barriers. The object would be to include in one's studies populations that exchange genes frequently, seldom, or not at all. One would expect to find that behavioral similarities between populations depended on the amount of gene flow between them. If one could show that this was actually the case, one would have done more than simply demonstrated that species tend to be behaviorally polytypic. One would have also demonstrated that behavioral polytypism is a function of the genetic segregation of a species.

The Belief in the Uniformity of Behavioral Adaptations

A corollary of the belief in species-typical behavior is the belief that behavioral adaptations are uniform throughout a species. Most psychologists would accept it as an article of faith that species are behaviorally adapted to their environment, but many psychologists would find it hard to illustrate the point with examples from their own work. As a rule, psychologists study animals in the laboratory, in an environment that bears little resemblance to the natural environment of the species. Often they study animals that are many generations removed from the wild state. What the environment of the ancestors of the laboratory animals looked like, and how these ancestors adapted to that environment, frequently becomes a matter of conjecture. Psychologists tend to simplify matters conceptually by believing in species-wide behavioral adaptations that fit a species to its "normal" environment. Some species are adapted to life in wooded areas, others only to life in environments in which water abounds. If a population of a species were found in the "wrong" environment, for example, a grassland species in a wooded area, most psychologists would attribute this to the "plasticity" of the species, meaning the capacity of its members to adapt to several environments, even wrong ones. Believing in the genetic uniformity of species, most psychologists would not consider the possibility that the population inhabiting the wrong environment differed genetically from populations inhabiting the "species-typical" environment. This obviously would be the possibility that population geneticists would consider first. Whether, in any specific instance, the view of psychologists or that of populationists should prevail can be decided only by careful population-genetic breeding experiments.

The case of "substrate races" (see Mayr, 1963) illustrates the issues involved. Many species are polytypic with regard to color, with local populations of such species usually matching the color of the substrate on which they live (e.g., Dice and Blossom, 1937). This is an obvious morphologic adaptation that protects the population by making its members less conspicuous to predators. In such cases it can be asked whether individuals are also behaviorally adapted to the substrate in the sense that they actively seek it when forcibly removed from it. If in experiments this is found to be the case, one can further question whether the observed behavioral adaptation is innate or acquired. The answer to this must be sought in breeding tests.

The phenomenon of substrate selection is only one of many that illustrate behavioral adaptations to local conditions. Most successful species occupy not one, but many different environments, each environment requiring special kinds of morphologic, physiologic, and behavioral adaptations. Indeed, the success of a species is measured by its ability to adapt to a variety of environments. This raises the question whether successful species colonize differing environments with highly adaptable but genetically identical populations, or whether such species consist of many genetically different populations, each adapted by natural selection to its own particular habitat. Here is a broad area for behavior-genetic research that has been barely touched upon by psychologists (see King, 1967; Mayr, 1963; Pittendrigh, 1958). Research in the area would affect the current belief in the species-wide uniformity of adaptive behavior; it would support, or it would negate that belief.

The Belief in the Relative Permanence of Species-Typical Behavior

The belief in species-typical behavior is often coupled with the belief in the relative immutability of behavior. Of course no psychologist doubts that behavior has evolved, but many psychologists believe that evolution occurs so slowly that for all practical purposes evolutionary changes can be disregarded. To be sure, psychologists are not unique in this respect. Not long ago this view was prevalent even among students of evolution (see Mayr, 1963). Much depends on one's understanding of evolution. The modern view is that evolutionary change consists of changes in the gene pool of a Mendelian population. If such changes are reflected in the behavior of individuals belonging to the population, behavioral evolution has occurred. If in the case of a quantitative behavioral trait, the change in the gene pool is expressed in a shift of the population mean from one value to another, the population has evolved behaviorally. Numerous experiments have shown that the mean expression of any behavioral trait can be changed in laboratory populations by selective breeding (see Fuller and Thompson, 1960). It would appear that the results of selective breeding experiments have a direct bearing on the matter in hand. They demonstrate that in small confined populations evolutionary changes of behavior can occur fast enough to be observable; behavior is by no means immutable. Yet, it may be argued that the conditions under which most selective breeding studies are being conducted are too artificial to be truly representative of what presumably happens in nature. In selective breeding experiments, human intervention tends to be very specific and rather "un-natural." Therefore an example from a series of studies using a more biologic approach shall be given here.

Lee Ehrman (1964a), an associate of Dobzhansky, was interested in the evolution of "behavioral isolating mechanisms" (see Dobzhansky, 1958; Spieth, 1958; Ehrman, 1964b). When two populations of a given species are separated from each other for a long time, they diverge genetically. They may diverge to a point of no return. When they meet again, they may be reproductively isolated for numerous reasons (Mayr, 1963, p. 92), one of them being psychologic in nature: the members of the two populations may not "recognize" each other, or they may not "attract" each other sexually. According to this theory, behavioral sexual isolation is a by-product of genetic divergence. There exists, however, another theory. When two populations meet after a period of isolation, the members of these populations may still mate but, because of genetic divergence, such interpopulation or heterogamic matings may result in reproductively inferior hybrids. Under these conditions, natural selection would favor in each population those individuals that are sexually attracted only by members of their own population. The frequency of genes carried by such individuals would increase in the gene pool of their population, while the frequency of genes carried by less sexually discriminating individuals would decrease. According to this theory, then, behavioral sexual isolation is in part due to natural selection; behavioral isolation evolves only after populations that had been allopatric become sympatric again. While the two theories are not mutually exclusive, Ehrman wanted to find out whether reproductive isolation can evolve solely as an incidental by-product of genetic divergence. She used for this purpose six experimental populations of *Drosophila pseudoobscura,* all originally descended from a single founder population. The populations, however, had been kept separated for almost 5 years during which they had been exposed to different temperatures. At this point, Ehrman carried out experimental interpopulation matings and found that, in fact, during the years of separation a degree of sexual isolation had evolved between some populations. In contacts between them, homogamic matings were more frequent than heterogamic matings. Prior to testing, the populations had not been exposed to each other. Thus, the behavioral isolation that had evolved could not be attributed to natural selection of homogamic mating behavior. Behavioral isolation had arisen as a by-product of genetic changes in populations that had become adapted to different environments, in particular to different temperatures.

That temperature affects the genetic constitution of fruitfly populations had been suggested by field observations and confirmed by laboratory experiments. In one representative study (see Sinnott et al., 1958, p. 260), it had been found that in a certain locality in California the chromosomal composition of *Drosophila pseudoobscura* populations remains constant during winter but changes during the warmer parts of the year. From March to June the percentage of *ST* chromosomes decreases and that of *CH* chromosomes increases; from July to October the trend reverses itself: the percentage of *ST* increases and that of *CH* decreases. One hypothesis to explain the phenomenon was that carriers of *ST* chromosomes were reproductively more fit during the hot season of the year than other chromosomal types. This hypothesis was tested in the laboratory using methods developed to perfection by Dobzhansky and his associates: Experimental populations of fruitflies are maintained in wooden "population cages" (see Sinnott et al., 1958, p. 260). A mixture of flies with known proportions of different chromosomes is introduced in such a cage and allowed to breed freely. The population is sampled at intervals and changes in the proportion of individuals with different chromosomes are determined by microscopic inspection of the salivary-gland cells of the flies. By this method it has been shown that in a population containing at first few *ST* but many *CH* chromosomes, the proportion of *ST* will rapidly rise while the proportions of *CH* will drop generation after generation, provided that the flies are kept at a temperature of 25°C. But at a temperature of 15°C the reproductive fitness of the carriers of either chromosome type is equal, thus their proportions remain constant; no changes occur, just as none occur during the winter in natural populations in California. The evolutionary changes brought about deliberately in the laboratory closely resemble those observed in natural populations of *D. pseudoobscura*. As environmental temperature changes with the seasons of the year, the chromosomal composition of fruitfly populations changes (see Sinnott et al., 1958, p. 276; Dobzhansky, 1952, p. 226). Temperature is clearly one of the selective agents that affects the genetic composition of fruitfly populations.

Returning to Ehrman's (1964a) experiment, we can now better appreciate the method she used. Her six experimental populations of *Drosophila pseudoobscura* descended from a single founder population. For almost 5 years the six populations were maintained at different temperatures: populations A and B were kept at 16°C, C and D at 25°C, and E and F at 27°C. This presumably affected their genetic constitutions differentially. The fact that gene flow between populations was prevented enhanced their genetic divergence. Ehrman waited almost 5 years before she tested the populations, so that their respective gene pools would have time to "settle down" and to gain internal cohesion. Ehrman's study illustrates well the intricate combination of field observation, laboratory experiment, and deduction from theory that is so characteristic of the research conducted by Dobzhansky and his associates. Ehrman's study is of particular interest to psychologists because it

shows that one need not select for a specific behavioral trait to affect behavior. In Ehrman's study, a nonspecific agent, temperature, changed the
genetic constitution of the population. This genetic change became manifest
in a behavioral phenotype, the sexual behavior of members of the population.

Evolutionary changes can be observed in nature and induced in the
laboratory. The replacement of white moths by melanic forms, the seasonal
changes in fruitfly populations, the changes in laboratory populations subjected to "rough" selective breeding or to the more subtle manipulations of
Ehrman, all illustrate evolutionary changes that occur fast enough to be
observed. The volume of research dealing with contemporary evolutionary
changes or morphologic traits is growing rapidly (see Mayr, 1963; Ford,
1964), but population-genetic research dealing with evolutionary changes
in behavior is only in its very beginnings. It has not yet completely dispelled
the belief in the relative immutability of behavior. As was already pointed
out, most psychologists today acknowledge the fact of hereditary variation
within species, but are still accustomed to think of such variation as distributed around a relatively fixed species-typical average value. That this
average itself can shift, and that such a shift constitutes evolutionary change,
is, among psychologists, an as yet unfamiliar notion. Psychologists continue
to think of evolution in terms of major qualitative steps rather than in terms
of small quantitative changes that eventually result in what appears to be
a qualitatively different behavior (see Manning, 1967, p. 52). More research dealing with the problem is needed. If nothing else, it would stimulate
continued discussion of these important problems.

Progress in the area of behavioral population genetics will depend on the
choice of animal species suitable for research. Much of what is known to date
has come from studies of insect populations, in particular populations of
Drosophila (see Ewing and Manning, 1967). Fruitflies are convenient in
many ways: They are small, so that sizable populations of them can be
easily maintained under seminatural conditions in the laboratory. Many
generations of fruitflies can be bred in a relatively short time. Local populations, geographic races, subspecies, allopatric and sympatric incipient species
and species of *Drosophila* exist in many regions of the world. Most important,
however, is the fact that *Drosophila* are very well known genetically. From
the point of view of many psychologists, the major shortcoming of fruitflies
as subjects for behavioral research is the fact that they are insects. In the
search for higher animals that would share certain of the advantages of
fruitflies, some investigators have turned to deer mice, and important work
has been done with this animal (see King, 1967). Here, I would like to
suggest that wild *Mus musculus,* the relatives of our laboratory mice, may
in many respects be the best available candidates for population-genetic
studies of behavior. In support of this suggestion, the following section
dealing with the races and world distribution of wild *Mus musculus* will be
presented.

WILD MICE *MUS MUSCULUS*

Every year all over the world millions of laboratory mice are carefully bred, raised, and studied by thousands of scientists; next to man, the laboratory mouse is the scientifically best known mammal. Every year in barns and fields uncounted more millions of mice, wild members of the species *Mus musculus,* reproduce, lead busy lives, and die, but relatively little scientific information exists about them. Until fairly recently the taxonomy of *Mus musculus* was in a state of chaos. Various authors listed well over 100 names of reputed "good species," subspecies, and local varieties. More recent authors, however, according to Grüneberg, "agree that all these names refer to a single species of world-wide distribution. No sterility barriers occur between the members of this group" (Grüneberg, 1952, p. 3).

The taxonomy of *Mus musculus* was considerably simplified by the investigations of E. Schwarz and H. Schwarz (1943, 1945), and their views will be presented below. First, however, it will be necessary to settle some questions of terminology: (1) the term mouse has been used to refer to many species. Here the term mouse will be used only to refer to *musculus,* a species belonging to the genus *Mus.* (2) We will distinguish between wild and domesticated mice. Wild mice are unconfined animals over whose reproduction man does not have any control. Domesticated mice are animals that have lived in captivity for many generations, and whose breeding was controlled by man. (3) We will carefully distinguish between domestic and domesticated mice. The term domestic mouse, or house mouse, is a synonym for *Mus musculus;* it refers to both wild and domesticated mice. The term domesticated mouse points to a history of breeding in captivity. (4) We will recognize several types of wild mice: (a) aboriginal mice: these are *Mus musculus* that, as far as we know, never associated with man; (b) commensal mice: these are the mice that followed man around the globe as scavengers; (c) feral mice: these are mice that once were commensals of man but reverted to a more feral existence.

Aboriginal Mice

Aboriginal *M. musculus* are mice that to our best knowledge have never lived in close association with man. They are genuinely wild in the sense this word is usually used. Schwarz and Schwarz (1943) distinguished four aboriginal races or subspecies of *Mus musculus* Linnaeus:

1. *M. m. wagneri* Eversmann. This mouse is indigenous to the dry areas of Central Asia from the east bank of the Volga River to the Yellow Sea north to about 42° to 44° northern latitude, and south to about 38° to 36° northern latitude.
2. *M. m. spicilegus* Petényi. This mouse is found in southern and south

central Russia west of the Volga River, Bulgaria, Roumania, Hungary, and possibly Austria. The western and northern limits of the range of this mouse have not been determined, but it is known that *M. m. spicilegus* does not occur south of the Caucasus Mountains.

3. *M. m. manchu* Thomas. This aboriginal mouse is known to inhabit the area of Yalu and upper Sungari rivers in southeast Manchuria. It was also found on the islands of Kyushu, Honshu, and Hokkaido of Japan, but the exact limits of its range were not known to Schwarz and Schwarz in 1943.

4. *M. m. spretus* Lataste. The aboriginal *M. m. spretus* inhabits the Iberian Peninsula, France immediately north of the Pyrenees, the Balearic Islands, and Northwest Africa, north of the Atlas mountains. The African *M. m. spretus* are the only aboriginal mice found outside of the Eurasian continent.

Commensal Mice

The first three races of aboriginal mice mentioned above have given rise to commensal forms; the fourth, *M. m. spretus,* has not.

1a. Commensal descendants of *M. m. wagneri:* Mediterranean mice. Descendants of aboriginal *M. m. wagneri* were carried by man from Central Asia into Persia, Iraq, Syria, Palestine, and from there into North Africa as far as the Atlantic. From North Africa these commensals reached Italy and Spain where they are known as *M. m. brevirostris.* We will refer to them as Mediterranean mice.

1b. Commensal descendants of *M. m. wagneri:* West European mice. The Mediterranean mice, *M. m. brevirostris,* spread into northwestern Europe; into Germany as far east as the Elbe River; into Switzerland, France, Belgium, Holland, and the western part of Denmark and Norway; into the British Isles, including the Orkney and Shetland groups, the Hebrides, Faroes, and Iceland. These descendants of Mediterranean mice are known as *M. musculus domesticus;* we will refer to them as West European mice.

1c. Commensal descendants of *M. m. wagneri* in America. The Mediterranean mouse, *M. m. brevirostris,* was carried by settlers into Latin America and those southern parts of the United States which formerly were under Spanish rule. The West European mouse, *M. m. domesticus,* was brought by settlers into North America, from the northern part of the Central States to Alaska, and also Hawaii. It was also introduced into those southern regions of North America that were settled by the French. Thus, in general, we find in America the Mediterranean mouse, *M. m. brevirostris,* in the South, and the West European mouse, *M. m. domesticus,* in the North. In many parts of North America the ranges of the Mediterranean and West European mice overlap. Let us note that all American mice descended from *M. m. wagneri,* the aboriginal mouse of Central Asia.

All commensal mice of Africa also descended from *M. m. wagneri,* as did the commensal mice of southern Asia, South China, Indochina, the Malay countries, Australia, and Polynesia.

2. Commensal descendants of *M. m. spicilegus:* East European mice. Eastern Europe and most of Russia was settled by *M. m. musculus,* a commensal descendant of *M. m. spicilegus,* the aboriginal mouse of southeastern Russia. *M. m. musculus,* to be called here East European mouse, is the commensal mouse of Austria, Germany up to the Elbe River, the eastern part of Denmark and Norway, Sweden, Poland, the Baltic countries, and most of northern Russia up to the White Sea. Its range is expanding eastwards along the Siberian railroad. At the western borders of its range the East European mouse encounters populations of the West European mouse (see Mayr, 1963, p. 379). In its southern range the East European mouse is sympatric with the aboriginal *M. m. spicilegus* from which it descended and which it replaces around human habitations. The East European house mouse has not been introduced into America (Schwarz, 1945).

3. Commensal descendants of *M. m. manchu:* Japanese mice. The commensal mice of Japan, *M. m. molossinus,* belong here. The Japanese mouse occurs also on the Kurile and Iki islands, and in Korea.

The fourth group of aboriginal mice, *M. m. spretus,* as was mentioned already, did not develop commensal forms.

The conclusions of Schwarz and Schwarz (1943) regarding the wild and commensal stocks of *Mus musculus* were based on an examination of many thousands of specimens in various museum collections, and on a critical evaluation of the existing literature. Schwarz and Schwarz have developed various morphologic criteria to distinguish between aboriginal and commensal mice. For example, according to these authors, characteristics common to all aboriginal forms include medium size and a tail that is always shorter than the length of the head and body. In all of them, the lower side of the animal is white, with or without dark underfur, and there is a distinct line of demarcation between the upper and lower sides. In all commensals certain trends of evolution are apparent. The tail becomes longer than the head and body. Color patterns change: the demarcation line between the upper and lower parts of the body becomes less distinct; the more recessive alleles of the agouti series become more common. And, "the skull of the commensals shows the well known changes characteristic of captive and domestic mammals, i.e., a reduction in bulk of the face, and of the size of the molar teeth; in certain highly advanced types there is a tendency to have the molar series shortened by the suppression of the last molar" (Schwarz and Schwarz,

1943, p. 60). Similar morphologic criteria are given by the authors to distinguish between the large number of mouse populations described in this paper.

Feral Mice

Populations of feral mice exist in many parts of the world, and numerous papers describing such populations have been published. For example, Koford (1968) found *Mus musculus* in desert areas of Peru. Breakey (1963) studied feral mice in the salt marshes of San Francisco Bay. Baker (1946) described feral mice inhabiting the open grass and brush land that is characteristic of the volcanic parts of Guam. Engels (1948) observed feral mice living in the dunes of some North Carolina coastal islands. Jameson (1898) communicated in 1897 to the Linnaean Society of London "a probable case of protective coloration in the house mouse." Jameson reported on mice he had found living in burrows in the sandhills on the North Bull in Dublin Bay. Feral populations of mice have been described in many more publications, but the small sample given here is sufficient. It shows that feral mice can be found in various regions of this world, and particularly in places that resemble the area inhabited by aboriginal mice. All aboriginal mice "are typical dry area animals occurring in savannahs, steppes, and even in desert country" (Schwarz and Schwarz, 1943, p. 71). Commensal mice can return to a feral existence where the climate is relatively dry.

Domesticated Mice

Although this article is devoted to wild mice, a few words about domesticated mice cannot be avoided; certain misconceptions about the origin of our inbred strains of laboratory mice must be briefly discussed. Sometimes, explicitly or implicitly, it is assumed that our inbred strains descend from individuals drawn at random from a natural population of mice (see Bruell, 1967; McClearn, 1967). Certain other assumptions follow. Stated in simple terms, this line of reasoning is often followed: Suppose a wild base population consisted of individuals carrying alleles *A* and *a* of a gene in homozygous and heterozygous combinations. Individuals are drawn from that wild base population, and they become the founders of a number of inbred strains. With inbreeding, heterozygotes disappear and homozygous *AA* and *aa* strains result. As a group, the strains will carry all the allelic forms of the *A* gene contained in the gene pool of the base population, but each strain carries only one allele of that gene. If we now, so the reasoning goes, instituted random breeding among our inbred strains, we would again obtain a population harboring all the alleles of the *A* gene in homozygous and heterozygous combinations. Thus, in an approximate way, we would have reconstituted the natural base population. This theoretical formulation is behind many diallel studies in which a number of inbred strains are completely intercrossed (see Bruell, 1964, 1967); the theory also underlies the use for research of four-way or eight-way crosses between strains of mice (see McClearn, 1967, p. 319; Green, 1966). Unfortunately, the basic assumption upon which this reasoning rests is wrong.

The mice that founded our inbred strains did not descend from any *one* natural Mendelian population. Their background was quite heterogeneous. Most of them were bought from pet dealers or were obtained from mouse fanciers (see Jay, 1963; Bruell, 1964; Staats, 1966). In the 19th century mice were bred by numerous mouse fanciers (Davies, 1896), and in Europe and the Orient the tradition of keeping color mutants reaches back into antiquity (Keeler, 1931). Our laboratory mice very probably descended from these fancy mice, that is, mice that had been domesticated for a very long time. The *BALB/c* strain was bred from animals bought from an "Ohio dealer" (Staats, 1966). The family of *C57* strains was derived from color mutants C. Little bought from a Miss Lathrop, a mouse fancier of Granby, Massachusetts (Staats, 1966). And so it goes for the other strains. The notion of a "natural base population from which the founders of our inbred strains were drawn" is a fiction. Thus, whatever the number of strains we would care to cross, the population we would produce in this way would not even in rough approximation reconstruct a natural population. Inbred and crossbred populations of laboratory mice, valuable as they may be for certain kinds of research (see McClearn, 1967), cannot serve as models for natural populations of mice.

POPULATION GENETICS OF MOUSE BEHAVIOR

The domesticated laboratory-bred mouse has played and continues to play an important role in mammalian genetics. For basic genetic research the domesticated mouse offers many advantages: It is prolific; it breeds well in the laboratory, and up to four generations can be obtained within a year. It is among the smallest mammalian species known; many individual mice can be maintained in relatively few square feet of laboratory space. It can be easily handled; during centuries of domestication all wildness has been bred out of it, presumably by a process of selection. The laboratory mouse is tame by nature, it need not be tamed. What made domesticated mice, however, particularly attractive to early geneticists was the fact that many coat-color mutations of the mouse were already well known at the turn of the century (Davies, 1896; Keeler, 1931). Even before the rediscovery of Mendel's work, investigators studied the inheritance of coat colors of mice (von Guaita, 1898). Their interpretations, of course, missed the mark, but immediately after Mendel's work became known, they were corrected by the first "Mendelians" (Bateson, 1903). From then on genetic research on the mouse never ceased. The process was self-accelerating. The better known the mouse became genetically, the easier it was to fit new findings into the broad scheme that was developing. Today, the inbred mouse of the laboratory is the genetically best known mammal, man excepted.

What domesticated mice have become for mammalian genetics, the wild *Mus musculus* could become for population genetics: the most important research animal. We have briefly reviewed what is known about the races and world distribution of mice. The species is ubiquitous, it is found wherever man lives. It consists of several distinguishable subspecies, many geographic races, and uncounted local demes and isolates. In addition, members of the species have developed various degrees of association with man: some avoid man, others are his commensals, still others have returned to a feral existence, and many are domesticated, live in captivity, and are completely dependent on man. *Mus musculus* is truly a polytypic species; few other species display as broad a spectrum of forms. This in itself would recommend wild *Mus musculus* for population-genetic research. The fact that wild mice share all the advantages of laboratory mice makes them doubly attractive. They reproduce fast, they are small, and most of the genetic knowledge gained in dealing with their domesticated relatives presumably applies to them. Mouse populations could be used to study most of the problems of behavioral population genetics, but they have seldom if at all been used for that purpose. Thus, in this section on "Population Genetics of Mouse Behavior" we will not be able to offer much in the way of substantial information; we will be limited to the listing of some possible research projects. Before we proceed, however, two more general topics must be taken up: We must speak about the necessity of specifying the mouse populations one works with, and we must discuss briefly the question of test selection.

The Systematic Position of Mouse Populations

In zoologic work it is customary to specify precisely the systematic position of the animals one studies. This custom has not yet been fully adopted by psychologists; in many cases psychologists deem it sufficient to note that cats, dogs, pigeons, or rats were used. But the custom is spreading, and in behavior-genetic research one now routinely specifies the strains of animals one works with. In population-genetic studies of mouse behavior it will be important to adopt an analogous procedure. The species *Mus musculus* is divided into many populations of varying interrelatedness. A pictorial representation of this family of populations would assume the form of a genealogic tree akin to the phylogenetic tree of evolutionists, or, more precisely, it would be a detailed representation of one branch of that tree. The branch would divide into four minor branches, representing the four subspecies or races recognized by Schwarz and Schwarz (1943). Each of these branches would divide again, and the littlest twigs on the branch would indicate small isolates of mice, perhaps the populations of mice on remote farms, or even single farm buildings (see Anderson, 1964). The relative position on the genealogic tree of any two mouse populations one wanted to compare would have to be considered and noted; of course, this is what evolutionists do when they compare species, genera, or families of animals. The classification of mouse populations given by Schwarz and Schwarz (1943), or any other system yet to be devised, would have to be used. Whatever the system, some systematic approach will be necessary for an orderly development of the behavior-genetic study of mouse populations.

Natural and Arbitrary Comparisons

There are innumerable ways in which mouse populations could differ behaviorally. Thus the comparing of populations presupposes a rational method of test selection. Mendelian populations can differ because they were exposed to different selection pressures, or they can differ because of chance. Whenever we have reason to believe that populations differ because of natural selection, we would want to compare them using "natural" tests, that is, tests that would measure the difference brought about by natural selection. For example, mouse populations living in a dry desert-like country presumably have evolved behavioral adjustments to their desert environment (see Koford, 1968). Meaningful natural comparisons of desert mice with those inhabiting regions amply supplied with water would include tests of drinking behavior, water conservation, or thirst motivation. In some cases, however, we will have no knowledge of the differing selection pressures to which the populations we compare have been exposed. In still other cases, we will have reasons to believe that the populations differ not because of "natural" causes but because of chance. In these instances, we will have to compare the populations using arbitrary measures. There is nothing wrong with that. One of the hypotheses we are testing is that Mendelian populations tend to diverge genetically. The gene pools of the populations become distinct, and their distinctness is reflected in all-pervasive phenotypic changes of the population. Thus, in the extreme case, any measure should do. A case in point is a measure used successfully by Ford (1964) in studies of populations of *Maniola jurtina,* a common butterfly of meadows and hayfields. This butterfly is polymorphic for spots it carries on the underside of its hind-wings: the spots may be absent, or present in any number up to five. Ford compared populations of *M. jurtina* by comparing the incidence of the various spot-morphs in each population. Ford had no specific hypotheses regarding the benefits, if any, the butterfly derived from the number of spots it carried. He was simply guided by the belief that polymorphism persists in populations only if it is adaptive, that adaptively neutral traits do not exist (see Mayr, 1963, p. 161), and that any trait that varies within and between populations reveals something about their genetic structure. From this point of view, any conveniently measured heritable phenotype will do for population-genetic research.

The use of measures that are both natural and convenient would be most desirable. However, it should be noted that arbitrary but convenient measures have been used successfully in much of population-genetic research. Spot-patterns in butterflies, banding patterns in snails, chromosome structures in *Drosophila,* and blood groups in man are only a few examples of traits that have been employed to characterize populations, although, at the time the traits were chosen for research, their adaptive significance was unknown (see Mayr, 1963, p. 161). Not only qualitative but also quantitative characters of unknown adaptive significance have been used. For example, Stalker and Carson (1948) compared thorax length, femur length, wing width, and wing length of *Drosophila robusta* populations occurring at different altitudes in the Great Smoky Mountains, Tennessee. Thus, in the absence of specific hypotheses about possible natural differences between the populations one compares, the past experience of population geneticists would seem to justify the use of any available easily administered test. This approach to test selection has also worked well for behavioral geneticists who, not knowing what to expect, compared strains of animals with any test at hand. The same pragmatic way of choosing tests for the behavioral comparison of populations could be used by students of behavioral population genetics. The rule should be to use "natural" measures when possible, and arbitrary measures when necessary.

Comparisons of Isolates, Demes, and Races

One of the ambitious goals of the population-genetic study of the mouse could be to obtain "behavioral profiles" of the various races and subraces of mice. Today we can fairly well characterize the behavior of each of the many breeds of dogs or strains of mice. Eventually, it should be possible to characterize in a similar way the behavior of the races of mice, if indeed the races differ behaviorally. But this goal lies in the distant future, and the population-genetic study of mice will have to start with less ambitious projects. One such project would compare the behavior of small isolated populations of mice. During the summers of 1962 and 1963, Anderson (1964) studied populations of mice on seven farms in the vicinity of Calgary, Alberta. His data suggest that mouse populations inhabiting even adjacent farm buildings remain reproductively isolated; immigration of mice into established populations appears to be rare (Selander and Yang, 1970; Crowcroft, 1966). Anderson found that adjacent groups of mice differed in genetic composition and that these differences persisted for 2 years despite gross disturbances incurred when males were removed for testing during the first year of Anderson's 2-year study. Anderson compared the populations with regard to certain genes they carried, but it would be quite feasible to study behavioral differences between such populations. The choice of behavioral tests used for the purpose would be arbitrary, since, obviously, one would have no specific hypotheses as to the nature of the differences one would encounter. Small isolated populations tend to be established by a few founders. The genes carried by the founders constitute a minute sample drawn from the gene pool of their base population and are subject to gross errors of sampling (Anderson, 1964). The latter fact would lead one to expect behavioral differences between isolates. But since chance governs the foundation of isolates, one cannot predict what differences will be observed. The situation is strictly analogous to that occurring when inbred strains are formed. The progenitors of the strains are chosen at random, a fact leading to unpredictable behavioral differences between strains.

From the study of panmictic isolates, one would go on to the study of demes, that is, Mendelian populations of mice in well-defined localities. Particularly useful for research would be the relatively isolated demes of mice existing on small islands (see Berry, 1964, 1967; Engels, 1948; Jameson, 1898). Such demes would differ in important ways from the isolates encountered in farm buildings or human dwellings. Although island populations presumably also often derive from a small number of founders, many such populations are given more time to diversify genetically than the very impermanent and constantly threatened populations existing close to man. Since, for all practical purposes, island populations are not exposed to gene flow from other mouse populations, they can evolve a cohesive gene pool. Given sufficient time, they can even develop special behavioral adaptations to their island environment (see Jameson, 1898). Mouse populations on uninhabited islands would be particularly desirable for population-genetic research since they could be used for experimental studies, as will be discussed later.

Comparative studies of isolates and demes would prepare one for the more ambitious undertaking of comparative research dealing with races of mice. Truly thorough racial studies of mice would perhaps start with samples of animals from each of the major races and subraces described by Schwarz and Schwarz (1943). But, on a more realistic scale, one would be satisfied with comparative studies of mice of West European and Mediterranean origin found in northern and southern parts of the United States (Schwarz, 1945). In Europe the choice of mice of different racial origin would be larger. While all mice in America belong to subraces of *Mus musculus wagneri,* populations of mice belonging to all major races, except *M. m. manchu,* exist in Europe. The proper, representative sampling of mouse populations for comparative studies of mouse races would pose a major problem. Even if, with the help of taxonomists, populations of mice could be clearly identified as belonging to one or another race, misinterpretations of data could readily occur. Samples from many mouse populations belonging to one racial group will have to be compared with an equally representative sample of mouse populations belonging to another racial group before any statements about racial differences are made. If too few population samples were included in one's study, one could easily mistake local behavioral adaptations of the populations for behavioral differences between races.

Behavioral Adaptations of Mouse Populations

Mouse populations could be used to study behavioral adaptations to the environment. Many such adaptations come to mind. We have spoken already about behavioral adaptations to desert-like environments (see Koford, 1968). Adaptations related to temperature regulation, such as nest building, could differ between mouse populations living in different climates (see King et al., 1964). Adaptations to the substrate could also exist. Jameson (1898), in addition to noting the "protective coloration" of mice inhabiting the sandhills of an island in Dublin Bay, reported that these mice had "acquired" the habit of digging deep burrows "to enable them the better to escape from their enemies" (Jameson, 1898, p. 468). Investigators conducting population-genetic studies will think of many other possible behavioral adaptations of populations, and we need not go into more detail. Two special kinds of adaptations, however, should be mentioned: the adaptation for life as a commensal of man, and the adaptation to life at the border joining the territories of two distinct racial groups.

The mouse followed man around the globe, a fact that in itself points to behavioral adaptations. The process which moves an animal gradually into dependence on man constitutes a major evolutionary phenomenon (see Hale, 1962). Darwin recognized it as such and accorded it a central place in his writings. The wild ancestors of most of our domesticated animals are extinct, but in the case of the mouse we possess a complete series of populations encompassing wild, commensal, and domesticated forms. This offers many opportunities for behavioral research. Interesting comparative studies could be conducted with aboriginal and commensal populations of mice that are sympatric. For example, in Spain and Northwest Africa commensal *Mus musculus breviostris* live in the same region as the aboriginal *M.m. spretus*. In many East European countries the aboriginal *M.m. spicilegus* is sympatric with the commensal *M.m. musculus*. The commensal animals must have evolved behavioral characteristics that permit them to live close to man. One can assume that they fear man less than the aboriginal mice inhabiting the same areas. On the other hand, one might expect that they have lost some behavioral adjustments that permit aboriginal forms to survive away from the shelter provided by human habitations. Commensal animals can breed and raise young in the homes of man and also in captivity. During the past year this writer has caught hundreds of commensal mice in Ohio and bred them in the same see-through plastic cages laboratory mice are being bred in. No special hiding places, such as turned around flowerpots or wooden breeding boxes, were provided. All animals bred well, numerous large litters were raised, and cannibalism, so common among inbred mice, was virtually absent. It would be interesting to find out whether aboriginal mice would breed equally well in captivity. Inability to breed close to man may be one factor keeping wild mice away from human habitation, fear of man may be another, but there must be other behavioral differences between aboriginal and commensal mice that should be studied. Parenthetically it should be noted that for the sake of simplicity we have spoken here in a rather "typologic" manner of aboriginal mice on the one hand, and commensal mice on the other. But it should be obvious that different populations of mice have retained various degrees of wildness, or reached various degrees of commensalism. This makes *Mus musculus* particularly attractive for the study of the behavioral adaptations that lead to the close association of species.

Aboriginal mice are sympatric with commensal mice, but their populations often remain segregated. In Spain, according to Schwarz and Schwarz (1943, p. 71), "the local wild form (*M. m. spretus*) has never developed a commensal because, at the time when human settlements became established, the immigrants themselves brought with them their own commensals, viz. *M. m. praetextus,* and *M. m. brevirostris.*" Here we have a case of populations that, after a period of allopatric evolution, became sympatric again. But now some form of isolating mechanism keeps them separated. The nature of the isolating mechanism should be explored. One may wonder whether in this case population hybrids were less fit than either of the parental stocks, and whether, in line with Dobzhansky's hypothesis, this favored the development of behavioral barriers to interpopulation matings. Similar questions could be asked about *M. m. domesticus* and *M. m. musculus.* Populations of these two commensals meet on a broad front in Europe, but only a narrow "hybrid belt" has developed at the place of contact between the groups. Mayr (1963, p. 378; see also Selander and Yang, 1969) has pointed to the situation:

> "One of the least understood aspects of hybrid belts is their width: some are very wide, others amazingly narrow. In hybrid belts that must have existed for thousands of years this narrowness is a great puzzle. One would expect either that reproductive isolation would be acquired as a result of an inferiority of the hybrids, or that gradual infiltration of the hybridizing genes would steadily widen the hybrid belt until it occupied the greater part of the ranges of the hybridizing populations.... In the house mouse *Mus musculus* in central Denmark, a dark-bellied southern race (*domesticus*)

meets a light-bellied northern race (*musculus*). The width of the total area of introgression is only 50 kilometers and the zone of truly intermediate populations is only a few kilometers wide."

In North America, the Mediterranean mouse (*M. m. brevirostris*) meets the West European mouse (*M. m. domesticus*) where territories formerly owned by Spain border former French and British possessions. What happens when Mediterranean and West European mouse populations meet has remained largely unexplored. In Spain, in Central Europe, in America, and in many other places where mouse populations of differing racial origins meet, there exist numerous opportunities for the study of behavioral isolating mechanisms that keep mammalian populations reproductively segregated.

Longitudinal Studies

Mendelian populations are dynamic systems that grow and change like organisms. One can study them as one studies groups of individuals: cross-sectionally or longitudinally. Several Mendelian populations can be observed and compared; this would be analogous to a cross-sectional study of a group of individuals. Or a single Mendelian population can be studied at successive points in time; this would be comparable to a longitudinal study of a developing organism. The studies considered so far were mainly of a "cross-sectional" nature. But observational and experimental longitudinal studies of mouse populations would also be entirely feasible.

One goal of longitudinal studies would be to learn about the behavioral stability of populations. We have mentioned Anderson's (1964) 2-year observational study of mouse populations inhabiting farm buildings. Using the relative frequencies of certain alleles as his measure, Anderson ascertainted that the genetic composition of the populations he observed did not change from one summer to another. Analogous studies, using behavioral measures, have not been conducted, and little is known about the behavioral stability of natural populations. To be concrete, consider the mouse population on Skokholm, a small island at the southwestern tip of Wales. Berry (1967) reported that the size of this population fluctuates between about 200 or 300 individuals in the spring to between 2,000 and 10,000 in the autumn; winter mortality is obviously very high. Suppose one studied for several years samples of Skokholm mice in fall and in spring, establishing the mean value and variance for some behavioral trait. The mean value could remain stable; or it could change predictably in one direction, for example, being higher in spring than it was in fall; or it could fluctuate unpredictably. Let us dismiss the last case as trivial: unpredictable fluctuations would point to a trait affected by forces other than the rigors of winter, to a poorly devised behavioral test, to poor sampling, or to all these things.

Let us consider the case of directional changes first: one could assume that a population that responded to recurring selection pressures, such as seasonal changes in temperature, with unidirectional changes of behavior, would become better and better adapted to its environment. We usually associate such directional changes with evolution. However, results of numerous selective breeding experiments indicate that this notion is too simple, and that it must be qualified. When, in breeding experiments, directional changes are forced upon a population too fast, the fitness of the population tends to decrease, and the selected lines often die out. When selection is relaxed, selected lines frequently return to the mean value they displayed before selection started. They thus manifest what Lerner (1954) has called "genetic homeostasis," or the tendency of a "population to equilibrate its genetic composition and to resist sudden changes" (see Mayr, 1963, p. 288). If, in our hypothetical case, we actually observed directional changes in the behavior of Skokholm mice following a severe winter, we would be interested to find out whether the population would return to a base value during the summer. In the long run, behavioral evolution involves change, but it presumably involves the development of homeostatic mechanisms that guarantee behavioral stability in the face of short-term changes in environmental conditions.

Turning to behavioral stability, we can think of several ways in which it could come about: it could be due to random, disruptive, or stabilizing selection (Mather, 1953). Winter could decimate the mouse population at random, that is, mice could succumb to winter conditions no matter what their behavior. If this were so, the mean value and the variance of behavioral traits would remain the same. Or, winter could eliminate those mice that scored close to the center of the distribution on our behavioral test, "disrupting" the distribution. In this case, the mean value of the trait would remain unchanged but its variance would increase. Finally, mice scoring at the extremes of the distribution could perish during the winter, while the "mediocre" and the "ordinary," survive. The latter case, referred to as "stabilizing selection," has been often observed in studies involving morphologic characters (see Mayr, 1963, p. 282). Berry (1967, p. 80) observed it in Skokholm mice: selection caused a reduction in skeletal variance of 20 to 30 percent over the winter. What happens to behavior in natural populations we do not know. Are such populations behaviorally stable or unstable? Is stability achieved by random, disruptive, or stabilizing selection; is instability kept in check by homeostatic mechanisms? These questions could be studied using natural populations of mice.

Small islands with resident mouse populations, or uninhabited islands on which mouse populations could be established, would be very useful for longitudinal population-genetic studies of mouse behavior, particularly if one wanted to conduct experimental studies. Many types of experimental research would be of great interest. For example, a group of mice, randomly assembled from several natural populations, could be released on an island. In such a heterogeneous group, one would expect, at first, that sample means for various behavioral traits would fluctuate. In time, however, as the gene pool of the population gained cohesion, sample means should stabilize. Along similar lines, one would expect a certain degree of behavioral stability in established mouse populations, but one should be able to upset such stability by "planned gene flow," that is, by adding "immigrants" to a resident population. Or, one could settle an island with two behaviorally distinct populations, each transplanted from its home territory to different areas of the island. Under such favorable conditions it is quite possible that the development of a hybrid belt, or even the emergence of behavioral isolating mechanisms could be observed.

It would be very easy to add many pages to this list of suggestions for research on population genetics of mouse behavior. But I will stop at this point. E. B. Ford, one of the founders of experimental population genetics, wrote in 1962 in the preface to his book on "Ecological Genetics" the memorable words: "This book was planned in 1928, and in considerable detail. At that period I believed it would be necessary for myself and others to work for a quarter of a century before it could be written. I was overoptimistic; more than thirty years were in fact needed" (Ford, 1964). I will benefit from Ford's experience and hazard no guess as to how long it will take to carry out some of the projects that I have suggested here. But I will venture one prediction: in the next decade behavioral population genetics will move into the position now occupied by behavioral genetics. It will assume that position because of its obvious implications for the issue of racial differences in human behavior (see Spuhler and Lindzey, 1967) and the wider problem of behavioral evolution. What animals will be used in prototypic studies of behavioral population genetics cannot be predicted with certainty, but is is clear that wild *Mus musculus* would serve the purpose well.

REFERENCES

Anderson, P. K. 1964. Lethal alleles in Mus musculus: Local distribution and evidence for isolation of demes. Science, 145:177–178.

Baker, R. H. 1946. A study of rodent populations on Guam, Mariana Islands. Ecol. Monogr., 16:393–408.

Bateson, W. 1903. The present state of knowledge of color-heredity in mice and rats. Proc. Zool. Soc. 2:71–99.

Berry, R. J. 1964. The evolution of an island population of the house mouse. Evolution, 18:468–483.

——— 1967. Genetical changes in mice and men. Eugen. Rev., 59:78–96.

Breakey, D. R. 1963. The breeding season and age structure of feral house mouse populations near San Francisco Bay, California. J. Mammal., 44:153–168.

Bruell, J. H. 1964. Inheritance of behavioral and physiological characters of mice and the problem of heterosis. Amer. Zool., 4:125–138.

——— 1967. Behavioral heterosis. *In* Hirsch, J., ed., Behavior-Genetic Analysis, New York, McGraw-Hill Book Co., pp. 270–286.

Crowcroft, P. 1966. Mice All Over, Chester Springs, Pa., Dufour Editions.

Davies, C. J. 1896. Fancy Mice . . . Illustrated, 4th ed. By an old fancier. With criticisms . . . on crossing, etc. . . . by Dr. C. Blake . . . and an appendix on fancy mice for exhibition by Mrs. Leslie Williams. London, Upcott Gill.

Dice, L. R., and P. M. Blossom. 1937. Studies of mammalian ecology in southwestern North America. Publications of the Carnegie Institute, pp. 458–485.

Dobzhansky, T. 1955. A review of some fundamental concepts and problems of population genetics. Cold Spring Harbor Symp. Quant. Biol., 20:1–15.

——— 1958. Species after Darwin. *In* Barnett, S. A., ed., A Century of Darwin. Cambridge, Harvard University Press, pp. 19–55.

——— 1959. Variation and evolution. Proc. Amer. Philosoph. Soc., 103:252–263.

——— 1962. Mankind Evolving: The Evolution of the Human Species, New Haven, Yale University Press.

——— 1967. On types, genotypes, and the genetic diversity in populations. *In* Spuhler, J. N., ed., Genetic Diversity and Human Behavior, Chicago, Aldine Publishing Co., pp. 1–18.

Ehrman, L. 1964a. Genetic divergence in M. Vetukhiv's experimental populations of Drosophila pseudoobscura. 1. Rudiments of sexual isolation. Genet. Res., 5:150–157.

——— 1946b. Courtship and mating behavior as a reproductive isolating mechanism in Drosophila. Amer. Zool., 4:147–153.

Engels, W. L. 1948. White-bellied house mice on some North Carolina coastal islands. J. Hered., 39:94–96.

Ewing, A. W., and A. Manning. 1967. The evolution and genetics of insect behavior. Ann. Rev. Entomol., 12:471–494.

Ford, E. B. 1964. Ecological Genetics, New York, John Wiley and Sons, Inc.

Fuller, J. L., and W. R. Thompson, 1960. Behavior Genetics, New York, John Wiley and Sons, Inc.

Green, E. L. 1966. Breeding systems. *In* Green, E. L., ed., Biology of the Laboratory Mouse, 2nd ed., New York, McGraw-Hill Book Co., pp. 11–22.

Grüneberg, H. 1952. The Genetics of the Mouse, 2nd ed., The Hague, Nijhoff.

Hale, E. B. 1962. Domestication and the evolution of behavior. *In* Hafez, E. S. E., ed., The Behavior of Domestic Animals, Baltimore, The Williams and Wilkins Co., pp. 21–53.

Hirsch, J. 1963. Behavior genetics and individuality understood. Science, 142:1436–1442.

Jameson, H. L. 1898. On a probable case of protective coloration in the house-mouse. (Mus musculus, Linn.) J. Linn. Soc. (Zool.), 26:465–473.

Jay, G. E., Jr. 1963. Genetic strains and stocks. *In* Burdette, W. J., ed., Methodology in Mamalian Genetics, San Francisco, Holden-Day, Inc., pp. 83–123.

Keeler, C. E. 1931. The Laboratory Mouse: Its Origin, Heredity, and Culture, Cambridge, Harvard University Press.

Kettlewell, H. B. D. 1959. Darwin's missing evidence. Sci. Am., March:48–53.

———— 1961. The phenomenon of industrial melanism in the Lepidoptera. Ann. Rev. Entomol., 6:245–262.

King. J. A. 1967. Behavioral modification of the gene pool. *In* Hirsch, J., ed., Behavior-Genetic Analysis, New York, McGraw-Hill Book Co., pp. 22–43.

———— D. Maas, and R. G. Weisman. 1964. Geographic variation in nest size among species of Peromyscus. Evolution, 18:230–234.

Koford, C. B. 1968. Peruvian desert mice: Water independence, competition and breeding cycle near the equator. Science, 160:552–553.

Lerner, I. M. 1954. Genetic Homeostasis, Edinburgh, Oliver and Boyd.

Manning, A. 1967. Genes and the evolution of insect behavior. *In* Hirsch, J., ed., Behavior-Genetic Analysis, New York, McGraw-Hill Book Co., pp. 44–60.

Mather, K. 1953. The genetical structure of populations. Symp. Soc. Exp. Biol., 7:66–95.

Mayr, E. 1959. Where are we? Cold Spring Harbor Symp. on Quant. Biol., 21:1–14.

———— 1963. Animal Species and Evolution, Cambridge, Harvard University Press.

McClearn, G. E. 1967. Genes, generality and behavior research. *In* Hirsch, J., ed., Behavior-Genetic Analysis, New York, McGraw-Hill Book Co., pp. 307–321.

Pittendrigh, C. S. 1958. Adaptation, natural selection, and behavior. *In* Roe, A., and G. G. Simpson, eds., Behavior and Evolution, New Haven, Yale University Press, pp. 390–416.

Schwarz, E. 1945. On North American house mice. J. Mammal., 26:315–316.

———— and H. Schwarz. 1943. The wild and commensal stocks of the house, Mus musculus, Linnaeus. J. Mammal., 24:59–72.

Selander, R. K., and S. Y. Yang. 1970. Biochemical genetics and behavior in wild house mouse populations. *In* Lindzey, G. and D. D. Thiessen, eds., Contributions to Behavior-Genetic Analysis: The Mouse as a Prototype, New York. Appleton-Century-Crofts.

———— W. G. Hunt and S. Y. Yang. 1969. Protein polymorphism and genetic heterozygosity in two European subspecies of the house mouse. Evolution, 23:379–390.

Simpson, G. G. 1958. The study of evolution: Methods and present status of theory. *In* Roe, A., and G. G. Simpson, eds., Behavior and Evolution. New Haven, Yale University Press, pp. 7–26.

Sinnott, E. W., L. C. Dunn, and T. Dobzhansky. 1958. Principles of Genetics, New York, McGraw-Hill Book Co.

Skinner, B. F. 1966. The phylogeny and ontogeny of behavior. Science, 153:1205–1213.

Spiess, E. B. 1968. Experimental population genetics. Ann. Rev. Genet., 2:165–208.

Spieth, H. T. 1958. Behavior and isolating mechanisms. *In* Roe, A., and

G. G. Simpson, eds., Behavior and Evolution, New Haven, Yale University Press, pp. 363–389.

Spuhler, J. N., and G. Lindzey. 1967. Racial difference in behavior. *In* Hirsch, J., ed., Behavior-Genetic Analysis, New York, McGraw-Hill Book Co., pp. 366–414.

Staats, J. 1966. The laboratory mouse. *In* Green, E. L., ed., Biology of the Laboratory Mouse, 2nd ed., New York, McGraw-Hill Book Co., pp. 1–9.

Stalker, H. D., and H. L. Carson. 1948. An altitudinal transect of Drosophila robusta. Evolution, 2:295–305.

Thompson, W. R. 1967. Some problems in the genetic study of personality and intelligence. *In* Hirsch, J., ed., Behavior-Genetic Analysis, New York, McGraw-Hill Book Co., pp. 344–365.

von Guaita, G. 1898. Versuche mit Kreuzungen von verschiedenen Rassen der Hausmaus. Ber. Naturf. Ges., Freiburg, 10:317–332.

11

Biochemical Genetics and Behavior in Wild House Mouse Populations[*]

ROBERT K. SELANDER AND SUH Y. YANG
*Department of Zoology,
The University of Texas at Austin*

INTRODUCTION

Although no mammal is better known genetically than the house mouse (*Mus musculus*), the available information pertains largely to the domestic, laboratory strains, the wild populations from which these were derived having remained essentially uninvestigated. Two years ago we initiated a long-term study of polymorphic variation in wild populations, concentrating our efforts primarily in North America. Our principal concern has not been with behavioral characters but, rather, with allelic variation at genetic loci controlling the synthesis of certain enzymes and other proteins of the blood, as revealed by starch gel zone electrophoresis. Through the use of biochemical characters, we seek to determine (1) the nature and extent of genic variation within populations; (2) the effects of population subdivision on patterns of genetic variation at the microgeographic level; (3) the extent of variation in allele frequencies at the macrogeographic level; (4) the selective factors responsible for the maintenance of polymorphisms; and (5) the degree to which the species has evolved in the New World since its introduction from stocks derived from Europe.

As our work progressed, it soon was apparent that we would have to consider the behavior and social system of the house mouse, for we found evidence of extremely fine genetic subdivision of populations, even within

[*] Our research is supported by grants from the Public Health Service (NIH GM-15769) and the National Science Foundation (GB-6662).

We wish to express our appreciation to R. Ralin for valuable assistance in preparation of the manuscript and to S. Stewart for assistance in computer programming and data handling procedures. We are also indebted to W. G. Hunt and L. Wheeler for contributing unpublished data from their own researches.

single barns and fields. Our findings supported previous suggestions that the panmictic unit (deme) in this species is very small indeed—perhaps on the order of 10 individuals—and that behavioral interactions among individuals are importantly involved in determining patterns of population subdivision, and, hence, indirectly, patterns of genetic variation among and within populations.

In this review, we intend to discuss several aspects of genetic variation in wild populations of the house mouse that will be of interest to behavior geneticists who are turning their attention from domestic, inbred strains to wild mice and the more complex problems of "behavioral population genetics," as defined and outlined by Bruell (Chapter 10). Our review touches upon certain aspects of (1) the systematics of the house mouse; (2) the geographic patterns of genetic variation in North America; (3) the nature of population subdivision on a local level, emphasizing the potential contribution of behavior studies to population genetics; and (4) the possible adaptive significance and evolutionary consequences of subdivision and resultant genetic drift in populations.

SYSTEMATICS

Although the house mouse is virtually world wide in distribution (Walker, 1964, pp. 921–922), it has not been well studied by mammalian systematists. A few modern, quantitative studies of geographic variation in parts of Europe are available (Zimmerman, 1949; Berry, 1964; Ursin, 1952), but most of the systematic literature, consisting of qualitative descriptions of morphologic subspecies, offers little of interest to population or behavior geneticists. The only comprehensive systematic review of the species is that of Schwarz and Schwarz (1943), in which populations are classified into 15 morphologically defined subspecies assigned to four subspecies groups. This classification is summarized by Bruell (Chapter 10) and need not be discussed at length here. However, we will take this opportunity to comment on the present state of systematic research dealing with the house mouse in an effort to anticipate some of the problems likely to be encountered by behavior geneticists and population geneticists using wild mice in their investigations.

It is current taxonomic opinion that the house mouse is a single biologic species. Supporting this conclusion, Grüneberg (1952, p. 3) notes that sterility barriers among nominal species and subspecies have not been demonstrated. However, because the biologic species definition (Simpson, 1961; Mayr, 1963; Selander, 1969) does not hinge upon sterility barriers, their absence does not preclude the possibility of the house mouse being divided into several partially or completely closed, coadapted genetic systems. And, indeed, we already have evidence suggesting that the house mouse, like many widespread, systematically "difficult" species, is a complex array of populations (semi-species) that are to varying degrees genetically isolated through behavioral, ecologic, or other mechanisms. For example, where *M. m. domesticus* of the *wagneri* subspecies group meets *M. m. musculus* of the *spicilegus* group in Europe, intergradation is confined to an extremely narrow, temporally stable zone of allopatric hybridization, through which little introgression of genes is occurring (Ursin, 1952), despite interbreeding of the parental types and backcrossing of the hybrids. Here, as in other similar cases of "suture-zone" hybridization (see review by Remington, 1968), it seems probable that introgression is restricted, in part at least, because potentially introgressant genes disrupt the parental coadapted complexes, thereby altering the adaptiveness of other genes by modifying the genetic environment in which they operate (Mayr, 1963; Yang and Selander, 1968).

The reported coexistence (sympatry) of morphologically distinctive "indoor" (commensal) and "outdoor" (aboriginal or feral) populations* of the house mouse is of considerable interest to evolutionists. According to Schwarz and Schwarz (1943, p. 71), two types of mice coexist on the Iberian Peninsula and in northwestern Africa, an aboriginal form, *M. m. spretus,* and an immigrant commensal form classified as *M. m. praetextus* or *M. m. brevirostris.* The nature of the isolating mechanisms maintaining this situation remains uninvestigated, but, clearly, the sympatry of supposed subspecies raises questions regarding species limits in the house mouse. In this connection, it is noteworthy that this case of sympatry involves subspecies belonging to different subspecies groups. However, sympatry of subspecies assigned to the same subspecies group is also reported (Schwarz and Schwarz, 1943), as, for example, that involving *M. m. urbanus* ("indoor") and *M. m. homourus* ("outdoor") of the *wagneri* group in northern India. Significantly, perhaps, Schwarz and Schwarz (1943, p. 63) note that, in this case, "segregation . . . is not always complete, and interbreeding occurs."

According to Schwarz and Schwarz (1943), North and South American mice were derived entirely from commensal European stocks, so that mice in the Western Hemisphere are, by their definition, either commensal or feral. Thus in this hemisphere there is no opportunity to study genetic, behavioral, or other differences between aboriginal and commensal forms in sympatry, although differences between commensal and feral populations

* We have adopted the terminology of Bruell (Chapter 10). Domestic mice are those living in captivity and whose breeding is controlled by man. Wild mice of several types are distinguished: aboriginal mice are those having no history of association with man or his structures, that is, mice living in native habitats; commensal mice are those associated with man and his structures and agriculture; feral mice are those that were originally commensal but have secondarily come to occupy native habitats.

could be investigated. In such an undertaking it will be important to select populations which are permanently either commensal or feral. We find that mice collected in agricultural fields are genetically and morphologically indistinguishable from those taken in barns or other man-made structures. Few truly feral populations existing in regions far removed from human settlements have been found in North America. Although Koford (1968) found house mice in the desert regions of Peru, Justice (1962) reported that populations are not maintained in unmodified desert habitats in Arizona.

Certain limitations of the subspecies concept in dealing effectively with the classification of populations in systematically complex species or semispecies groups will be apparent from our discussion thus far. Further inadequacies, arising, ultimately, from the inherent typologic basis of the concept (Wilson and Brown, 1953), are revealed when the systematist attempts to describe geographic patterns of character variation in terms of subspecies. Because such patterns tend to be discordant, the grouping of populations into subspecies on the basis of geographic considerations or similarities in mean values of expression of one or a few characters often fails to contribute to an understanding of character variation, and, indeed, may obscure the actual patterns of variation. Once the subspecies concept is applied, all populations of a species must be subspecifically assigned, even though this may be entirely arbitrary on the basis of existing patterns of character variation. This difficulty frequently is reflected in the systematic review by Schwarz and Schwarz (1943). For example, on page 66, the *"Differential diagnosis"* of the subspecies *M. m. domesticus* states that "The *typical* form is distinguished from *brevirostris* by the back being overlaid, instead of suffused, with black, by the flank line being absent, by the tail being dark brown, the underside distinctly speckled, and the feet having only the end phalanges white, instead of the whole of the fingers and toes. [italics ours]" Use of the term *"typical"* is understood when we read (p. 65) that some populations are assigned to *M. m. domesticus* on geographic grounds, although, actually, they "approach the *brevirostris* style." Here we also learn that a "highly specialized type with the back bluish grey, the belly buff, with a sharp line of demarcation, and pale feet is found in certain areas of western France (Poitou) and on several islands in the Outer Hebrides. . . ." It seems likely that many populations presently assigned to one subspecies will prove to be genetically more similar to populations assigned to other subspecies.

Schwarz and Schwarz (1943), followed by Hall and Kelson (1959, p. 770), claim that two subspecies of *Mus musculus* occur in North America. The northwestern European form *M. m. domesticus,* introduced primarily from England, France, and Germany, ranges from Alaska to the "northern part of the central states" (and also occurs in Hawaii), while the Mediterranean subspecies *M. m. brevirostris,* introduced into South and Central America by Spainsh colonizers along shipping lanes from the Iberian Peninsula, now ranges to the southern part of the United States. The ranges of these subspecies supposedly meet in the central states, where, according to Schwarz and Schwarz (1943, p. 70), there is "a mixed population, that to a certain extent is biologically segregated." Details of intergradation between the two subspecies are not given; nor are the distributions outlined in any detail.

From this account, the nonsystematist might conclude that mouse populations in North America assigned to *M. m. brevirostris* or to *M. m. domesticus* by Schwarz and Schwarz (1943) are identical to those of the same subspecies occurring in Europe. This would be a mistake. Use of these names for populations in North America may actually mean nothing more than that most of the mice introduced into North America are believed to have come originally from parts of Europe within the ranges of these subspecies. But, in fact, it seems unlikely that mice from Asia and from other parts of Europe and the Western Hemisphere have not been introduced into North America, or that there has been no regional subintroduction within North America. More importantly, the notion, implicit in these subspecific designations, that mice in North America have remained genetically static since introduction, is certainly erroneous. Studies of the house mouse and other introduced or colonizing species, notably the house sparrow (*Passer domesticus*), which was also introduced to North and South America from Europe, demonstrate that populations rapidly evolve in adaptation to the environmental conditions in new areas (Johnston and Selander, 1964; see also Mayr, 1963; Baker and Stebbins, 1965). House mice introduced from Europe to the Faeroe Islands in the North Atlantic within historical times, and perhaps less than 300 years ago, are now so distinctive morphologically that they have been regarded as a separate subspecies or, even, as a separate species, *M. faeroensis* (Miller, 1912; Barrett-Hamilton and Hinton, 1921; Evans and Vevers, 1938). Berry (1964) has shown that, although mice have been on the Welsh island of Skokholm for only 70 years, they are now distinct from mainland mice both in overall size and in skeletal characteristics. House mice have been in North America for several hundred years and have had ample time to evolve gene complexes adapted to local conditions. Evidence of extensive geographic variation in gene frequencies in Texas is presented below, and we may note that mice from Jalapa, Mexico, and from the Trinidad Mountains of Cuba have been formally described as subspecies (Allen and Chapman, 1897; Moulthrop, 1942). The full extent to which North American house mouse populations differ inter se and from the European populations from which they were largely derived remains to be determined. We have not yet examined sufficient material from Europe to permit close comparison of introduced and native stocks. However, for North American populations our studies clearly do not support the systematic interpretation of Schwarz and Schwarz (1943). In particular, we find no evidence of either the coexistence or hybridization of subspecies in the central United States (Selander, Yang, and Hunt, 1969).

In concluding, we emphasize the need for intensive investigation of distribution and variation in the house mouse. The impressionistic, typologic account of morphologic variation and the classification presented by Schwarz and Schwarz (1943) serve merely to identify some of the major systematic problems requiring attention. Whether the subspecies concept, as employed by these authors, has merit in dealing with geographic variation and problems of species limits in populations currently designated as *Mus musculus* remains to be seen when modern, quantitative analyses of character variation become avaliable.

GENETIC VARIATION IN WILD POPULATIONS

Morphologic Characters

Systematic studies of the house mouse deal almost entirely with quantitative, polygenic morphologic characters such as body size, relative body proportions, and pelage color, all of which have large environmental components of variance and may also be subject to age and sexual variation. Interpreting the underlying genetic basis of interpopulation and intrapopulation variation is difficult for a variety of reasons. For example, studies of domestic and wild mice demonstrate that phenotypic variability of quantitative traits may be unrelated, or even inversely related, to genetic variability (Biggers et al., 1958; Bader, 1956; Bader and Lehmann, 1965).

Of the quantitative characters studied in mice, teeth dimensions offer the best opportunity for genetic analysis. A partitioning of the variance in width of the three mandibular molars in wild populations by Bader (1965), demonstrated that the additive genetic component is relatively high, on the order of 40 percent. In mice from various parts of the mainland of England and Wales, Van Valen (1965) studied the effects of phenotypic selection on the width of the first upper molar, M^1, which is an age-invariant character. Surprisingly, he found evidence of both directional and destabilizing selection, older individuals being on the average larger and more variable in molar width than younger individuals. The populations studied were heterogeneous for intensity of selection and for selection on the mean, but not on the variance. Food or structure of the ricks in which the mice lived proved to be a major source of heterogeneity, but the functional significance of the selection remains unknown.

Somewhat intermediate between the quantitative characters of the systematist and the true genetic polymorphisms preferred for population-genetic analysis are the "epigenetic polymorphisms" (Berry and Searle, 1963) the "quasi-continuous" variants of Grüneberg) of the skeleton studied in wild populations by Berry (1963), Weber (1950), Doel (1958), and others. These characters would not seem to be well suited for population genetics or systematics because they are polygenically determined in complex fashion and are heavily influenced by environmental factors.

Apart from the biochemical polymorphisms herein reported, the only polymorphism that has been intensively studied in wild populations of the house mouse is that at the T locus controlling certain aspects of development of axial structures in the caudal region. Thirteen prenatal lethal and three viable recessive alleles have been detected in wild populations in the United States (Dunn et al., 1960, p. 224). Heterozygotes $(+/t)$ are detected by breeding wild-caught mice with laboratory mice heterozygous for the dominant brachyury allele $(T/+)$.

Homozygotes for certain t alleles are either unconditional lethals or male-sterile when homozygous. However, selection against these alleles is counterbalanced by their increase from an abnormal transmission ratio in heterozygous males, the effective sperm pool of which contains between 85 and 99 percent t-bearing sperm. A deterministic model for the balance between selection and abnormal segregation (Bruck, 1957) predicts that from 60 to 95 percent of adults in natural populations should be heterozygous for t, whereas actual frequencies of heterozygotes range from 35 to 50 percent (the mean frequency of t alleles in populations is 0.174) (Lewontin and Dunn, 1960). More accurate predictions were obtained by a stochastic model, developed by Lewontin and Dunn (1960) for lethal alleles and later applied by Lewontin (1962) to male-sterile alleles, based on the premise that populations are subdivided into small family groups in which there is loss of the t allele due to genetic drift. Empiric support for the Lewontin-Dunn model was supplied by Anderson (1964), Anderson et al. (1964), and Petras (1967b). Levin et al. (1964) extended the model to consider certain effects of interdeme migration, and Petras (1967b) has shown that Bruck's deterministic model can predict t allele frequencies consistent with empirical frequencies, provided an inbreeding coefficient is incorporated. Following a report by Dunn and Bennett (1966) that the "viable" male-sterile t alleles in fact reduce viability in homozygotes from 45 to 90 percent, Lewontin (1968), using both stochastic and deterministic models, demonstrated that the evolutionary dynamics of these alleles are essentially independent of viability.

Because Dunn et al. (1958) suggest the possibility of heterosis for lethal t alleles, it appears that the complete story of the evolutionary dynamics of this locus is not as yet available.

Several coat color polymorphisms are known in wild house mouse populations (Dunn et al., 1960), but no comprehensive investigation of gene frequencies has been undertaken.

BELLY COLOR (Agouti Locus).—Petras (1967c) investigated a polymorphism involving A^w and A^+ alleles at the agouti locus in mice from five farms in Washtenaw County, Michigan, and also summarized data on the frequency of the A^w allele reported in the literature. A^w is dominant, producing a creamy white to white phenotype, while the recessive homozygote (A^+A^+) has a grey belly.

Polymorphism at the agouti locus is widespread in wild populations in the United States. For the Michigan samples, the overall frequency of A^w was 0.150 ± 0.011, and there was heterogeneity in frequency among the five farms sampled. In a sample of 200 mice from Davis, California, Evans and Storer (1944) found the frequency of A^w to be 0.13 ± 0.018, and Engels (1948) found a frequency of A^w of 0.227 ± 0.055 in a sample of 30 mice from the southeastern United States. Belly color polymorphism of this type was recorded in samples from many localities in the United States by Dunn et al. (1960, p. 226), and Grüneberg (1952, pp. 37–38) also reported this polymorphism in wild populations. In our work, we find it very widespread; many local populations appear to be fixed for the A^+ allele, but we have yet to sample a large population monomorphic for the A^w allele.

The study of variation in allele frequencies at the agouti locus using the frequency of white bellied phenotypes in samples is complicated by the fact that genes at other loci may similarly influence belly color (e.g., "snowy belly" detected in a wild mouse taken at Arlington, Virginia, which is due to multiple alleles, not to an allele at the agouti locus [Eaton and Schwarz, 1946; Falconer, 1947]).

PIED WHITE SPOTTING ON VENTRUM.—This character, known to be under polygenic control (Dunn, 1942), is widespread (see literature cited by Petras, 1967c, p. 294). In North America, Dunn et al. (1960) found it in 6 of 13 samples examined, and Petras (1967c, p. 294) noted it in 28 (4.8 percent) of 580 mice examined from the Ann Arbor region. We also have found it in most regions collected.

OTHER COAT COLOR VARIATION.—Other coat color variants such as unusually light or dark agouti colors are not uncommon in wild mouse populations (Dunn et al., 1960, p. 227). Although Petras (1967c) failed to uncover certain recessive coat color mutant alleles in heterozygous condition in mice from Washtenaw County, Michigan, by mating wild mice with homozygous laboratory mice, such alleles have been recorded elsewhere (Grüneberg, 1952). For example, Brown (1965) studied a population of between 25 and 69 mice living in a granary on the McCrosky Farm, Greene County, Missouri (Table 1). In April, 1962, 28 percent of the population possessed pale yellow pelage and pink eyes, a condition due to a recessive allele similar to, if not identical with, pink-eyed dilution (p) at the P locus of link-

TABLE 1

**Frequency of Pelage Types and Alleles at the *P* Locus in Mice
Collected at McCrosky Farm, Missouri[a]**

Date	Number of mice trapped	Percent with pale yellow pelage and pink eyes	Allele frequency[b]	
			P	*p*
1962				
April	32	28.1	0.46	0.54
August	44	40.9	0.36	0.64
December	58	46.6	0.32	0.68
1963				
April	22	0.0	<1.00	>0.00
August	29	0.0	<1.00	>0.00
December	37	5.0	0.75	0.25

[a]Adapted from Brown (1965, p. 463, Table 1).
[b]We have calculated allele frequencies, using Haldane's (1956) method.

age group I (Grüneberg, 1952, p. 54). The frequency of the p allele increased from 0.54 to 0.68 from April to December, 1962, in which period population size also increased and the mice were free from predation by domestic cats living on the farm. In January, 1963, the cats were given free access to the granary, and by April, 1963, they had reduced the population to 22 mice, all of which were agouti color. (It is assumed that the pale-colored mice were more easily seen by the cats against the dark background of the granary.) However, the p allele remained in the population, as indicated by the capture of two homozygous mutant mice in December, 1963, 3 months after the cats were again excluded from the granary.

Biochemical Polymorphism

Electrophoresis, combined with histochemical staining for the demonstration of multiple molecular forms of proteins, provides a tool opening new areas of investigation and promoting rapid advance in many conventional areas of population genetics (see review by Lewontin, 1967) and systematics. The technique detects variation in the mobility of proteins dependent upon net charges of molecules arising from the addition, deletion, or substitution of one or more amino acids in polypeptides encoded by single genes. For descriptions of electrophoretic techniques, the reader is referred to Smith (1968). Lush (1967) has summarized biochemical genetic variation in non-human vertebrates, and the nature and genetic control of multiple molecular forms of molecules are discussed by Markert (1968), Ogita (1968), Kaplan (1968), Watts (1968), and Shaw (1969). Because allelic variants are distinguishable by their mobility on gels (electrophorograms), and environmental influences on the phenotypes are, for practical purposes, nil, discrete phenotypic differences among individuals are translatable into genotypic differences, and allele frequencies for samples are obtained directly, without genetic analysis through breeding. Isoallelic variation, as demonstrated by electrophoresis, is particulary useful in observational and experimental studies of selection, which, in the past, have usually involved morphologically visible, deleterious mutant alleles (MacIntyre and Wright, 1966, pp. 371–372).

At present, nothing is known of the physiologic adaptive functions of the polymorphisms studied in the house mouse, and selection coefficients cannot be attached to the alleles. However, this is true for almost all morphologic or other polymorphisms (Lewontin, 1967).

Many of the polymorphic loci described in the house mouse were first detected in inbred strains (see, for example, Ruddle and Roderick, 1966, 1968; Ruddle and Harrington, 1967), where polymorphism invariably is manifested as interstrain variation, all individuals of a given inbred strain being uniformly homozygous. Recently, investigation has extended to wild populations, with our own program and the work of Petras (1967a) on the *Hbb* and *Es-2* loci in southeastern Michigan. Although we have detected 17 polymorphic loci in wild populations (Selander and Yang, 1969; Selander, Hunt, and Yang, 1969), our research has largely been concerned with polymorphism in three esterases and hemoglobin occurring in the blood. The common alleles at these loci are indicated in Table 2, and various phenotypes are illustrated in Fig. 1. The variation is independent of sex and age. Methods of tissue preparation, electrophoresis, and staining are given by Selander, Yang, and Hunt (1969).

PLASMA ESTERASE 1 (*Es-1*).—*Es-1* is an autosomal locus closely linked with *Es-2* and oligosyndactyly (*Os*) in linkage group XVIII (Popp, 1965; Petras and Biddle, 1967). Two alleles (*Es-1^a* and *Es-1^b*) are known in inbred mice (Popp and Popp, 1962) and both occur in wild populations. On our electrophorograms, two bands appear in the heterozygote *Es-1^a/Es-1^b*, and homozygotes show a single band (Fig. 1).

PLASMA ESTERASE 2 (*Es-2*).—Seven alleles are known at the *Es-2* locus, which was described from the plasma. Recently, we have found that the *Es-2* system stains intensely in liver and kidney extracts, where it is located in the V esterase zone designated by Ruddle and Roderick (1966). Petras (1963) showed that most inbred strains are homozygous for *Es-2^b* and described the *Es-2^a* allele from wild populations in Michigan. We have discovered two additional common alleles in wild populations (Table 2, Fig. 1). Because of the known or suspected presence of the "silent" ("null") allele *Es-2°* in various wild populations, allele frequencies cannot accurately be estimated by direct translation of phenotypes. Instead, we have used a maximum likelihood

TABLE 2

Allelic Variation at Four Biochemical Loci in the House Mouse[a]

Allele	Phenotype	Occurrence
	Plasma esterase *Es-1*	
Es-1ᵃ	Fast-migrating band	*C57BL, C57L/R1*, and *C57BR/cdj;* wild populations
Es-1ᵇ	Slow-migrating band	Most inbred strains; wild populations
	Plasma esterase *Es-2*	
Es-2ᵃ	Band absent or very faint with mobility similar to *Es-2ᶜ*	*RFM/Un* and *RF/AL*; wild populations
Es-2ᵇ	Fast-migrating band	Inbred strains; wild populations
Es-2ᶜ	Medium-mobility band	*PL/J*; wild populations
Es-2ᵈ	Slow-migrating band	Wild populations in southwestern United States and Ohio
	Erythrocyte esterase *Es-3*	
Es-3ᵃ	Faint band with mobility intermediate between *Es-3ᵇ* and *Es-3ᶜ*	*C57BL, BALB/cJ, B10·D2/Sn*, and others; unrecorded in wild populations
Es-3ᵇ	Band with mobility intermediate between *Es-3ᶜ* and *Es-3ᵈ*	*RF, YBR/He*, and *RFM/Un;* wild populations
Es-3ᶜ	Slow-migrating band	Most inbred strains; wild populations
Es-3ᵈ	Fast-migrating band	Wild populations in southwestern United States
	Hemoglobin *Hbb*	
Hbbˢ	Single fast-migrating band	*C57BL* and other inbred strains; wild populations
Hbbᵈ	Fast- and slow-migrating bands	Most inbred strains; wild populations

[a]Only those alleles occurring commonly in inbred strains or wild populations have been included; rare alleles are listed in Selander, Yang, and Hunt (1969).

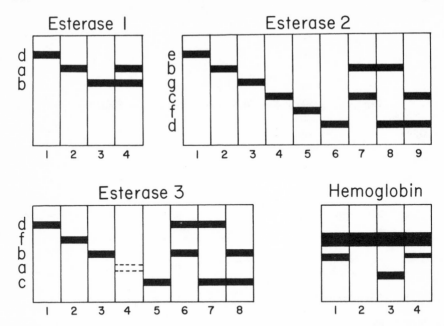

FIG. 1. Diagrammatic representation of phenotypes of four polymorphic proteins in house mouse, showing homozygous patterns of nonsilent alleles and heterozygous patterns commonly found in wild populations. Esterase 1: Samples 1–3, patterns of mice homozygous for *Es-1ᵈ*, *Es-1ᵃ*, and *Es-1ᵇ* alleles, respectively; sample 4, pattern of *Es-1ᵃ/Es-1ᵇ* heterozygote. Esterase 2: Samples 1–6, homozygous patterns; samples 7–9, heterozygous patterns. Esterase 3: Samples 1–5, homozygous patterns; samples 6–8, heterozygous patterns. Hemoglobin: Samples 1–3, patterns of mice homozygous for *Hbbᵈ*, *Hbbˢ*, and *Hbbᵖ* alleles, respectively; sample 4, pattern of *Hbbᵈ/Hbbˢ* heterozygote.

method based on the relative proportions of the "silent" homozygote (Es-2^a/Es-2^a) and other phenotypes.

ERYTHROCYTE ESTERASE 3 (Es-3).—Variation at the Es-3 locus controlling an esterase in the erythrocytes was described by Hunter and Strachan (1961) and Pelzer (1965), who detected the alleles Es-3^a and Es-3^c in inbred strains. Two additional common alleles, Es-3^b and Es-3^d, were discovered by us in wild populations, and the former has also been found in the RF/J and YBR/He strains (Popp, 1966). To date we have found three alleles occurring commonly in wild populations, but most polymorphic populations are segregating for only two of them, Es-3^b and Es-3^c. Homozygotes show a single dark band on the gels, while two lighter bands are represented in heterozygotes (Fig. 1).

HEMOGLOBIN TYPE (Hbb).—In inbred strains (Russell and Gerald, 1958) and in wild populations, two alleles are present (Fig. 1). The term "diffuse" has been used to describe the Hbb^d/Hbb^d and Hbb^d/Hbb^s phenotypes because of their polymerization characteristics (Bonaventura and Riggs, 1967). Using our material, Bonaventura (1969) demonstrated that the creative cysteinyl residue ($\beta13$) responsible for polymerization is in the same position on the β-chain in wild mice as in the inbred strain *BALB/c*. A third allele, Hbb^p, described from the domestic strain *Aa/Ss* by Morton (1962, 1966), has not been recorded in wild populations. The *Hbb* locus shows close linkage with the albino locus on linkage group I (Popp and St. Amand, 1960, 1964; see also Hutton et al., 1962).

Geographic Variation in Biochemical Characters

Our survey of variation in wild populations is based on 10,000 mice collected in 300 barns and agricultural fields. For most purposes, we have used the single barn or field as the unit of sampling, and for each unit, gene frequencies and expected numbers of the genotypes have been calculated by the exact formula for small samples (Levene, 1949).

Much of our material comes from Texas, where we have systematically sampled populations in all parts of the state. In addition, we have sampled populations in California, Arizona, New Mexico, Utah, Kansas, Missouri, Minnesota, Wisconsin, Illinois, Florida, Jamaica, and Venezuela. Comparable studies are being pursued in our laboratory for mice in the Hawaiian Islands by L. Wheeler and in Denmark by W. G. Hunt.

Our discussion will center on variation in Texas, but we will comment on certain aspects of variation elsewhere. Sample localities in Texas have been grouped into sample regions, as shown in Fig. 2. For each locus presumed regional equilibrium allele frequencies have been estimated by computing unweighted mean frequencies for all barns and fields represented by samples larger than nine mice. Unweighted means were used in preference to pooled means because of marked interbarn heterogeneity within regions (see below). Sample sizes by region are indicated in Table 3 and geographic patterns of variation in allele frequencies for three loci are shown in Fig. 3 through 5. An analysis of variation in North America, together with complete genetic data for all samples, has been presented elsewhere (Selander, Yang, and Hunt, 1969).

Es-1.—All wild mice from Texas and from most other localities in North America, South America, and the West Indies show only the slow-migrating band of the genotype $Es-1^b/Es-1^b$. Similarly, wild mice from Washtenaw County, Michigan, examined by Petras (1967a, p. 265) were homozygous for $Es-1^b$. But $Es-1^a$ occurs in low frequency in southern California and southern Arizona. In view of the monomorphism of most North American populations, it is noteworthy that mice from the Hawaiian Islands are polymorphic, with

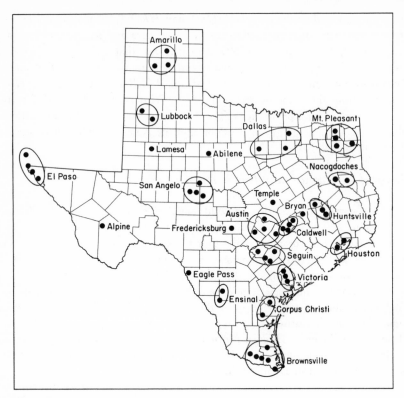

FIG. 2. Sample localities and regions in Texas. (From Selander, Yang, and Hunt, 1969. *Univ. Texas Publ.*, 6918:271.)

Es-1ᵃ and *Es-1ᵇ* in approximately equal frequencies (L. Wheeler). An analysis of the hybridizing subspecies *M. m. musculus* and *M. m. domesticus* in Denmark indicates that populations of the former are fixed for *Es-1ᵃ* and those of the latter for *Es-1ᵇ*, heterozygotes being confined to the zone of hybridization (Selander, Hunt, and Yang, 1969).

Es-2.—In most wild populations *Es-2ᵇ* is the most common allele, and, not infrequently, it is locally fixed; the only exception is in *M. m. musculus* of northern Denmark, in which *Es-2ᶜ*, an allele also represented in most regions, appears to be fixed (Selander, Hunt, and Yang, 1969). The "silent" allele *Es-2ᵃ* is also widespread but is generally in low frequency. *Es-2ᵈ* has so far been recorded only in southern Texas, California, and Arizona.

Within Texas, the pattern of geographic variation in allele frequency at the *Es-2* locus is complex (Fig. 3). Populations in central and northern Texas are relatively homogeneous, with *Es-2ᵇ* and *Es-2ᶜ* at frequencies of about 0.90 and 0.10, respectively. Locally, the *Es-2ᵃ* allele is present in low

TABLE 3

Samples of Wild House Mice Collected in Texas for Analysis of Biochemical Polymorphism

Region	Farms	Barns or fields		Individuals	Individuals used to calculate mean allele frequencies
		Total	Total with $n > 9$		
Dallas	3	7	7	184	184
Temple	2	2	1	37	34
Austin	25	32	15	2339	2188
Fredericksburg	2	2	1	14	13
Abilene	1	1	1	32	32
San Angelo	7	7	5	178	157
Seguin	8	17	8	286	257
San Antonio	2	2	0	8	–
Houston	5	12	3	59	32
Ensinal	3	13	13	765	765
Brownsville	10	20	12	551	515
Alpine	3	3	3	116	116
Victoria	3	3	3	62	62
Corpus Christi	3	9	6	158	147
Nacogdoches	8	14	7	165	123
Bryan	4	10	7	501	488
Huntsville	3	4	2	34	32
Caldwell	4	6	2	96	84
Mt. Pleasant	6	12	5	259	235
Lamesa	1	3	2	37	37
Lubbock	3	11	9	604	588
Amarillo	3	5	5	188	188
El Paso	6	10	1	37	10
Eagle Pass	1	3	2	81	80
Totals	116	208	120	6791	6367

frequency. Eastward from central Texas the frequency of *Es-2ᵃ* increases clinally, reaching 0.24 in eastern Texas, where *Es-2ᶜ* is absent. To the north, *Es-2ᵇ* approaches fixation. In southern Texas, *Es-2ᵈ* occurs at a frequency of 0.15, together with *Es-2ᵇ* and *Es-2ᶜ*. A clinal pattern of variation between central Texas and the Rio Grande Valley of extreme southern Texas is suggested by allele frequencies in the Corpus Christi region.

Es-3.—Populations from all regions sampled are segregating for *Es-3ᵇ* and *Es-3ᶜ*. In Texas the frequency of *Es-3ᵇ* varies from about 0.70 in the western part (including the Panhandle) to about 0.50 in the central part of the state (Fig. 4). Then eastward from the Austin region, the frequency of *Es-3ᵇ* increases clinally, reaching 0.90 in eastern Texas. The Brownsville region is distinctive in having *Es-3ᵈ* present. Because the overall frequency of this allele is 0.02, its occurrence might be attributed solely to mutation, but this hypothesis would not account for its absence in other regions. Possibly it occurs at higher frequencies to the south in Mexico.

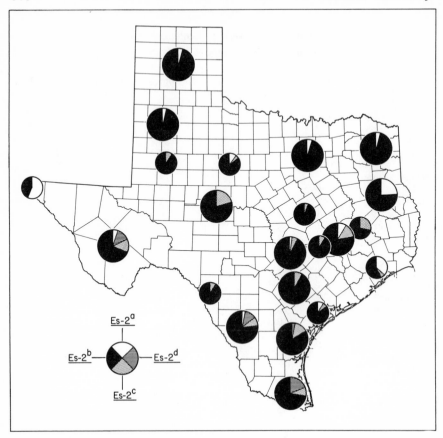

FIG. 3. Geographic variation in allele frequencies at Esterase 2 locus in Texas. Indicated areas of circles are proportional to mean regional frequencies. Small diagrams represent small samples. (From Selander, Yang, and Hunt, 1969. *Univ. Texas Publ.*, 6918:271.)

Hʙ.—In most regions, populations are polymorphic for *Hbb*[s] and *Hbb*[d], but the former allele tends to be more common and may be fixed locally. *Hbb*[s] approaches fixation in the southeastern part of Texas, and the two alleles are equally common in the Brownsville region (Fig. 5). In terms of number of alleles, *Hbb* is the least variable locus investigated.

POPULATION SUBDIVISION

Lewontin (1967, p. 47) states the central problem of theoretical population genetics "as the prediction of the genetic composition of a population at any time, t_1, given the composition at some previous time, t_0, and given

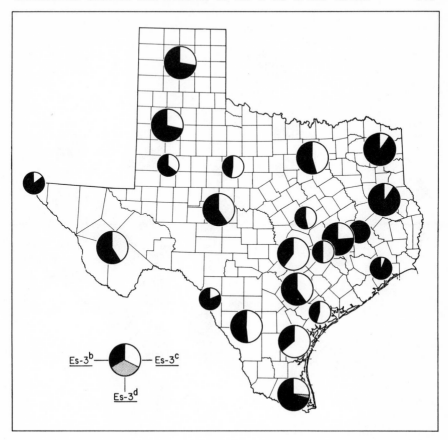

FIG. 4. Geographic variation in allele frequencies at Esterase 3 locus in Texas. (From Selander, Yang, and Hunt, 1969. *Univ. Texas Publ.,* 6918:271.)

the genetic mechanisms, the breeding structure of the population, the fitness of the genotypes, [and] the amount of mutation and migration during the intervening time." For wild house mice we have essentially no information on genotypic fitness or mutation rates, but we are able to examine some aspects of population structure and migration as they affect the genetic composition of populations.

For the house mouse and other rodents, there exists a considerable literature on movements and home ranges, but, as noted by Lewontin (1967, p. 57), information of this type tells us "virtually nothing quantitative ... about gene flow from one part of a population to another." Two parameters of potential interest, but in practice difficult to estimate, are N_e, the effective population size, and m, the migration rate. Wright (1951) worked out the relationship between the coefficient of inbreeding (F) and N_e and m, so that,

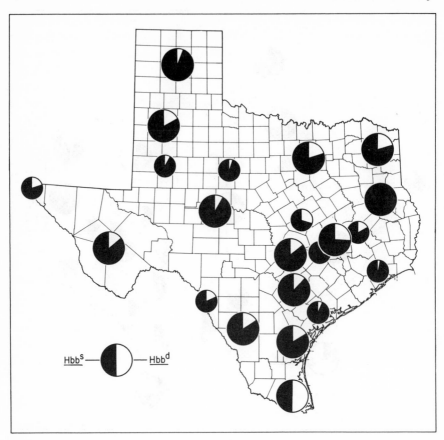

FIG. 5. Geographic variation in allele frequencies at *Hbb* hemoglobin locus in Texas. (From Selander, Yang, and Hunt, 1969. *Univ. Texas. Publ.,* 6918:271.)

given F and N_e or m, it is possible to estimate N_e or m (Petras, 1967a), but it may be impossible to obtain reliable estimates of F in wild populations, owing to heterotic and other effects (see below).

Extremes of population subdivision are described by Wright's two models. In Model I (Isolation by Distance) the population is continuously distributed, with varying density and varying distance between individuals, but no clear physical division of the population into subgroups. The limits of randomly mating units depend largely on the average dispersal distance from birth to mating. Populations of certain species of *Drosophila* may exemplify this model (Lewontin, 1967, p. 58). In Model II (Island Model) a species population is divided into small units (demes), within which random breeding occurs; these are almost completely isolated from one another by physical barriers or by territoriality or other social behavior. Even in continuously distributed populations, insular subdivision may be achieved if potential immigrants are killed or driven from territories by resident members of demes.

Territoriality as a factor effecting an insular pattern of population subdivision has received relatively little attention from population geneticists, who generally think of insularity in terms of pockets of habitat separated by areas of unsuitable habitat. But the genetic consequences are the same, whether migration is restricted by physical "barriers" or by behavioral "barriers" in the form of territoriality. Where animals occupy isolated pockets of suitable habitat, both types of barriers may be involved, together with "isolation by distance." However, the effects of territoriality alone may be observed within continuously distributed populations.

With this background, we may now consider the size of demes in wild populations of the house mouse and the mechanisms underlying population division. Several lines of evidence are available.

The behavioral and ecologic investigations of Crowcroft and Rowe (1963), Anderson and Hill (1965), Southwick (1955a, 1955b), and Brown (1953) demonstrate that wild house mice are strongly territorial (notwithstanding previous claims to the contrary; e.g., see Blair, 1953, p. 24) and that the territorial behavior of individuals organized into tribes (family groups or endogenous households) results in the subdivision of populations into small breeding units (demes), several of which may exist within single barns or other buildings. While neither the actual sizes of these demes nor the areal extent of their distributions is adequately known, there is reason to believe that they may be very small. One line of evidence comes from measurements of the distances moved by marked individuals within rooms or barns and the frequency of movement between barns or corn ricks. For example, Young et al. (1950), studying two populations in a room 170 by 100 feet, found that the average distance moved by a mouse between trappings was 12 feet, with 90 percent of the mice moving less than 30 feet. The estimated home range of the mouse in corn ricks in England is 50 square feet on a single level, with little vertical movement (Southern and Laurie, 1946). Surveying the ecologic literature, Petras (1967a, p. 269) concluded that the best estimate of physical migration between barns or other buildings is 5 percent, but the level of genetically effective migration is probably much lower.

In the absence of data sufficient for direct determination of deme size, indirect approaches have been applied to the problem. From ecologic data on the movements of mice into two corn ricks (A and B) in England (Rowe et al., 1963), Petras (1967a, p. 266), employing methods of estimation presented by Li (1955), has calculated the genetically effective population size (N_e) of the founder populations and the genetically effective size of the whole population $(\overline{N}_e)$ of a rick over several generations. If calculations are based on all immigrants (mice entering the ricks in the 10-month period between establishment of the ricks and their thrashing), N_e values of the founder populations are 70 for rick A and 77 for rick B.

The most convincing evidence of an insular pattern of subdivision comes from a recent study by Reimer and Petras (1967) of mice in "population cages" consisting of series of nest boxes connected by narrow runways. Despite the artificiality of the experimental situation, their results probably reflect the situation in more natural populations. Briefly, their results may be summarized as follows.

Populations of house mice become divided into tribes composed of a dominant male, several females, and several subordinate males. The formation of tribal units is due primarily to male territoriality, but females also contribute to the defense of the tribal territory. The tribes are stable over long periods (at least several generations) in spite of high densities, and, in older tribes, the dominant male is replaced by one of his offspring. Intertribe migration is rare and involves only females, there being effectively no movement of males between tribes. The genetically effective size of these units is five or less if the subordinate males do not contribute to the gene pool, or 10 or more if they do. Unfortunately, no information on the breeding status of subordinate males is available.

If wild populations of the house mouse are, in fact, subdivided into small tribes, with very limited intertribe migration, intertribe variation in gene frequencies resulting from random fluctuation is inevitable, and it should be possible to demonstrate genetic heterogeneity between barns or even within single barns large enough to support several tribes (Selander, 1970).

Interbarn and Interfarm Variation

Among large samples of mice from two or more barns on a single farm, it is often possible to demonstrate significant heterogeneity in allele frequencies at one or more loci. Representative data are presented in Tables 4 and 5 for Nieschwietz Farm, in the Ensinal Region, where 10 small chicken barns were sampled on October 22 (barns 1 to 5 and 7) and November 3 (barns 8 to 11), 1967. The observed and expected numbers of *Es-3* phenotypes are included in Table 4 to demonstrate the close agreement of these values that characterizes all of our samples.

Another example of marked interbarn variation is provided by data for collections from four chicken barns at Hildreth Farm, near Dripping Springs, central Texas (Table 6). These barns, which house laying hens held in three longitudinal rows of cages suspended above the earth floor, are 192 feet long and 24 feet wide and are set in a row, side-by-side, at intervals of 80 feet. Mark-and-recapture studies in July and August, 1967, indicated an average population size of 3,000 mice per barn, or a density of 5.9 mice per square yard. Interbarn frequencies are heterogeneous for all three loci. Sampling

TABLE 4

Allele Frequencies at *Es-3*[a] Locus in Samples from Nieschwietz Farm, Ensinal, Texas

Barn	Number of mice	Genotypes and phenotypes: Observed (expected)			Allele frequency	
		Es-3b / Es-3b MM	*Es-3c / Es-3c* SS	*Es-3b / Es-3c* MS	*Es-3b* M	*Es-3c* S
1	49	4 (5.4)	20 (21.4)	25 (22.1)	0.34	0.66
2	86	12 (11.4)	35 (34.4)	39 (40.2)	0.37	0.63
3	101	22 (20.8)	31 (29.8)	48 (50.3)	0.45	0.55
4	25	6 (3.5)	12 (9.5)	7 (12.0)	0.38	0.62
5	50	5 (5.0)	23 (23.0)	22 (22.0)	0.32	0.68
7	106	21 (19.8)	35 (33.8)	50 (52.3)	0.43	0.57
8	90	26 (23.9)	23 (20.9)	41 (45.2)	0.52	0.48
9	31	13 (13.4)	3 (3.4)	15 (14.1)	0.66	0.34
10	47	5 (3.5)	26 (24.5)	16 (19.0)	0.28	0.72
11	31	8 (5.3)	13 (10.3)	10 (15.3)	0.42	0.58
Pooled	616	122 (108.4)	221 (207.3)	273 (300.3)	0.420	0.580

[a]Heterogeneity $\chi^2_{(9)} = 40.12**$.

TABLE 5

Allele Frequencies at *Es-2*[a] and *Hbb*[b] Loci in Samples From Nieschwietz Farm, Ensinal, Texas

Barn	Number of mice	*Es-2*				*Hbb*	
		Es-2a "Silent"	*Es-2b* F	*Es-2c* M	*Es-2d* S	*Hbbd* Diffuse	*Hbbs* Single
1	49	0.00	0.77	0.03	0.20	0.17	0.83
2	86	0.00	0.84	0.01	0.15	0.23	0.77
3	101	0.06	0.78	0.01	0.15	0.17	0.83
4	25	0.00	0.68	0.10	0.22	0.22	0.78
5	50	0.13	0.63	0.08	0.16	0.10	0.90
7	106	0.00	0.78	0.04	0.18	0.17	0.83
8	90	0.09	0.76	0.03	0.12	0.11	0.89
9	31	0.00	0.73	0.11	0.16	0.13	0.87
10	47	0.00	0.89	0.01	0.10	0.22	0.78
11	31	0.00	0.86	0.03	0.11	0.11	0.89
Pooled	616	0.034	0.778	0.035	0.153	0.165	0.835

[a]Heterogeneity $\chi^2_{(27)} = 119.17**$ for *Es-2*, calculating numbers of alleles from maximum likelihood estimates of frequencies.
[b]Heterogeneity $\chi^2_{(9)} = 19.92*$ for *Hbb*.

TABLE 6

Allele Frequencies at Three Loci in Samples from Hildreth Farm, Austin Region, Texas

Sample	Date of collection	Number of mice	Es-2			Es-3		Hbb	
			Es-2a "Silent"	Es-2b F	Es-2c M	Es-3b M	Es-3c S	Hbbd Diffuse	Hbbs Single
Barn 1									
Whole Barn	April 1967	54	0.000	0.907	0.093	0.167	0.833	0.171	0.829
Whole Barn	July 1967	108	0.000	0.912	0.088	0.144	0.856	0.134	0.866
West Side	July 1967	54	0.000	0.981	0.019	0.148	0.852	0.128	0.872
East Side	July 1967	54	0.000	0.843	0.157	0.139	0.861	0.139	0.861
Whole Barn	December 1967	186	0.000	0.946	0.054	0.188	0.812	0.167	0.833
West Side	December 1967	48	0.000	0.969	0.031	0.292	0.708	0.187	0.813
East Side	December 1967	27	0.000	0.981	0.019	0.185	0.815	0.148	0.852
East Side	June 1968	583	0.000	0.880	0.120	0.122	0.878	0.166	0.834
Barn 2									
Whole Barn	August 1967	108	0.135	0.832	0.033	0.407	0.593	0.413	0.587
West Side	August 1967	54	0.086	0.895	0.019	0.472	0.528	0.470	0.530
East Side	August 1967	54	0.166	0.788	0.046	0.343	0.657	0.360	0.640
Whole Barn	December 1967	310	0.064	0.897	0.039	0.385	0.615	0.329	0.671
West Side	December 1967	101	0.092	0.860	0.048	0.446	0.554	0.396	0.604
East Side	December 1967	74	0.000	0.986	0.014	0.284	0.716	0.270	0.730
Whole Barn	May 1968	802	0.082	0.876	0.042	0.393	0.607	0.353	0.647
West Side	May 1968	423	0.094	0.871	0.036	0.433	0.567	0.402	0.598
East Side	May 1968	379	0.068	0.882	0.050	0.350	0.650	0.298	0.702

314

TABLE 6 continued

Sample	Date of collection	Number of mice	Es-2			Es-3		Hbb	
			Es-2a "Silent"	Es-2b F	Es-2c M	Es-3b M	Es-3c S	Hbbd Diffuse	Hbbs Single
Barn 3									
Whole Barn	August 1967	108	0.089	0.897	0.014	0.407	0.593	0.218	0.782
West Side	August 1967	54	0.000	0.972	0.028	0.435	0.565	0.194	0.806
East Side	August 1967	54	0.152	0.848	0.000	0.380	0.620	0.241	0.759
Whole Barn	December 1967	80	0.000	0.981	0.019	0.394	0.606	0.194	0.806
Barn 4									
Whole Barn	August 1967	108	0.060	0.856	0.084	0.343	0.657	0.189	0.811
West Side	August 1967	54	0.041	0.880	0.079	0.333	0.667	0.183	0.817
East Side	August 1967	54	0.075	0.834	0.090	0.352	0.648	0.194	0.806
Whole Barn	December 1967	198	0.046	0.859	0.095	0.407	0.593	0.220	0.780

through the year has revealed no seasonal variation in allele frequency for any of the loci studied. To date we have found no evidence of seasonal variation in any mouse population studied in Texas, and we have yet to detect short-term secular changes in barns sampled over the 2-year period of our study.

Additional examples of interbarn heterogeneity at one or more loci are given in Table 7.

To obtain an approximate measure of interbarn allele frequency variation within farms, we computed for the $Es-3^b$ and Hbb^s alleles weighted interbarn variances of arcsin transformations of frequencies for each of 26 farms, then pooled these variances (weighted by degrees of freedom) to provide an esti- mate of mean interbarn variance.* A similar analysis of pooled frequencies for farms within 18 regions of the state yielded an estimate of mean interfarm variance within regions. Representative interfarm variation is shown in Table 8, and the estimates of interbarn and interfarm variance are presented in Table 9. For both the *Es-3* and *Hbb* loci, the interfarm variance is about half again as large as the interbarn variance. Despite the fact that there appears to be greater geographic variation in Texas in allele frequencies at the *Es-3* locus than at the *Hbb* locus, the two loci are equally variable among farms within regions.

Intrabarn Variation

At the outset of our investigation, we assumed, as did Petras (1967a), that populations within barns are essentially panmictic. However, spatial sub-sampling within barns revealed significant variation in gene frequency. Repre- sentative data are presented in Table 10 for a large chicken barn at Lay Egg Ranch, where, for sampling purposes, the west half of the barn was divided into four equal areas of approximately 108 square yards each. The $Es-2^b$ allele is fixed in this barn, but allele frequencies for the *Es-3* and *Hb* loci are heterogeneous among the four subsamples.

In an effort to obtain a detailed picture of intrabarn variation, mice were live-trapped in a 1.6-foot grid-pattern throughout barn 2 and on the east side of barn 1 at Hildreth Farm (Table 6). A total of 870 traps set in the grid-pattern on the west side of barn 2 on the night of May 10, 1968, yielded 443 mice, and a similar trapping effort on the east side of barn 2 on the night of May 15, 1968, yielded 383 mice. On the night of June 3, 1968, 400 mice were caught in 870 traps set on the east side of barn 1; unsuccessful traps were left in position on the grid over the night of June 4 and yielded an

---

* Because the samples used in these calculations have only two alleles at each of these loci, similar variance estimates are obtained using the alternate alleles $Es-3^c$ and Hbb^d. However, because three or more alleles are represented at the $Es-2$ locus, variance estimates differ for the various alleles and, thus, are not directly comparable to those for the *Es-3* and *Hbb* loci.

TABLE 7

Interbarn Variation in Allele Frequency for Several Farms in Texas

Farm and Barn	Number of mice	Es-2				Es-3			Hbb	
		Es-2a "Silent"	Es-2b F	Es-2c M	Es-2d S	Es-3b M	Es-3c S	Es-3d F	Hbbd Diffuse	Hbbs Single
Barton Farm[a]										
Barn 1	71	0.13	0.87	0.00	0.00	0.75	0.25	0.00	0.25	0.75
Barn 3	126	0.00	1.00	0.00	0.00	0.68	0.32	0.00	0.14	0.86
Barn 10	90	0.00	1.00	0.00	0.00	0.74	0.26	0.00	0.40	0.60
Barn 11	51	0.00	1.00	0.00	0.00	0.83	0.17	0.00	0.33	0.67
Barn 12	106	0.00	1.00	0.00	0.00	0.77	0.23	0.00	0.26	0.74
Pickard Farm[b]										
Barn 1	75	0.00	0.75	0.09	0.16	0.43	0.57	0.00	0.36	0.64
Barn 2	60	0.00	0.55	0.24	0.21	0.72	0.24	0.04	0.48	0.52
Barn 4	60	0.00	0.61	0.17	0.22	0.78	0.22	0.00	0.62	0.62
Barn 5	61	0.00	0.61	0.10	0.29	0.81	0.18	0.01	0.62	0.38
Steep Hollow Farm[c]										
Barn 1	72	0.05	0.90	0.05	0.00	0.79	0.21	0.00	0.35	0.65
Barn 5	67	0.15	0.78	0.07	0.00	0.81	0.19	0.00	0.20	0.80
Barn 6	61	0.23	0.68	0.09	0.00	0.77	0.23	0.00	0.36	0.64

TABLE 7 continued

Farm and Barn	Number of mice	Es-2				Es-3			Hbb	
		Es-2a "Silent"	Es-2b F	Es-2c M	Es-2d S	Es-3b M	Es-3c S	Es-3d F	Hbbd Diffuse	Hbbs Single
Chessher Farm										
Barn R3	5	0.00	1.00	0.00	0.00	0.80	0.20	0.00	0.30	0.70
Barn R4	11	0.00	0.82	0.18	0.00	0.73	0.27	0.00	0.05	0.95
Barn R5	5	0.00	1.00	0.00	0.00	0.70	0.30	0.00	0.20	0.80
Barn I1	42	0.00	0.92	0.08	0.00	0.74	0.26	0.00	0.23	0.77
Barn I2	54	0.00	1.00	0.00	0.00	0.52	0.48	0.00	0.03	0.97
Barn I3	24	0.00	1.00	0.00	0.00	0.73	0.27	0.00	0.10	0.90
Barn I4	51	0.00	0.99	0.01	0.00	0.56	0.44	0.00	0.20	0.80
Barn I5	2	0.00	1.00	0.00	0.00	0.75	0.25	0.00	0.25	0.75

[a]Heterogeneity χ^2 = 98.71** (4 d.f.) for *Es-2*, 10.25* (4 d.f.) for *Es-3*, 39.31** (4 d.f.) for *Hbb*.
[b]Heterogeneity χ^2 = 24.35** (5 d.f.) for *Es-2*, 76.12** (5 d.f.) for *Es-3*, 25.55** (3 d.f.) for *Hbb*.
[c]Heterogeneity χ^2 = 21.21** (4 d.f.) for *Es-2*, 0.49 (2 d.f.) for *Es-3*, 10.20** (2 d.f.) for *Hbb*.
[d]Heterogeneity χ^2 = 31.67** (7 d.f.) for *Es-2*, 16.78* (7 d.f.) for *Es-3*, 24.23** (7 d.f.) for *Hbb*.

TABLE 8
Interfarm Variation in Allele Frequency for Several Regions in Texas

Region and Farm	Number of barns	Number of mice	Es-2				Es-3		Hbb	
			$Es\text{-}2^a$ "Silent"	$Es\text{-}2^b$ F	$Es\text{-}2^c$ M	$Es\text{-}2^d$ S	$Es\text{-}3^b$ M	$Es\text{-}3^c$ S	Hbb^d Diffuse	Hbb^s Single
Dallas Region[a]										
Littlebrook Farm	1	78	0.00	1.00	0.00	0.00	0.31	0.69	0.13	0.87
Mayhard Farm	4	73	0.05	0.92[b]	0.00	0.00	0.69	0.31	0.21	0.79
Rockwell Farm	2	33	0.00	0.98	0.02	0.00	0.42	0.58	0.23	0.77
Austin Region[c]										
Cook-Synoett Farm	1	105	0.00	0.99	0.01	0.00	0.31	0.69	0.13	0.87
Garfield Farm	1	12, 97[d]	0.00	0.98	0.02	0.00	0.25	0.75	0.18	0.82
Herbert Farm	1	65, 90[e]	0.00	0.94	0.06	0.00	0.46	0.54	0.22	0.78
Norco Farm	1	14	0.00	0.94	0.06	0.00	0.57	0.43	0.22	0.78
Robinson Farm	1	18, 52[f]	0.00	1.00	0.00	0.00	0.42	0.58	0.12	0.88
Lay Farm	1	324	0.00	1.00	0.00	0.00	0.40	0.60	0.25	0.75
Empire Farm	1	81, 83[g]	0.00	0.99	0.01	0.00	0.28	0.72	0.03	0.97
Stanley Farm	4	68	0.00	0.54	0.46	0.00	0.52	0.48	0.00	1.00
Hildreth Farm	4	432	0.07	0.87	0.06	0.00	0.32	0.68	0.24	0.76
Lubbock Region[h]										
Barton Farm	5	444	0.02	0.98	0.00	0.00	0.74	0.26	0.26	0.74
Gary Farm	4	139	0.06	0.94	0.00	0.00	0.70	0.30	0.08	0.92
Vance Farm	2	21	0.19	0.81	0.00	0.00	0.60	0.40	0.00	1.00

TABLE 8 continued

Region and Farm	Number of barns	Number of mice	Es-2				Es-3		Hbb	
			Es-2ᵃ "Silent"	Es-2ᵇ F	Es-2ᶜ M	Es-2ᵈ S	Es-3ᵇ M	Es-3ᶜ S	Hbbᵈ Diffuse	Hbbˢ Single
Bryan Region[l]										
Reliance Farm	4	248	0.00	0.73	0.27	0.00	0.69	0.31	0.27	0.73
Bullock Farm	1	42	0.15	0.84	0.01	0.00	0.84	0.16	0.10	0.90
Steep Hollow Farm	3	200	0.14	0.79	0.07	0.00	0.79	0.21	0.31	0.69
Syptac Farm	2	11	0.18	0.66	0.16	0.00	0.64	0.36	0.36	0.64
Ensinal Region[j]										
Nieschwietz Farm	10	616	0.03	0.78	0.04	0.15	0.42	0.58	0.17	0.83
Burkholder Farm	2	128	0.00	0.63	0.37	0.00	0.84	0.16	0.06	0.94

[a]Heterogeneity $\chi^2 = 18.11$** (4 d.f.) for *Es-2*, 43.95** (2 d.f.) for *Es-3*, 4.71 (2 d.f.) for *Hbb*.

[b]The *Es-2ᵉ* allele is also represented, at a frequency of 0.03.

[c]Heterogeneity $\chi^2 = 629.46$** (16 d.f.) for *Es-2*, 46.55** (8 d.f.) for *Es-3*, 90.65** (8 d.f.) for *Hbb*.

[d]12 mice for *Es-3* and *Hb*, 97 mice for *Es-2*.

[e]90 for *Es-3* and *Hb*, 65 for *Es-2*.

[f]18 for *Es-3* and *Hb*, 52 for *Es-2*.

[g]81 for *Es-3* and *Hb*, 83 for *Es-2*.

[h]Heterogeneity $\chi^2 = 40.85$** (2 d.f.) for *Es-2*, 5.78 (2 d.f.) for *Es-3*, 51.71** (2 d.f.) for *Hbb*.

[i]Heterogeneity $\chi^2 = 140.75$** (6 d.f.) for *Es-2*, 16.91** (3 d.f.) for *Es-3*, 16.41** (3 d.f.) for *Hbb*.

[j]Heterogeneity $\chi^2 = 291.50$** (3 d.f.) for *Es-2*, 146.97** (1 d.f.) for *Es-3*, 19.33** (1 d.f.) for *Hbb*.

TABLE 9

Pooled Interbarn and Interfarm Variances in Allele Frequency (Arcsin Transformation) for Three Loci in Samples from Texas

Source of variation	Number of farms or regions pooled	Arcsin variance (in degrees)	
		Es-3^b	Hbb^s
Interbarn within farms	26	52.39	43.81
Interfarm within regions	18	71.38	72.91

additional 183 mice. Positions on the grid at which mice were caught in barn 2, together with the genotypes for the *Es-3* locus are shown in Fig. 6.

Barn 2 was selected for detailed study because sampling in August, 1967, and December, 1967, had shown significant differences between the west and east sides in frequency of alleles at both the *Es-3* and *Hbb* loci (Table 6). These differences were again evident in May, and thus are known to have persisted for at least 8 months. The samples collected in May also demonstrate significant east-west variation in age structure and fecundity within barn 2, with higher percentages of juveniles and pregnant females on the east side (Selander, 1970).

Fig. 6 demonstrates that spatial variation in allele frequencies within barn 2 is not simply clinal from east to west; rather, each locus shows a complex mosaic pattern, with areas of high and low frequency of particular alleles on either side of the barn. The degree to which the spatial distribution of alleles departs from random on either side of a barn was determined by variance tests for homogeneity of the binomial distribution, employing counts

TABLE 10

Allele Frequencies at Two Loci in Samples from a Barn at Lay Ranch, Austin Region, Texas[a]

Sample	Number of mice	Es-3		Hbb	
		Es-3^b M	Es-3^c S	Hbb^d Diffuse	Hbb^s Single
West 1	96	0.542	0.458	0.255	0.745
West 2	75	0.520	0.480	0.120	0.880
West 3	70	0.314	0.686	0.250	0.750
West 4	59	0.119	0.881	0.424	0.576

[a]Heterogeneity $\chi^2_{(3)}$ = 68.26** for *Es-3*, 32.22** for *Hbb*.

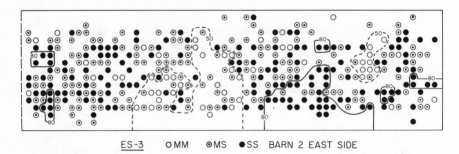

ES-3   O MM   ⊙ MS   ● SS   BARN 2 EAST SIDE

FIG. 6. Esterase 3 genotypes of mice collected on east side of barn 2 at Hildreth Farm. Circle: $Es-3^b/Es-3^b$ (MM). Dot in circle: $Es-3^b/Es-3^c$ (MS). Dot: $Es-3^c/Es-3^c$ (SS). M and S refer to electrophoretic bands of medium and slow mobilities, respectively. Distance between capture points is 1.6 feet. 50 percent and 80 percent isofrequency lines are indicated. (From Selander, 1970. *Amer. Zool.,* in press.)

of alleles from mice taken on 22 nonoverlapping "quadrats" of 25 grid-points. Two series of tests were made, one employing both adults and juveniles, and another using only adult mice. Results of these tests (Table 11) demonstrate a nonrandom component, with clumping of similar alleles. Only for the *Es-2* locus for adults on the east side of barn 2 does the x-square value fall below the 5 percent probability level, and, when the juvenile mice are included, a highly significant value is obtained.

To aid in visualizing patterns of regional variation in gene frequency within barn 2, the genotypes $Es-3^b/Es-3^b$, $Es-3^b/Es-3^c$, and $Es-3^c/Es-3^c$ were scored 1, 2, and 3, respectively, and for each grid-point a frequency estimate

TABLE 11

Tests of Heterogeneity in Distribution of Alleles among 22 "Quadrats" in Two Barns at Hildreth Farm

Barn and side	$\chi^2_{(21)}$ values for indicated locus		
	Es-2	*Es-3*	*Hbb*
Adults and Juveniles			
Barn 2			
West Side	63.26**	60.65**	33.93*
East Side	48.33**	65.59**	48.86**
Barn 1			
East Side	38.96**	50.37**	32.13*
Adults			
Barn 2			
West Side	61.44**	53.54**	36.01*
East Side	28.55	66.70**	41.78**

of the *Es-3ᶜ* allele was obtained by computing a two-dimensional running average of scores, in which the score of a mouse taken at a grid-point was weighted 1.0 and the scores of mice taken at grid-points in the five adjacent rows and columns on all sides of the grid-point were weighted by the inverse square of their distance from the grid-point. Similar estimates were obtained for the *Hbbᵈ* and *Es-2ᶜ* alleles.

With frequency estimates available for each grid-point, isofrequency contours can be drawn. Many of the smaller regions of high or low allele frequency reflect the chance association of several mice of similar genotype, but the major "peaks" and "valleys," as shown in Fig. 6, are believed to correspond to tribes or to groups of genetically related tribes occupying neighboring territories. Note that on the east side of barn 2 areas of frequencies of 0.50 and 0.80 for the *Es-3ᶜ* allele occur less than 10 feet apart.

HETEROZYGOTE DEFICIENCY AS AN INDEX TO SUBDIVISION.—Petras (1967a, p. 266), following Rasmussen (1964), has used an inbreeding coefficient (F) calculated from observed deficiencies of heterozygotes in samples as a measure of the degree of departure of populations from panmixia through subdivision. The coefficient adopted by Petras is $F = (H_e - H_o)/H_e$, where H_e is the proportion of heterozygotes expected on the basis of Hardy-Weinberg equilibrium, and H_o is the proportion observed. We prefer to express heterozygote deviation as $D = (H_o - H_e)/H_e$, since observed deficiencies will then be indicated by negative values.

From data on the *Es-2* locus in 296 mice collected over a 2- to 4-year period in numerous small buildings on five adjacant farms in Michigan, Petras (1967a) determined that $F = 0.1804 \pm 0.0626$ ($D = -0.1804$ or, by small sample formulas, $D = -0.1819 \pm 0.0601$.) However, the validity of the inbreeding coefficient approach to the problem of measuring subdivision may be questioned on several grounds. First, any temporal variation in allele frequency will be interpreted as spatial heterogeneity due to inbreeding. And we note that Petras' data (1967a, p. 262) suggest an increase in frequency of the *Es-2ᵃ* allele from 1960 to 1962. Second, the effect of subdivision may be partially or completely masked by the effects of heterosis or directional selection. For example, for the *Es-3* locus in samples from Texas D is inversely related to the number of mice trapped in a barn on a single night, which is, to some degree, positively correlated with population size and, in turn, with barn size (Fig. 7). On the average, D is positive in small barns in which only one or a few demes would be present, and negative in very large barns in which numerous demes would exist. Possibly there is heterosis at this locus, which in larger barns is masked by a heterozygote deficiency resulting from subdivision and inbreeding, but, in any event, a meaningful measurement of inbreeding cannot be obtained from this data.

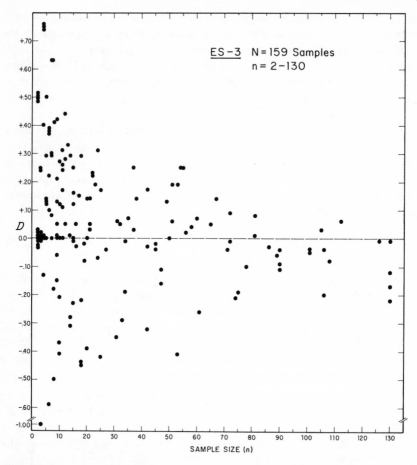

FIG. 7. Relationship of heterozygote deviation (D) and sample size (n) for Esterase 3 locus in samples from Texas. Equation for linear regression of D on n, with D values weighted by reciprocals of their variances: $D = 0.148 - 0.00208\,n$ $(F = 42.79^{**}$, with 1 and 157 d.f.). (From Selander, 1970. *Amer. Zool.,* in press.)

DISCUSSION AND CONCLUSIONS

Genetic Variation in Wild Populations

An analysis of genetic variation in biochemical characters clearly points up the great differences between inbred strains and wild populations of the house mouse. Whereas individuals of a given inbred strain are uniformly homozygous, all wild populations are highly polymorphic. For many polymorphic loci studied, more alleles are represented in wild populations in Texas than in all of the inbred strains combined. For example, only three alleles are known at the *Es-2* locus in inbred strains, yet at least five occur in Texas. Only two alleles ($Es-3^a$ and Hbb^p) for two loci described from inbred strains have yet to be detected in wild populations. It is apparent that the genetic variation represented in inbred strains is only a fraction of the total variation in the species as a whole.

In wild populations, many of the alleles at the various loci are very widespread, perhaps worldwide, in distribution, while others are relatively restricted geographically. Within a single subregion of continental North America, as, for example, within Texas, there is conspicuous geographic variation in allele frequencies at all polymorphic loci studied. Geographic patterns are clinal for the most part, and those of different loci are not strongly concordant. In general, the most distinctive allele frequencies are found in the pine-woods region of eastern Texas and in the Rio Grande Valley region, both of which have distinctive climatologic, floral, and faunal features. The patterns of variation at these biochemical loci in Texas are not unlike those shown by morphologic or other characters of native animals, a fact which suggests adaptive response to geographic patterns of environmental factors.

Our investigations of variation in allele frequencies at biochemical loci have demonstrated heterogeneity at all levels of population structure: intrabarn, interbarn, interfarm, and geographic (regional). At the lowest level, our findings confirm the hypothesis, derived from theoretic considerations and from ecologic and behavioral studies of mice in enclosures and under natural conditions, that populations are subdivided into units (family groups, endogenous households, or tribes) with an effective breeding size well below that at which random fluctuations and inbreeding play a major role in determining gene frequencies. On the basis of the available evidence, we presume that territorial behavior of tribe members, especially males, is the principal factor reducing migration among tribes and thus creating an insular pattern of subdivision.

The effects on population structure of interdemic distance and interspersed areas of unsuitable habitat cannot at this time be properly evaluated. However, it is likely that the physical isolation imposed by discontinuities in habitat between barns within farms and between farms within geographic regions contributes significantly to the total genetic variance in the species. Petras' (1967a) conclusion that "population discontinuity observed between farms because of unsuitable habitat appears to contribute very little to the inbreeding coefficient [i.e., to population subdivision]" may be applicable to his own data derived from populations on adjacent farms, but, as our analysis of interfarm variance has shown, it does not hold generally.

The existence of heterogeneity in gene frequencies among barns on the same farm is indicative of very restricted migration among barns, even where they are situated only a few yards apart. While the possibility of variation in selective factors acting on mice in different barns cannot be entirely ruled out, we doubt that selection is a major cause of interbarn heterogeneity. On many of the farms, the barns are identical in structure, and we have been unable to identify any environmental factor in which the barns differ. With regard to interfarm variation in gene frequencies within regions, the hypothesis of differential selection is more plausible, but even at this level of population structure stochastic elements involved in the establishment of farm populations by small numbers of founders may be of overriding importance.

Evolutionary Significance of Genetic Drift

That genetic drift—the tendency toward random fixation of alleles— occurs in small, isolated demes is indisputable theoretically and is convincingly demonstrated by studies of house mouse populations. The central question, however, is: What is the evolutionary significance of random fluctuations of genes in local populations (Mayr, 1963, p. 204)? Wright (1951) has himself noted that "fluctuations in gene frequency in small, completely isolated communities rarely if ever contribute to evolutionary advances, but merely to trivial differentiation. . . ." Because small, isolated demes are only temporarily withdrawn from the free gene exchange of the species, it is probable, as stressed by Mayr (1963, pp. 210–211), that random fixation is of negligible evolutionary importance. In most wild populations of the house mouse, the average "life-span" of demes may be only a few generations; with depletion of the temporarily available food supplies exploited by local populations, dispersal occurs, with heavy mortality and the eventual founding of new demes. Hence, the tendency toward fixation of alleles is periodically reversed. In populations inhabiting chicken barns or other structures in which the food supply is not subject to periodic depletion, demes may persist for several years. Yet eventually the panmictic effects of dispersal and reorganization of population structure are experienced.

The fact that genes of *Mus musculus* are subject to genetic drift cannot be advanced as evidence of their selective neutrality. The fact that geographic variation in allele frequency at all of the polymorphic loci studied is clinal provides strong circumstantial evidence against selective neutrality (Mayr, 1963, p. 207). Moreover, for the *Es-3* and *Hbb* loci an excess of heterozygotes in samples from small local populations suggests a balanced polymorphism mediated by heterozygote advantage (Selander, 1970). Thus, our findings do not support Bruell's (1967, p. 279) suggestion that there exist in wild house mouse populations gene pools "untested and unchallenged" by selection and, hence, filled with "a random assortment of genes," many of which are selectively neutral.

Social Systems and Genetic Variation

It is perhaps worthwhile to speculate on the adaptive advantage of the polygamous, territorial social system of the house mouse that is in large part responsible for the genetic subdivision of wild populations. The house mouse is a species which, in its commensal form, evolved to exploit grain and other agricultural products of man that are abundant at seasonal periods and are often stored in small areas. Under these conditions, there would seem to be a selective premium on a rapid population growth rate to permit rapid exploitation of food supplies once they are located by mice. Moreover, it may be advantageous for males to hold territories in or near the food supply to insure cover, food, and nesting areas, and to prevent interference with reproduction. Because the male does not participate in parental care, there is no loss in fitness to females which settle in a territory already occupied by one or more females. That young males remain in the parental territory is at first sight puzzling but may be explained as follows. If the chances of a young male dispersing to a new supply of food and establishing a territory are very poor, it may be advantageous for the male to remain in or near a food supply, on the chance of taking over the parental territory or an adjacent one with the death of a dominant male. It may also be possible that young males get occasional matings within their parental territory. With depletion of the food supply, dispersal occurs, a few individuals find unexploited food supplies, and the cycle continues.

There are several reports indicating that the interpretation of social structure derived from studies of commensal populations may not apply to populations in certain other habitat types. For example, Berry (1968, pp. 466–467) suggests that "the whole population on Skokholm [Island] can be regarded as a panmictic unit." Skokholm Island has an area of 244 acres, and estimates of population size range from 150 to 400 mice in spring, following extremely high mortality in winter, to 2,500 to 5,000 mice in the fall. Berry (1968, p. 467) believes that the Skokholm mice do not have a "harem" system, but admits that little is known of the social system. He cites a nearly equal sex ratio as evidence of a different social system, noting that females tend to predominate in buildings and corn ricks (Evans, 1949; Southwick, 1958; Rowe et al., 1963). He also notes that, whereas commensal mice have home ranges on the order of 15 m^2 (Young et al., 1950; Brown, 1953; Southern, 1954), feral mice generally have a much wider range (Baker, 1946; Fitch, 1958; Justice, 1961).

Against the idea that feral populations inhabiting small islands are panmictic is evidence from a study by Anderson et al. (1964), in which males heterozygous for the recessive allele t^{w11} were released in 1956 and 1957 on Great Gull Island, Long Island Sound, New York, with an area of 18 acres supporting a population of between 60 and 600 mice. By 1961, the introduced allele had reached high frequency in the immediate area of the introduction but was spreading very slowly.

Amount of Genic Variation in Natural Populations

The electrophorogram method permits for the first time assessment of the amount of genic variation in natural populations. For 18 randomly chosen loci in samples of natural populations of *Drosophila pseudoobscura*, Lewontin and Hubby (1966; see also Hubby and Lewontin, 1966) obtained the following estimates: 39 percent of loci in the genome are polymorphic (two or more alleles) over the whole species; the average population is polymorphic for 30 percent of all loci; and, for the individual genome, between 8 and 15 percent of the loci will be in heterozygous state. Similarly, Stone et al. (1968) estimate that at least 45 percent of the enzyme "systems" studied in populations of *D. nasuta* and *D. ananassae* in Fiji and Samoa are polymorphic. For human populations, an estimate of 25 percent is provided by Harris' (1966; see also Harris et al., 1968) work on 12 randomly chosen enzymes. Our estimate of the minimal proportion of polymorphic loci in the house mouse species as a whole, as indicated by samples from North America, and Denmark, is 41 percent. The average population is polymorphic at 26 percent of loci, and the average individual heterozygosity is 8.5 percent (Selander, Hunt, and Yang, 1969).

Current estimates of genic heterozygosity based on variation in electrophoretic mobility are minimal, since the technique detects only a part of the variation present. Lewontin (1967, p. 40) estimates that only about 26 percent of mutations result in a net change in mobility of proteins, in which case, essentially every locus is polymorphic and the proportion of heterozygosity per individual is about 50 percent.

For alleles with frequencies of only a few percent, a balance between mutation and selection against the heterozygotes provides a sufficient explanation for their maintenance in populations. However, the maintenance of alleles at higher frequencies requires mechanisms involving some form of selection, the nature of which constitutes a central problem in population genetics and evolutionary theory. Because the maintenance of the unexpectedly large amount of genic variation by heterotic selection (balanced polymorphism) would seem to involve intolerable genetic loads in populations (Lewontin and Hubby, 1966; but see also Sved et al., 1967; Milkman, 1967; and J. L. King, 1967), additional selective mechanisms are sought. For example, Kojima and Yarbrough (1967) propose that frequency-dependent selection, with selective neutrality of alleles (or inversions) at equilibrium frequencies (and, hence, minimal load), may be a general mechanism maintaining a large proportion of genic polymorphism in natural populations. This type of selection has been demonstrated in laboratory populations for inversion systems of the second and third chromosomes in *Drosophila ananassae* (Tobari and Kojima, 1967) and for the esterase 6 locus in *D. melanogaster* (Kojima and Yarbrough, 1967; Yarbrough and Kojima, 1967), and has been implicated in mating success of males of *Drosophila* species (Ehrman, 1966).

Final Comment

Bruell (Chapter 10) sets as the major goal of behavioral population genetics the elimination of out-dated "typologic thinking" still prevalent among psychologists. We will conclude our contribution to this volume by emphasizing another potentially beneficial effect of interaction between behavior geneticists and population geneticists. While population behavior geneticists are to concern themselves with the intrapopulation and interpopulation variation in behavior and its genetic basis, population geneticists are interested in the ways in which behavior affects the genetic structure of populations. J. A. King (1967) has discussed the present state of the "ecological genetics of behavior," or the study of the relationships between behavior and changes in gene frequencies of populations. Although the genetic consequences of behavior—including, importantly, those behaviors affecting migration, dispersion, and mating systems—have been worked out mathematically by R. A. Fisher, S. Wright, and others, and have been discussed by ecologists (e.g., Blair, 1953), we have few data bearing on this subject, which, thus, can only be discussed in generalities (J. A. King, 1967, p. 33). Hopefully, research directed toward an understanding of the interaction of genetics and behavior in natural populations of the house mouse and other species will establish a better balance between theory and empiric evidence.

REFERENCES

Allen, J. A., and F. M. Chapman. 1897. On a collection of mammals from Jalapa and Las Vigas, state of Vera Cruz, Mexico. Bull. Amer. Mus. Nat. Hist., 9:197–208.

Anderson, P. K. 1964. Lethal alleles in *Mus musculus:* Local distribution and evidence for isolation of demes. Science, 145:177–178.

——— L. C. Dunn, and A. B. Beasley. 1964. Introduction of a lethal allele into a feral house mouse population. Amer. Nat., 48:57–64.

——— and J. L. Hill. 1965. *Mus musculus:* Experimental induction of territory formation. Science, 148:1753–1755.

Bader, R. S. 1956. Variability in wild and inbred mammalian populations. Quart. J. Florida Acad. Sci., 19:14–34.

——— 1965. A partition of variance in dental traits of the house mouse. J. Mammal., 46:384–388.

——— and W. H. Lehmann. 1965. Phenotypic and genotypic variation in odontometric traits of the house mouse. Amer. Midl. Nat., 74:28–38.

Baker, H. G., and G. L. Stebbins, eds. 1965. The Genetics of Colonizing Species, New York, Academic Press, Inc.

Baker, R. H. 1946. A study of rodent populations on Guam, Mariana Islands. Ecol. Monogr., 16:393–408.

Barrett-Hamilton, G. E. H., and M. A. C. Hinton. 1921. A history of British mammals. Part 20, London. Gurney and Jackson.

Berry, R. J. 1963. Epigenetic polymorphism in wild populations of *Mus musculus*. Genet. Res. (Camb.), 4:193–220.

———— 1964. The evolution of an island population of the house mouse. Evolution, 18:468–483.

———— 1968. The ecology of an island population of the house mouse. J. Anim. Ecol., 37:445–470.

———— and A. G. Searle. 1963. Epigenetic polymorphism of the rodent skeleton. Proc. Zool. Soc. Lond., 140:577–615.

Biggers, J. D., A. McLaren, and D. Michie. 1958. Variance control in the animal house. Nature, 182:77–80.

Blair, W. F. 1953. Population dynamics of rodents and other small mammals. Advances Genet., 5:1–41.

Bonaventura, J. 1969. Polymeric hemoglobins of the house mouse (*Mus musculus* L.): Isolation of cysteinyl peptides. Biochem. Genet. 3:239–247.

———— and A. Riggs. 1967. Polymerization of hemoglobins of mouse and man: Structural basis. Science, 158:800–802.

Brown, L. N. 1965. Selection in a population of house mice containing mutant individuals. J. Mammal., 46:461–465.

Brown, R. Z. 1953. Social behavior, reproduction and population changes in the house mouse (*Mus musculus* L.). Ecol. Monogr., 23:217–240.

Bruck, D. 1957. Male segregation ratio advantage as a factor in maintaining lethal alleles in wild populations of house mice. Proc. Nat. Acad. Sci. U.S.A., 43:152–158.

Bruell, J. H. 1967. Behavioral heterosis, *In* Hirsch, J., ed., Behavior-Genetic Analysis, New York, McGraw-Hill Book Co., pp. 270–286.

Crowcroft, P., and F. P. Rowe. 1963. Social organization and territorial behaviour in the wild house mouse (*Mus musculus* L.). Proc. Zool. Soc. Lond., 140:517–531.

Doel, M. S. 1958. Genetical studies on the skeleton of the mouse. XXIV. Further data on skeletal variation in wild populations. J. Embryol. Exp. Morphol., 6:569–574.

Dunn, L. C. 1942. Studies on spotting patterns. V. Further analysis of minor spotting genes in the house mouse. Genetics, 27:258–267.

———— A. B. Beasley, and H. Tinker. 1958. Relative fitness of wild house mice heterozygous for a lethal allele. Amer. Nat., 92:215–220.

———— A. B. Beasley, and H. Tinker. 1960. Polymorphisms in populations of wild house mice. J. Mammal., 41:220–229.

———— and D. Bennett. 1966. Report on viability. Mouse News Letter, 34:20–22.

Eaton, O. N., and E. Schwarz. 1946. The "snowy belly" mouse. J. Hered., 37:31–32.

Ehrman, L. 1966. Mating success and genotype frequency in *Drosophila*. Anim. Behav., 14:332–339.

Engels, W. L. 1948. White-bellied house mice. J. Hered., 39:94–96.

Evans, F. C. 1949. A population study of house mice (*Mus musculus*) following a period of local abundance. J. Mammal., 30:351–363.

———— and T. I. Storer. 1944. Abundance of house mice at Davis, California, in 1941–42. J. Mammal., 25:89–90.

———— and H. G. Vevers. 1938. Notes on the biology of the Faeroe mouse (*Mus musculus faeroensis*). J. Anim. Ecol., 7:290–297.

Falconer, D. S. 1947. Genetics of "snowy belly" in the house-mouse. J. Hered., 38:215–219.

Fitch, H. S. 1958. Home ranges, territories and seasonal movements of vertebrates on the natural history reservation. Univ. Kansas Publ. Mus. Nat. Hist., 11:63–326.

Grüneberg, H. 1952. The Genetics of the Mouse, 2nd ed., The Hague, Nijhoff.

Haldane, J. B. S. 1956. The estimation and significance of the logarithm of a ratio of frequencies. Ann. Hum. Genet., 20:309–311.

Hall, E. R., and K. R. Kelson. 1959. The Mammals of North America, New York, The Ronald Press Co., Vol. 2.

Harris, H. 1966. Enzyme polymorphisms in man. Proc. Roy. Soc. Lond., B164:298–310.

———— D. A. Hopkinson, and J. Luffman. 1968. Enzyme diversity in human populations. Ann. N.Y. Acad. Sci., 151:232–242.

Hubby, J. L., and R. C. Lewontin. 1966. A molecular approach to the study of genic heterozygosity in natural populations. I. The number of alleles at different loci in *Drosophila pseudoobscura*. Genetics, 54:577–594.

Hunter, R. L., and D. S. Strachan. 1961. The esterases of mouse blood. Ann. N.Y. Acad. Sci., 94:861–867.

Hutton, J. J., J. Bishop, R. Schweet, and E. S. Russell. 1962. Hemoglobin inheritance in inbred mouse strains, I. Structural differences. Proc. Nat. Acad. Sci. U.S.A., 48:1505–1513.

Johnston, R. F., and R. K. Selander. 1964. House sparrows: Rapid evolution of races in North America. Science, 144:548–550.

Justice, K. E. 1961. A new method for measuring home ranges of small mammals. J. Mammal., 42:462–470.

———— 1962. Ecological and Genetical Studies of Evolutionary Forces Acting on Desert Populations of *Mus musculus,* Tucson, Arizona-Sonora Desert Museum.

Kaplan, N. O. 1968. Nature of multiple molecular forms of enzymes. Ann. N.Y. Acad. Sci., 151:382–399.

King, J. A. 1967. Behavioral modification of the gene pool. *In* Hirsch, J., ed., Behavior-Genetic Analysis, New York, McGraw-Hill Book Co., pp. 22–43.

King, J. L. 1967. Continuously distributed factors affecting fitness. Genetics, 55:483–492.

Koford, C. B. 1968. Peruvian desert mice: Water independence, competition, and breeding cycle near the equator. Science, 160:552–553.

Kojima, K., and K. M. Yarbrough. 1967. Frequency-dependent selection at the esterase 6 locus in *Drosophila melanogaster*. Proc. Nat. Acad. Sci. U.S.A., 57:645–649.

Levene, H. 1949. On a matching problem arising in genetics. Ann. Math. Statist., 20:91–94.

Levin, B., M. L. Petras, and D. I. Rasmussen. 1964. The effect of migration in maintaining a polymorphism in the house mouse. Genetics, 50:264–265.

Lewontin, R. C. 1962. Interdeme selection controlling a polymorphism in the house mouse. Amer. Nat., 96:65–78.

——— 1967. Population genetics. Annual Reviews of Genetics, Palo Alto, Calif., Annual Reviews, Inc., Vol. 1, pp. 37–70.

——— 1968. The effects of differential viability on the population dynamics of *t* alleles in the house mouse. Evolution, 22:262–273.

——— and L. C. Dunn. 1960. The evolutionary dynamics of a polymorphism in the house mouse. Genetics, 45:705–722.

——— and J. L. Hubby. 1966. A molecular approach to the study of genic heterozygosity in natural populations. II. Amount of variation and degree of heterozygosity in natural populations of *Drosophila pseudoobscura*. Genetics, 54:595–609.

Li, C. C. 1955. Population Genetics, Chicago, Univ. Chicago Press.

Lush, I. E. 1967. The Biochemical Genetics of Vertebrates Except Man, Amsterdam, North-Holland Publishing Co.

MacIntyre, R. J., and T. R. F. Wright. 1966. Response of esterase 6 alleles of *Drosophila melanogaster* and *D. simulans* to selection in experimental populations. Genetics, 53:371–387.

Markert, C. L. 1968. The molecular basis for isozymes. Ann. N. Y. Acad. Sci., 151:14–40.

Mayr, E. 1963. Animal Species and Evolution, Cambridge, Mass., Harvard Univ. Press.

Milkman, R. D. 1967. Heterosis as a major cause of heterozygosis in nature. Genetics, 55:493–495.

Miller, G. S. 1912. Catalogue of the mammals of western Europe (Europe exclusive of Russia) in the collection of the British Museum, London.

Morton, J. R. 1962. Starch-gel electrophoresis of mouse haemoglobins. Nature, 194:383–384.

——— 1966. The multiple electrophoretic bands of mouse haemoglobins. Genet. Res. (Camb.), 7:76–85.

Moulthrop, P. N. 1942. Description of a new house mouse from Cuba. Sci. Publ. Cleveland Mus. Nat. Hist., 5:79–82.

Ogita, Z. 1968. Genetic control of enzymes. Ann. N.Y. Acad. Sci., 151:243–262.

Pelzer, C. F. 1965. Genetic control of erythrocyte esterase forms in *Mus musculus*. Genetics, 52:819–828.

Petras, M. L. 1963. Genetic control of a serum esterase component in *Mus musculus*. Proc. Nat. Acad. Sci. U.S.A., 50:112–116.

——— 1967c. Studies of natural populations of *Mus*. III. Coat color polymorphisms and their bearing on breeding structure. Evolution, 21:259–274.

——— 1967b. Studies of natural populations of *Mus*. II. Polymorphism at the *T* locus. Evolution, 21:466–478.

——— 1967c. Studies of natural populations of *Mus*. III. Coat color polymorphisms. Canad. J. Genet. Cytol., 9:287–296.

——— and F. G. Biddle. 1967. Serum esterases in the house mouse, *Mus musculus*. Canad. J. Genet. Cytol., 9:704–710.

Popp, R. A. 1965. Loci linkage of serum esterase patterns and oligosyndactyly. J. Hered., 56:107–108.

————— 1966. Inheritance of an erythrocyte and kidney esterase in the mouse. J. Hered., 57:197–201.

————— and D. M. Popp. 1962. Inheritance of serum esterases having different electrophoretic patterns. J. Hered., 53:111–114.

————— and W. St. Amand. 1960. Studies on mouse hemoglobin loci. I. Identification of hemoglobin types and linkage of hemoglobin with albinism. J. Hered., 51:141–144.

————— and W. St. Amand. 1964. A sex difference in recombination frequency in the albinism-hemoglobin interval of linkage group I in the mouse. J. Hered., 55:101–103.

Rasmussen, D. I. 1964. Blood group polymorphism and inbreeding in natural populations of the deer mouse *Peromyscus maniculatus gracilis*. Evolution, 18:219–229.

Reimer, J. D., and M. L. Petras. 1967. Breeding structure of the house mouse, *Mus musculus,* in a population cage. J. Mammal., 48:88–99.

Remington, C. L. 1968. Suture-zone hybridization. *In* Dobzhansky, T., Hecht, M., and Steere, W., eds., Evolutionary Biology, New York, Appleton-Century-Crofts, Vol. 2, pp. 321–428.

Rowe, F. P., E. J. Taylor, and H. J. Chudley. 1963. The numbers and movements of house-mice (*Mus musculus*) in the vicinity of four corn ricks. J. Anim. Ecol., 32:87–97.

Ruddle, F. H., and L. Harrington. 1967. Tissue specific esterase isozymes of the mouse (*Mus musculus*). J. Exp. Zool., 166:51–64.

————— and T. H. Roderick. 1966. The genetic control of two types of esterases in inbred strains of the mouse. Genetics, 54:191–202.

————— and T. H. Roderick. 1968. Allelically-determined isozyme polymorphisms in laboratory populations of mice. Ann. N.Y. Acad. Sci., 151:531–539.

Russell, E. S., and P. S. Gerald. 1958. Inherited electrophoretic hemoglobin patterns among 20 inbred strains of mice. Science, 128:1569–1570.

Schwarz, E., and H. K. Schwarz. 1943. The wild and commensal stocks of the house mouse, *Mus musculus* Linnaeus. J. Mammal., 24:59–72.

Selander, R. K. 1969. The ecological aspects of the systematics of animals. *In* Stevens, R. B., ed., Systematic Biology. Proceedings of an International Conference, Washington, National Academy of Sciences Publ. 1962, pp. 213–239.

————— 1970. Behavior and genetic variation in natural poulations. Amer. Zool., in press.

————— W. G. Hunt, and S. Y. Yang. 1969. Protein polymorphism and genic heterozygosity in two European subspecies of the house mouse. Evolution, 23:379–390.

————— and S. Y. Yang. 1969. Protein polymorphism and genic heterozygosity in a wild population of the house mouse (*Mus musculus*). Genetics, in press.

————— S. Y. Yang, and W. G. Hunt. 1969. Polymorphism in esterases and hemoglobin in wild populations of the house mouse (*Mus musculus*). Studies in Genetics V. Univ. Texas Publ. 6918:271–338.

Shaw, C. R. 1969. Isozymes: classification, frequency, and significance. Internat. Rev. Cytol. 25:297–332.

Simpson, G. G. 1961. Principles of Animal Taxonomy, New York, Columbia University Press.

Smith, I. 1968. Chromatographic and Electrophoretic Techniques, New York, Interscience Pubs., John Wiley and Sons, Inc.

Southern, H. N., ed. 1954. Control of Rats and Mice, House mice, Oxford, Clarendon Press, Vol. 3.

—— and E. M. O. Laurie. 1946. The house mouse (*Mus musculus*) in corn ricks. J. Anim. Ecol., 7:134–149.

Southwick, C. H. 1955a. The population dynamics of confined house mice supplied with unlimited food. Ecology, 36:212–225.

—— 1955b. Regulatory mechanisms of house mouse populations: Social behavior affecting litter survival. Ecology, 36:627–634.

—— 1958. Population characteristics of house mice living in English corn ricks: Density relationships. Proc. Zool. Soc. Lond., 131:163–175.

Stone, W. S., M. R. Wheeler, F. M. Johnson, and K. Kojima. 1968. Genetic variation in natural island populations of members of the *Drosophila nasuta* and *Drosophila ananassae* subgroups. Proc. Nat. Acad. Sci. U.S.A., 59:102–109.

Sved, J. A., T. E. Reed, and W. F. Bodmer. 1967. The number of balanced polymorphisms that can be maintained in a natural population. Genetics, 55:469–481.

Tobari, Y. N., and K. Kojima. 1967. Selective modes associated with inversion karyotypes in *Drosophila ananassae*. I. Frequency-dependent selection. Genetics, 57:179–188.

Ursin, E. 1952. Occurrence of voles, mice, and rats (Muridae) in Denmark, with a special note on a zone intergradation between two subspecies of the house mouse (*Mus musculus* L.). Medd. Dansk. Naturhist. Foren. 114:217–244.

Van Valen, L. 1965. Selection in natural populations. IV. British house mice (*Mus musculus*). Genetica, 36:119–134.

Walker, E. P. 1964. Mammals of the World, Baltimore, Johns Hopkins Press, Vol. 2.

Watts, D. C. 1968. Variation in enzyme structure and function: The guidelines of evolution. *In* Lowenstein, O., ed., Advances in Comparative Physiology and Biochemistry, New York, Academic Press, Inc., pp. 1–114.

Weber, W. 1950. Genetical studies on the skeleton of the mouse. III. Skeletal variation in wild populations. J. Genet., 50:174–178.

Wilson, E. O., and W. L. Brown. 1953. The subspecific concept and its taxonomic application. Syst. Zool., 2:97–111.

Wright, S. 1951. The genetical structure of populations. Ann. Eugen., 15:323–354.

Yang, S. Y., and R. K. Selander. 1968. Hybridization in the grackle *Quiscalus quiscula* in Louisiana. Syst. Zool., 17:107–143.

Yarbrough, K. M., and K. Kojima. 1967. The mode of selection at the polymorphic esterase 6 locus in cage populations of *Drosophila melanogaster*. Genetics, 57:677–686.

Young, H., R. L. Strecker, and J. T. Emlen, Jr. 1950. Localization of activity in two indoor populations of house mice, *Mus musculus*. J. Mammal., 31:403–410.

Zimmerman, K. 1949. Zur Kenntnis der mitteleuropaeischen Hausmäuse. Zool. Jahrb. (Abt. f. Syst., Oekol. u. Geogr. d. Tiere), 78:301–322.

Index

Aboriginal mice 275, 284, 295
Activity 13ff, 24, 31ff, 40, 105, 188, 190, 191, 197
Albinism and behavior 48ff, 76, 144, 170, 185ff
Alcohol preference 215
Audiogenic seizure susceptibility 219ff

Behavioral tests 10ff, 31, 58, 73, 96, 116, 188, 222
Biochemical mechanisms 146, 155, 219ff
Biochemical polymorphism 301ff
Biometric analyses 67ff, 77ff, 97
Blood proteins 301ff

Chromosome mapping 161ff
Coadapted gene pools 266, 294
Coat color and behavior 48ff, 76, 144ff, 170, 176ff. See also Albinism and behavior.
Commensal mice 276, 284, 295
Convulsions 219ff, 225. See also Audiogenic seizure susceptibility.
Critical periods in development 163, 213, 225
Crossbreeding 7, 30, 61, 96, 224

Defecation 14, 23ff
Diallel cross method 6, 30, 75, 96
Domesticated mice 278

Dominance 26, 32, 62
Drosophila 29, 164, 272

Electroconvulsive shock 225
Electrophoresis 293ff
Emotionality 23ff, 100
Endocrine system 169, 212, 213
Environmental interactions 32, 67, 82, 99ff, 131, 141, 148ff, 195, 211, 305
Enzymes 163, 170, 211, 212, 221ff, 301ff
Esterase 302ff

Feral mice 278, 295

GABA 243
Gene pair model 126
Genetic correlation 16, 29, 39, 46, 123
Genetic drift 305, 308, 326

Handedness 115ff
Hemoglobin 302ff
Heritability 16, 26, 35, 38, 45, 82, 98
Heterogeneous population 5, 265
Heterosis 62, 83
Hoarding 91ff
Human handedness 122, 165

Inbred mice
  A   7ff
  A/Crgl   52
  AKR   7ff
  AKR/J   107
  BALB/c   7ff, 100
  BALB/cJ   30, 52, 68
  B6D2F   148
  C3H   95, 100, 216
  C3H/2   7ff
  C3H/HeJ   102
  C57   96
  C57BL   7ff, 216
  C57BL/10   100
  C57BL/6J   30, 48, 52, 61ff, 107,
    116, 144, 148, 155, 184ff, 224ff
  C57BL/Crgl   52
  DBA/1J   221
  DBA/2   7ff
  DBA/2J   61ff, 116, 148, 221,
    224ff
  Is/Bi   7ff
  JK   96, 102
  RIII   7ff
  SJL/J   155
  SWR/J   101
  YS/ChWf   155
Inbred strains, limitations of   4
Inbreeding depression   323
Intraclass correlation   38
Isolating mechanisms   272, 294

Learning   195, 216

Major gene effects   48, 76, 139ff,
  162ff
Maternal effects   32, 76
Mather's criteria   35, 80, 96
Mechanism-specific behavior   147,
  207ff
Metrazol seizure susceptibility   225
Mutants   139ff

Natural selection and evolution   93,
  149, 264, 284, 293ff, 326

Nervous system   213, 220, 227ff
Norepinephrine   222, 227

Open-field behavior   23ff, 100, 188

Parent-offspring regression   28, 37
Phenylketonuria and seizure suscepti-
    bility   163, 221, 233
Pleiotropism   144, 164
Polygenic effects   14, 32, 61, 78ff, 98
Polymorphism   266, 298ff
Population genetics   262ff, 298ff
Pyridoxine   247

Quantitative genetic analyses   25ff,
  298ff

Reproduction and behavior   57ff,
  165, 272
Reserpine   229

Selection   40
Sensory-motor performance   184ff
Serotonin (5-HT)   222, 227ff
Sex differences   103, 107, 192, 194
Sex-linkage   75
Sib correlations   37
Social behavior   311ff, 327
Species-typical behavior   268, 271
Systematics of the mouse   280, 294ff

Territorial behavior   311ff
Tetrabenazene   229
Tyron effect   34

Variance analyses   4, 16, 25ff, 34,
  51, 81, 312, 316